AF358693

New Trends in Production and Operations Management

New Trends in Production and Operations Management

Editor

Panagiotis Tsarouhas

Basel • Beijing • Wuhan • Barcelona • Belgrade • Novi Sad • Cluj • Manchester

Editor
Panagiotis Tsarouhas
AInternational Hellenic
University
Katerini
Greece

Editorial Office
MDPI
St. Alban-Anlage 66
4052 Basel, Switzerland

This is a reprint of articles from the Special Issue published online in the open access journal *Applied Sciences* (ISSN 2076-3417) (available at: https://www.mdpi.com/journal/applsci/special_issues/Production_Operations).

For citation purposes, cite each article independently as indicated on the article page online and as indicated below:

Lastname, A.A.; Lastname, B.B. Article Title. *Journal Name* **Year**, *Volume Number*, Page Range.

ISBN 978-3-7258-0417-7 (Hbk)
ISBN 978-3-7258-0418-4 (PDF)
doi.org/10.3390/books978-3-7258-0418-4

Contents

About the Editor

Panagiotis Tsarouhas

Panagiotis Tsarouhas, Dipl. Eng., Ph.D., is a Professor in the Department of Supply Chain Management (Logistics) at the International Hellenic University (Greece) and a tutor for the Postgraduate Quality Management and Technology course at Hellenic Open University. He received his Diploma in Mechanical Engineering from Universita degli Studi di Napoli Federico II (Italy) and his Ph.D. and M.Sc. from the Department of Mechanical and Industrial Engineering at the University of Thessaly in Volos, Greece. Tsarouhas served as the Head of Production and Maintenance Operations for the Technical Department of the Greek food industry's "Chipita International SA" for approximately ten years. He also has approximately 32 years of research/teaching experience and more than 90 research papers in international journals, book chapters, and conference proceedings. Tsarouhas has been involved in many research and practical projects in reliability analysis and maintenance engineering. His areas of interest are reliability maintenance engineering, quality engineering, Total Quality Management (TQM), Six Sigma, and supply chain management.

Editorial

New Trends in Production and Operations Management

Panagiotis Tsarouhas

Department of Supply Chain Management (Logistics), International Hellenic University, 60100 Katerini, Greece; ptsarouhas@ihu.gr

1. Introduction

Operations Management includes the management of all company activities that support the input–output cycle. Initially, production or manufacturing sections were the only places where the phrase "operations management" was used. The system, nevertheless, has developed over time and is now often used to refer to the administration of daily business operations of all units that ultimately direct toward the final service or product. Operations management seeks to maximize the efficiency of both the manufacturing process and broader corporate operations. Ensuring that a company's expenses and costs are reflected in its income is one of operations management's primary responsibilities. Thus, profit maximization and business expansion are ensured by effective operations management.

New trends in production and operations management require every action that aims to boost productivity and maximize profitability. To maintain market competitiveness, businesses continuously adopt new trends and technological breakthroughs (i.e., e-commerce markets, last mile logistics [1], human–machine systems reliability [2], humans and robots systems [3], supply chain management [4], etc.). Therefore, it might take the form of cost-cutting initiatives, the automation of repetitive processes, or the elimination of pointless tasks and extra fees (i.e., reduction in the electricity costs [5], etc.). Current management efficiencies and trends are always evolving with a focus on company efficiency. The most recent trends and advancements in production and operations management are covered in this Special Issue.

2. Strategies for New Trends in Production and Operations Management

The aim of this Special Issue is to collect the latest research on relevant topics, which are related to the interests and concerns of managers who manage the operations, design and supply chains of products and processes. Ten papers have been published in this Special Issue. Na et al. [1] focused on both operational and technological aspects and offered a thorough and organized assessment of current studies related to last mile logistics (LML). Amaya-Toral et al. [2] present a method for assessing and enhancing the man–machine system's dependability in a workshop for machine tools by taking into account system features, notably those provided by the machine shops of Chihuahua city. This sector of metal mechanics employs a low-volume manufacturing strategy and high-mix batches in their workshops.

In another study, a model for selective harvesting based on fruit maturity was created by Harel et al. [3]. Harvesting may be performed by humans or robots, and each type of harvester has a unique capacity for determining ripeness. Numerical experiments utilizing sweet pepper harvesting as a case study illustrate the model development and analysis. In their research, Zeng and Yang [4] examined a two-echelon closed-loop supply chain (CLSC) that includes a risk-neutral producer, a risk-averse fairness-neutral retailer, and a risk-neutral retailer who is concerned about fairness. A centralized, a decentralized, and three partially allied models are employed in cooperative game analysis to describe equilibrium conditions under five different circumstances. Roth et al. [5] provided a method for deciding which risk treatment approaches to use and how to classify and include the

Citation: Tsarouhas, P. New Trends in Production and Operations Management. *Appl. Sci.* **2023**, *13*, 9071. https://doi.org/10.3390/app13169071

Received: 1 August 2023
Accepted: 5 August 2023
Published: 8 August 2023

most appropriate measures into the production schedule. The decision-making process for choosing the measures is based on a hybrid multi-criteria approach, in which the three pertinent criteria—cost, energy flexibility, and risk reduction—are weighted using both the analytic hierarchy process and entropy, and then prioritized using the multi-attribute utility theory. The strategy was put into practice in MATLAB and validated with a case study in the foundry industry.

For the Hospital Pharmacy of the Centro Hospitalar de Vila Nova de Gaia/Espinho, E.P.E., Lima et al. [6] studied a system that can concurrently scan several Data Matrix codes and autonomously incorporate them into an authentication database. The trial findings were promising, and it is anticipated that the system may be utilized as a true resource for pharmacists with improvements like real-time feedback of the code's validation and greater hardware system stability. Wen [7] made an effort to research the alliance because certain members are frequently in low supply while others are experiencing an overstock. For instance, there is a greater demand than supply for hospitals in rural regions or for the central hospitals in the medical alliance. Large demand (HD) will shift some of their demand to large supply (HS), which might increase HS's effectiveness.

Mateus et al. [8] described evaluations based on many neural network models that were tested to predict global steel output. To generate a prediction for a nine-year period, the primary objective was to identify the best machine learning model that matches the global data on steel output. The study is crucial for comprehending how convolutional LSTM and GRU recurrent neural networks behave and respond to hyperparameters. The outcomes demonstrated that the GRU model performs better and is simpler to train for long-term prediction. Sawicki and Sawicka [9] concentrated on the problem of constructing stacks of at least two stackable pallet load components. In addition, this study focused on the portion of the distribution network's product flow that is prepared at the site of first assembly in the form of palletized loading units intended for the ultimate receiver. Such a unit does not go above the permitted weight or height restrictions. The purpose of the article's single-criteria binary programming model is to reduce the amount of pallet space needed to hold the built units. The savings from using the best design for the stacked palletized cargo units were shown through the tests carried out, and the model generated was tested on a test dataset.

Through a survey of the academic literature, Zhu et al. [10] examined the objectives to comprehend the categorization of idea management tools and their efficacy. A total of 38 journal publications ($n = 38$) from 2010 to 2020 were found after searching electronic databases (Scopus, ACM Digital Library, Web of Science Core Index, Elsevier ScienceDirect, and SpringerLink). The 30 distinct kinds of concept management tools that we found were used by stakeholders, software designers, hardware designers, and digital tool designers ($n = 21$), guidelines ($n = 5$), and frameworks ($n = 4$). The tools mentioned may help with several phases of idea management, including gathering, producing, implementing, monitoring, refining, retrieving, choosing, and sharing. Therefore, it is crucial to provide management tools for ideas that would enable users, designers, and other stakeholders to minimize bias in selecting and prioritizing ideas.

3. Future Trends on Production and Operations Management

Production and operations management are evolving rapidly because of the industrial revolution 4.0: the following are the future trends for this Special Issue that should be explored more in-depth in future research [1–5,8–10]:

- Automating traditional methods (e.g., manual processes).
- Focus on the safety, health, and wellness of employees.
- Improvement in communication (i.e., an integrated communications system is necessary to ensure collaboration across teams and departments).
- Agile organization solutions are required, due to market pressures and quickly changing consumer demand.

- Collection and analysis of data (e.g., failure/repair data of the equipment) are important for the decision made of the system's operations management.
- Focus on customers as referred in Total Quality Management (TQM) principles.

4. Conclusions

In conclusion, production and operations management is the connecting link between many departments. The most recent trends and advancements may support a variety of organizational and company success factors. By incorporating these trends and advances into corporate operations, operations management specialists might significantly contribute to increasing productivity and profit. By automating time-consuming and repetitive tasks, simplifying communication channels, and connecting front-line staff, businesses may make the most of their current personnel. By maintaining a healthy balance between people and technology, these developments will also enhance procedures.

Funding: This research received no external funding.

Conflicts of Interest: The author declares no conflict of interest.

References

1. Na, H.S.; Kweon, S.J.; Park, K. Characterization and Design for Last Mile Logistics: A Review of the State of the Art and Future Directions. *Appl. Sci.* **2022**, *12*, 118. [CrossRef]
2. Amaya-Toral, R.M.; Piña-Monarrez, M.R.; Reyes-Martínez, R.M.; de la Riva-Rodríguez, J.; Poblano-Ojinaga, E.R.; Sánchez-Leal, J.; Arredondo-Soto, K.C. Human–Machine Systems Reliability: A Series—Parallel Approach for Evaluation and Improvement in the Field of Machine Tools. *Appl. Sci.* **2022**, *12*, 1681. [CrossRef]
3. Harel, B.; Edan, Y.; Perlman, Y. Optimization Model for Selective Harvest Planning Performed by Humans and Robots. *Appl. Sci.* **2022**, *12*, 2507. [CrossRef]
4. Zeng, T.; Yang, T. Unfair and Risky? Profit Allocation in Closed-Loop Supply Chains by Cooperative Game Approaches. *Appl. Sci.* **2022**, *12*, 6245. [CrossRef]
5. Roth, S.; Huber, M.; Schilp, J.; Reinhart, G. Risk Treatment for Energy-Oriented Production Plans through the Selection, Classification, and Integration of Suitable Measures. *Appl. Sci.* **2022**, *12*, 6410. [CrossRef]
6. Lima, J.; Rocha, C.; Rocha, L.; Costa, P. Data Matrix Based Low Cost Autonomous Detection of Medicine Packages. *Appl. Sci.* **2022**, *12*, 9866. [CrossRef]
7. Wen, J. A Stochastic Queueing Model for the Pricing of Time-Sensitive Services in the Demand-Sharing Alliance. *Appl. Sci.* **2022**, *12*, 12121. [CrossRef]
8. Mateus, B.C.; Mendes, M.; Farinha, J.T.; Cardoso, A.J.M.; Assis, R.; da Costa, L.M. Forecasting Steel Production in theWorld—Assessments Based on Shallow and Deep Neural Networks. *Appl. Sci.* **2023**, *13*, 178. [CrossRef]
9. Sawicki, P.; Sawicka, H. Design Optimization of Stacked Pallet Load Units. *Appl. Sci.* **2023**, *13*, 2153. [CrossRef]
10. Zhu, D.; Al Mahmud, A.; Liu, W. A Taxonomy of Idea Management Tools for Supporting Front-End Innovation. *Appl. Sci.* **2023**, *13*, 3570. [CrossRef]

applied
sciences

Article

Design Optimization of Stacked Pallet Load Units

Piotr Sawicki [1,*] and Hanna Sawicka [2]

[1] Division of Transport Systems, Poznan University of Technology, ul. Piotrowo 3, 61-138 Poznań, Poland
[2] Division of Rail Transport, Poznan University of Technology, ul. Piotrowo 3, 61-138 Poznań, Poland
* Correspondence: piotr.sawicki@put.poznan.pl

Abstract: The article deals with the problem of building stacked pallet load units consisting of at least two stackable pallet load units. Moreover, this article concerns the part of the flow of goods in distribution networks that is prepared at the place of initial assembly in the form of palletized loading units designed for the final receiver. Such a unit does not exceed the limits of permissible weight or height. The article proposes a single-criteria binary programming model in which the goal is to minimize the pallet spaces required to accommodate the constructed units. In addition to the classical parameters of acceptable weight and height of the units, the constraints also take into account the fragility of the goods placed on each unit, filling the top layer of each unit, and its height homogeneity. The model developed was verified on a test dataset, and the savings from the use of optimum construction of the stacked palletized cargo units were demonstrated through the conducted experiments.

Keywords: sustainable logistics; sustainable transportation issues; pallet stacking; stacked pallet load unit problem (SPLUP)

Citation: Sawicki, P.; Sawicka, H. Design Optimization of Stacked Pallet Load Units. *Appl. Sci.* **2023**, *13*, 2153. https://doi.org/10.3390/app13042153

Academic Editor: Panagiotis Tsarouhas

Received: 28 December 2022
Revised: 1 February 2023
Accepted: 2 February 2023
Published: 7 February 2023

1. Introduction

1.1. The Essence of Constructing Pallet Load Units

Forming palletized load units—PLU is an issue closely related to the functioning of distribution networks and supply chains. It concerns the proper placement of products on the smallest possible number of carriers. In practice, there are two basic types of palletized load units, i.e., homogeneous and heterogeneous. In the first case, products of one type are loaded on a pallet, which is mostly performed by product manufacturers. In the other case, the loaded products differ in basic characteristics such as weight, length, width and height, and product type, and this case mainly occurs at distributors and dealers.

The arrangement of products in PLUs has a number of practical consequences. Firstly, it affects the stability of the pallet units, which, on the one hand, may involve the risk of product damage during basic logistics operations such as transportation, storage, and handling. Product safety depends on their placement on the pallet as they need to be layered to prevent crushing the ones located on lower and intermediate layers. Secondly, product arrangement affects the cost of logistics operations. The smaller the number of PLUs, the less space necessary for transportation and storage, the less energy required, and the lower the cost of performing these operations. The comprehensive research on the distribution network structure and its effectiveness related to the type of packages, including PLUs and parcels are presented in the previous work of the authors; see Sawicki and Sawicka [1].

In general, the essence of the problem of forming palletized load units involves the proper arrangement of products on the pallet to ensure its stability and compliance with basic constraints such as the limited weight and height of the load unit. An overview of the current state-of-the-art in this area, including the categories of decision-making problems, the way these problems are modeled, the procedures for solving them as well as the results obtained, are presented in the following subsection.

Based on the results of the literature review and the authors' empirical experience, this publication focuses on the issue of forming a specific type of load unit, which is a stacked pallet load unit—SPLU, commonly referred to as a sandwich-pallet. This issue involves creating a single load unit composed of several elementary PLUs previously arranged on separate pallets. The need to build this type of unit is a direct result of the increasingly noticeable phenomenon within the distribution network of the formation of PLUs whose weight and/or height significantly deviate from the limit values, i.e., the available space on the pallets is not fully utilized. This results from the requirements of purchasers, mainly large retail chains, who order large volumes of goods versus a variety of products and expect that the PLUs will be prepared in such a way that the goods reaching their transshipment warehouses will not require any additional operations, except for the redirection of dedicated PLUs for shipment to the final receivers. Thus, creating a few collective units instead of many smaller PLUs is becoming a necessity. In practice, such operations are undertaken on an ongoing basis, i.e., whilst loading, the forklift operator subjectively assesses whether it is feasible to combine the available PLUs into SPLUs. The FLT operator uses his own experience and intuition, yet whether such an operation is correct, both in terms of the durability of product packaging on individual PLUs and the created weight and height of the SPLU, cannot be guaranteed. The work undertaken in this area by the authors of this article provides a proposal to solve this problem by developing and verifying a procedure to optimize the process of creating SPLUs.

1.2. The State-of-the-Art Design of Palletized Load Units

In the literature, the problem of planning pallet load units is called *bin packing problem—BPP*. To solve the problem, one needs to define which items of different sizes need to be packed into a finite number of bins or containers, each of a given and fixed capacity so as to minimize the number of bins used. In the literature, the same problem, depending on the authors, is referred to as pallet loading problem—PLP, e.g., Dell'Amico and Magnani [2], Morabito et al. [3], pallet building problem—PBP, e.g., Alonso et al. [4], container loading problem—CLP, e.g., Lim and Zhang [5], or packing problem—PP, e.g., Ali et al. [6].

Since the packing problem is an NP-hard decision problem, it can also be analyzed, with the exception of its principal constraint, i.e., the weight of the product, as a soft version of two dimensions (2D), i.e., width x length of the products to be packed, e.g., G and Kang [7], or a more complex problem of three dimensions (3D), i.e., width x length x height, e.g., Dell'Amico and Magnani [1], Morabito et al. [3], or Ali et al. [6].

From the perspective of the location, the packing problem is analyzed and planned with respect to homogenous items by some of the researchers who refer to it as manufacturer's pallet loading—MPL, e.g., Marabito et al. [3]. When packaging involves heterogeneous items, it is called distributor's pallet loading—DPL; see Akkaya et al. [8].

In general, the bin packing problem can be analyzed as a stand-alone packing problem or as a part of a more complex problem, i.e., in combination with vehicle loading. As far as the stand-alone approach is concerned, it has been extensively analyzed as a decision problem for around 50 years. In recent years several papers have been published to review or compare different approaches to this problem [6,9]. Silva et al. [9] reviewed the papers with respect to the methods proposed for the solution of the problem. In conclusion, a group of the most challenging methods was identified. Ali et al. [6] analyzed different approaches to the decision-making problem concerning packing. They differentiated 3D off-line vs. on-line streams of packing problems. Off-line packing problems can happen when full knowledge about items is available beforehand. The on-line problem, i.e., real-time problem, is when items arrive one by one, and the packing decision should be made immediately without prior knowledge about the items. According to Ali et al., most of the packing problems described in the literature are of the off-line type. Contemporary examples of the on-line bin packing problem can be found in Lin et al. [10], where a pattern-based adaptive heuristics for the on-line bin packing problem is proposed. The distribution of items may be predicted based on the packed items, and the pattern is next applied in the packing of the items that arrive later.

Some new research into the bin packing problem was published in the last few years. Gzara et al. [11] analyzed a wide spectrum of practical constraints, including vertical support, load bearing, planogram sequencing, and weight limits. The authors performed extensive numerical tests to prove the ability of the approach to find high-quality solutions for industrial-size instances within a short computational time. There are also some recent papers on the application of more efficient computational procedures whose aim is to achieve computation results in the shortest time possible. Tresca et al. [12] published a paper on the 3D bin packing problem (3D-BPP) where they proposed a model oriented on pallets configurations to satisfy the practical requirements of item grouping by logistic features such as load bearing, stability, height homogeneity, overhang as well as weight limits, and robotized layer picking. The complex problem is solved with metaheuristics, which combines MILP formulation and layer-building heuristics. Moreover, 3D-BPP is analyzed in the work of Zuo et al. [13]. The authors addressed a novel 3D-BPP variant in which the shape-changing factor of non-rectangular and deformable items was incorporated into the model. In the research of Elhedhli et al. [14], another practical variant of 3D-BPP is addressed. The authors analyzed the mixed-case palletization problem, where item support with the presence of different sizes of items is considered. Elhedhli et al. proposed a novel problem formulation as well as a column-generation solution approach. El-Ashmawi and Elminaam [15] concentrated on the design of an approximate algorithm of BPP applicable to solve large-scale instances within a reasonable time. They proposed a modified version of the squirrel search algorithm (SSA) for solving the 1D bin packing problem. In the experiments, hard class instances of up to 200 items were tested, and the obtained result was compared against other approximate algorithms such as particle swarm optimization (PSO), African buffalo optimization (ABO), and crow search algorithm (CSA).

A typical complex decision problem is usually a combination of packing and vehicle loading problems [3,4,15]. In the work of Morabito et al. [3], the optimal solution to the problem is presented. However, the mathematical model is not presented in detail. This concept is applied to solve the combined problem of pallet and vehicle loading, and the size of unit packages and the size of pallets and vehicles are also discussed. A two-phase approach with a packing problem is also considered by Moura and Bortfeld [16]. In this approach, however, the main objective is to guarantee sufficient utilization of the truck loading space. Alonso et al. [4] considered the problem of building and placing pallets on the truck at the same time, i.e., pallet loading as the first phase and truck loading as the second phase. The authors implemented several extended constraints, including the total weight of the load, the maximum weight supported by each vehicle's axle, and the distribution of the load inside the vehicle. During the truck loading phase, it was allowed to stack one pallet on top of another. The model was constructed and solved with the GRASP algorithm, and the experiments were performed in the domain of the number of instances, vehicles, and computation time. Dell'Amico and Magnani [2] have also proposed a two-phase procedure. In the first phase, 2D layers were defined, while in the second phase, the layers were combined on the minimum number of pallets. The authors constructed a specialized metaheuristic with a MILP model of a 3D problem, which was subsequently solved by the Gurobi solver. The experiments were performed in the domain of the number of instances and computation time. Contrary to the previous research on BPP and vehicle loading problems, Moura Santos et al. [17] proposed the research on BPP with compatible categories, i.e., products that cannot be transported together. In this approach, the bin is the fleet of vehicles, and the problem is to allocate products with respect to the product categories. Moura Santos et al. concentrated on large instances of 200–1000 items, and they solved them with a variable neighborhood search (VNS) procedure. The experiments were also compared with alternative procedures.

There are several conclusions resulting from the state-of-the-art of the bin packing problem that should be highlighted, i.e.,

- Numerous papers on BPP focus on the analysis of a wide spectrum of constraints that are either theoretical or practical. There is hardly any research on small-size pallets, i.e., substantially below the maximum height or weight of the PLU. It is a typical situation in contemporary logistics.
- The BPP is an NP-hard decision problem, which is why a substantial number of studies have been carried out on the effectiveness of the optimization procedures.
- The case when PLUs can be stacked one on top of another is referred to as a combined problem of pallet and vehicle loading. It is not, however, a widely discussed topic in the literature.
- One specific way of formulating the BPP problem can be regarded as the inspiration for the formulation of the problem of stacking pallet load units, i.e., every single PLU can be treated as equivalent to a single layer considered in the PLU packing planning process.

1.3. Formulation of the Decision Problem

The article considers the problem of building stacked pallet load units (SPLUs) composed of elementary pallet load units (PLUs) whose height and gross weight parameters are significantly below the permissible values. Therefore, a mechanism is sought to plan the stacking of a set of PLUs in a way that will reduce the amount of space required on the load bed or in racking slots, while the parameters of the SPLUs built in this way should be within the permissible values of weight and height. It is assumed that the elementary PLUs are created in advance as a result of the adaptation to the specific requirements of the ordering party and may not be subject to content modification. It is also assumed that individual PLUs may contain homogeneous or mixed products, and the ordering party agrees to stack PLUs to form SPLUs.

The basic characteristics parameterizing the susceptibility of elementary PLUs to the formation of SPLUs include such parameters as gross weight and gross height, i.e., including the pallet itself, the fragility class of the stacked products, and the status of the PLUs resulting from the filling of its upper layers. Figure 1a,b show elementary PLUs with a total height well below the limit value; the PLUs in Figure 1a have a base status on which it is possible to stack more PLUs, and in Figure 1b, a *top* status: on this type of unit it is impossible to stack more PLUs. Figure 1c,d show SPLUs built from elementary PLUs. In the case of (c), all PLUs have *base* status, and in the case of (d), the highest layer is a *top*-type PLU, which makes it impossible to build further layers on it.

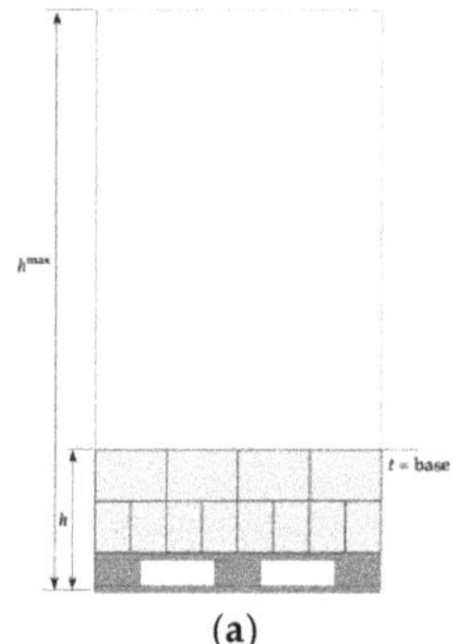

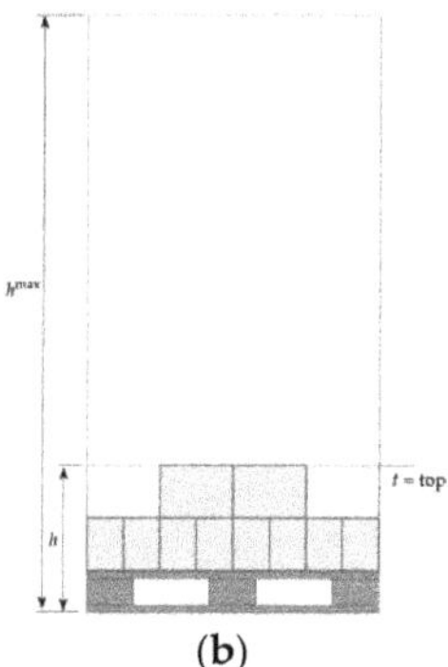

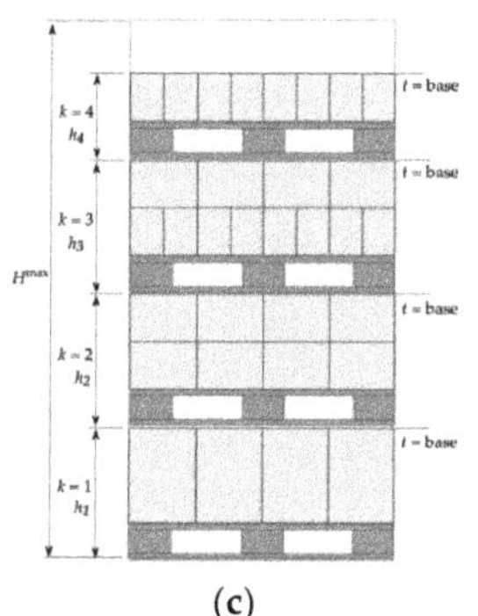

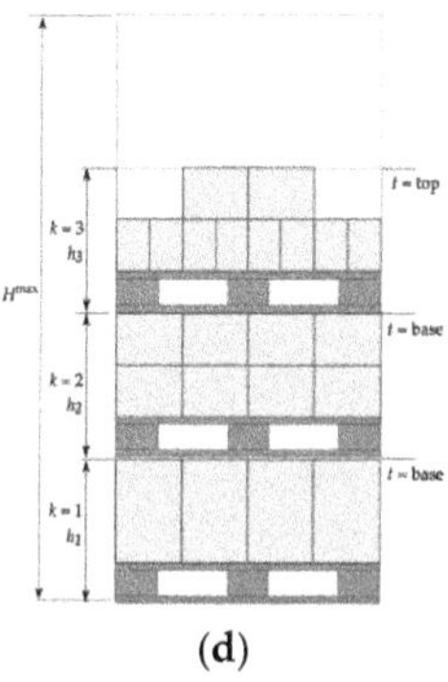

Figure 1. The considered forms of PLUs: (**a**) Elementary PLU of the *base* type, which has the necessary features for the construction of SPLUs; (**b**) Elementary PLU of the *top* type, which has limited susceptibility for the construction of SPLUs; (**c**) SPLU built only from PLUs of the *base* type; (**d**) SPLU built from SPLs of the *base* and *top* type.

1.4. Purpose and Scope of the Research

Based on the conclusions resulting from the state-of-the-art, see Section 1.2, and based on the formulation of the research problem, see Section 1.3, the purpose of the work was defined. The objective of the study is to develop a procedure to optimize the construction process of stacked pallet load units (SPLUs) built from elementary PLUs, taking into account, in addition to the classical limitations of the weight and height of the unit, the susceptibility of PLUs to the construction of successive layers, including the fragility of products, the flatness of the top layer of products contained on the PLU, and the filling of the top layer.

The article contains four key sections. In Section 1, the need for PLU formation as a cost driver in transportation and warehousing processes is discussed, the decision problem is defined, and the research objective is formulated. Section 2 presents the research methodology, including the definition of the objective function and a set of constraints; it also presents the assumptions for the proposed methodology, i.e., its applicability limits. Section 3 presents the results of computational experiments; the methodology is verified based on a test dataset. Section 4 is a summary of the article, where the obtained results are described with reference to the formulated purpose of the work, and the directions for further work are defined. Finally, a list of the literature used is presented.

2. The Research Methodology

2.1. Notations

In the presented research, some indexes, decision variables, and parameters are defined. The complete list is presented in the following tables; see Tables 1–3.

Table 1. The list of indexes in alphabetical order.

Symbol	Definition
i	an ordinal number of pallet load unit to be stacked; $i = 1, 2, \ldots, I$;
j	an ordinal number of stacked pallet load unit; $j = 1, 2, \ldots, J$;
k	an order of pallet load unit setting in the built stacked pallet load unit; $k = 1, 2, \ldots, K$, i.e., $k = 1$ is the lowest layer, and $k = K$ is the top layer.

Table 2. The list of variables in alphabetical order.

Symbol	Definition
x_{ijk}	a binary variable specifying which i-PLU should be assigned to the k-layer on the j-SPLU.
y_j	a binary variable specifying whether the j-SPLU is formed;

Table 3. The list of parameters in alphabetical order.

Symbol	Definition
f_i	a fragility class of the products on the i-PLU (-), $f = 1, 2, \ldots, n$
h_i	a gross height of i-PLU, including height of pallet expressed in (mm)
H_j^{max}	the maximum height of j-SPLU, expressed in (mm)
t_i	a parameter defining the upper layer of i-PLU (-); $t_i \in \{0; 1\}$, where: $t_i = 0$ if upper layer is flat and its filling is higher than 70% (*base* status); $t_i = 1$ otherwise (*top* status).
w_i	a gross weight of i-PLU, including weight of pallet, expressed in (kg)
W_j^{max}	the maximum weight of j-SPLU, expressed in (kg)

2.2. Key Assumptions

The following assumptions are made in the decision problem:

- Individual i-PLUs are built according to the final receiver's order. They contain one type of product, i.e., homogeneous PLU, as well as a mix of products, i.e., heterogeneous PLU, the so-called mix.
- In the case of heterogeneous i-PLU, containing products with different fragility classes, a representative fragility class is determined, i.e., the lowest fragility class among the products on the i-PLU.
- If the top layer of i-PLU is flat, that is, the total height of products accumulated on each layer of i-PLU is the same, then such a PLU can serve as the base for the construction of j-SPLU and its intermediate layers.
- If there is a difference in the height of the products from the point of view of the last layer on the PLU, i.e., the top layer is not flat, then such an i-PLU can only be the last layer of the built SPLU. The height of such an i-PLU is the maximum value for all the accumulated products on the i-PLU.
- If the last layer of i-PLU is filled up to at least 70%, such a PLU can serve as a base for the construction of SPLU and its intermediate layers.
- When the last layer of i-PLU is filled up to less than 70%, then such a unit can only be the last layer of the built SPLU.

2.3. Decision Variables and Objective Function

The mathematical model of the considered decision problem is formulated as a binary model and is presented in this section. The minimized objective function F is the number of j-SPLU necessary to handle all elementary i-PLUs included in the order. It is represented by the following formulas, see (1)–(11):

$$F = \min \sum_{j=1}^{J} y_j \tag{1}$$

where

$$y_j = \begin{cases} 1 \text{ if } \sum_{i=1}^{I} \sum_{k=1}^{K} x_{ijk} > 0 \\ 0 \text{ if } \sum_{i=1}^{I} \sum_{k=1}^{K} x_{ijk} = 0 \end{cases}, \forall j = 1, 2, \ldots, J \tag{2}$$

The unit of j-SPLU can be planned if i-PLU is assigned at any of the available k-layers of j-SPLU. Otherwise, the j-SPLU unit is not planned.

2.4. Constraints

The optimization model of the design the stacked pallet load units problem is constructed with nine constraints; see Formulas (3)–(11) below. Constraint (3) indicates that each i-PLU is assigned exactly to one of the j-SPLU within the considered order. The next constraint (4) indicates that on a k-layer of the j-SPLU, only one i-PLU can be located. Two other constraints say that the gross weight (5) and gross height (6) of the j-SPLU result from the requirements of the j-SPLU user, i.e., $W_j^{\max}$ and $H_j^{\max}$. Based on the constraint (7), successive layers on a j-SPLU are planned while preserving the fragility class of individual i-PLU on a k-layer. It means that the k-layer may contain an i-PLU of the same or higher fragility class than layer $k - 1$. Fragility class $f = 1$ means the lowest fragility, i.e., the product has a significant compression resistance, while $f = n$ means the highest fragility, i.e., the product has the lowest compression resistance. The constraints (8) and (9) result from the classification of i-PLUs into the *base* and *top* status. Status *base* means that on i-PLU, another k-layer of j-SPLU can be considered, and status *top*, otherwise. The unit i-PLU of the status *top* can be on the top of j-SPLU, exclusively. Constraint (8) guarantees that, at maximum, one i-PLU of *top* status can be allocated to a single j-SPLU. However, constraint (9) ensures that the i-PLU of the *top* status will not be located below the i-PLU of the *base* status. The last two constraints, i.e., (10) and (11), indicate that both decision variables y_j and x_{ijk} are of binary nature.

$$\sum_{j=1}^{J} \sum_{k=1}^{K} x_{ijk} = 1, \forall i = 1, 2, \ldots, I \tag{3}$$

$$\sum_{i=1}^{I} x_{ijk} \leq 1, \forall j = 1, 2, \ldots, J; \forall k = 1, 2, \ldots, K \tag{4}$$

$$\sum_{i=1}^{I} \sum_{k=1}^{K} w_i \cdot x_{ijk} \leq W_j^{\max}, \forall j = 1, 2, \ldots, J \tag{5}$$

$$\sum_{i=1}^{I} \sum_{k=1}^{K} h_i \cdot x_{ijk} \leq H_j^{\max}, \forall j = 1, 2, \ldots, J \tag{6}$$

$$\sum_{i=1}^{I} f_i \cdot \left(x_{ijk} - x_{ij(k-1)} \right) \geq 0, \forall j = 1, 2, \ldots, J; \forall k = 2, \ldots, K \tag{7}$$

$$\sum_{k=1}^{K} \sum_{i=1}^{I} t_i \cdot \left(x_{ijk} - x_{ij(k-1)} \right) \leq 1, \forall j = 1, 2, \ldots, J \tag{8}$$

$$\sum_{i=1}^{I} t_i \cdot \left(x_{ijk} - x_{ij(k-1)} \right) \geq 0, \forall j = 1, 2, \ldots, J; \forall k = 2, \ldots, K \tag{9}$$

$$x_{ijk} = \begin{cases} 1 \text{ if } i\text{-PLU is located at } k\text{-layer in the } j\text{-SPLU,} \\ 0 \text{ otherwise} \end{cases} \tag{10}$$

$$y_j = \begin{cases} 1 \text{ if } j\text{-SPLU is planned to be designed,} \\ 0 \text{ otherwise} \end{cases} \tag{11}$$

Fragility is the parameter that defines the compressive strength of a single package of a product exerted by products of the same type and piled on it. Products, usually homogenous, with a certain fragility of a full load unit, i.e., filled up to the maximum weight, are characterized by the height of this PLU, i.e., $h_i^{\max}$. It should also be assumed that the weight of a full pallet load unit of products with lower fragility f_i is substantially heavier than the weight of a full pallet load unit of products with higher fragility f_{i+1}, i.e., $w_i(f_i) > w_{i+1}(f_{i+1})$. Thus, in case of stacking incomplete pallet load units, i.e., $h_i < h_i^{\max}$, with different fragilities and maintaining the fragility ranking from the lowest to the highest $f_i > f_{i+1} > f_{i+2}$ etc., starting from the background of SPLU, the total height $H_j^{\max}$ of this SPLU may exceed the maximum height of each of the pallet full load unit of products, i.e., $H_j^{\max} > h_i^{\max}$.

2.5. Implementation Procedure

The optimization of the design of the stacked pallet load units is based on the formal assumptions and the mathematical model presented in Sections 3.1 and 3.2. The practical application of this approach is shown as a BPMN notation procedure, see Figure 2.

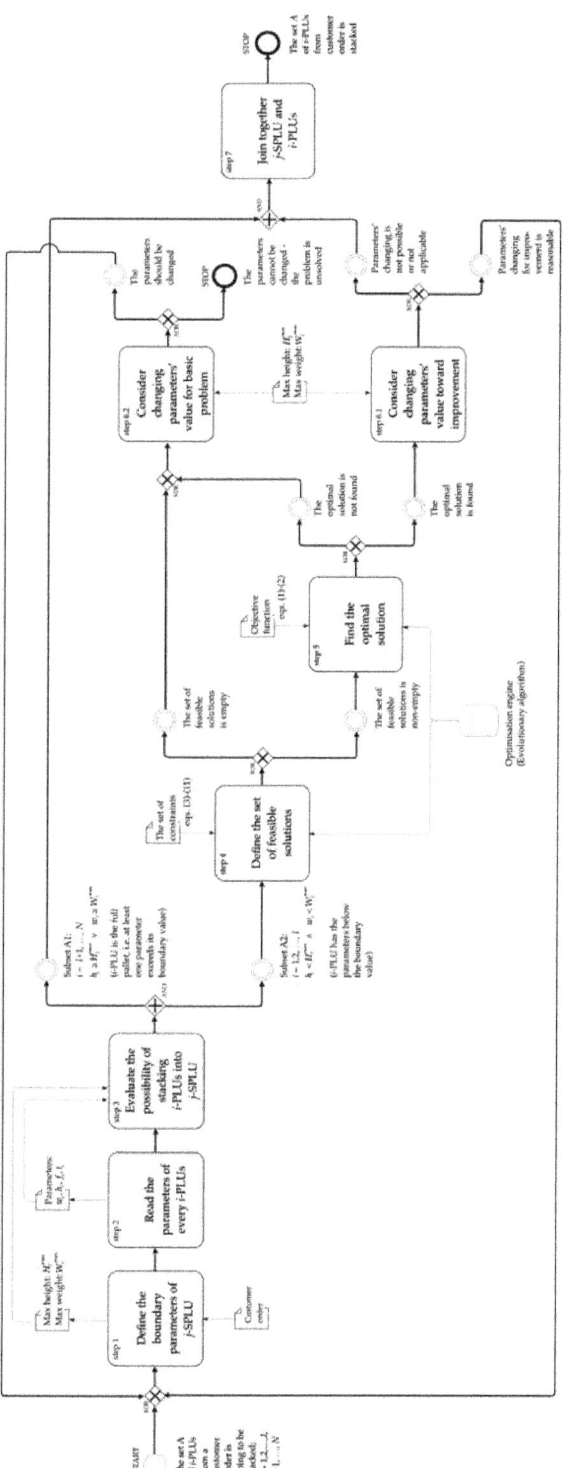

Figure 2. The implementation of stacking pallet load units optimization procedure.

The starting point is set A of i-PLUs, where $i = 1, 2, \ldots, I, I + 1, \ldots, N$, which constitutes all pallets prepared upon customer order. Initially, two boundary parameters of j-SPLU that should be designed on A, i.e., W_j^{max} and H_j^{max}, are defined (step 1). Then, each i-PLUs from the set A: $i = 1, \ldots, N$ is identified in terms of basic parameters (step 2), such as weight w_i, height h_i, fragility class f_i, and the degree of filling of the top layer t_i. Based on the identified parameters of each i-PLU and the boundary parameters of SPLU to be constructed, in the next step of the procedure (step 3), the assessment of the possibility of building stacked j-SPLUs from available i-PLUs is carried out. As a result, set A is divided into two subsets, i.e., $A1$: $i = I + 1, \ldots, N$, which includes i-PLUs that do not meet at least one boundary condition of W_j^{max} or H_j^{max}, and subset $A2$: $i = 1, 2, \ldots, I$, which includes those i-PLUs that do not exceed any of the boundary parameters. In step 4, a set of feasible solutions is built on $A2$; for this purpose, all the constraints described by Formulas (3)–(11) are applied. If, as a result of this operation, the set of feasible solutions is non-empty, in step 5, the optimal solution is searched; at this stage, the Formulas (1)–(2) of the objective function are applied. The computations within steps 4 and 5 are performed by optimization engines of the evolutionary algorithm. If, as a result of step 5, an optimal solution is found, in step 6.1, an analysis of whether it is necessary or applicable to modify boundary parameters for further improvement should be carried out. If not, then step 7 is performed, which consists of joining the j-SPLUs constructed in steps 4 and 5 with the i-PLUs that were separated in step 3 due to unmet boundary conditions; see subset $A1$ of step 3. The result of the procedure at this stage is a reduced number of pallet spaces that are necessary for the transport or storage of i-PLUs because some of them can be transformed to SPLUs, upon boundary conditions of W_j^{max} or H_j^{max}.

If, as a result of step 6.1, the change of boundary parameters is applicable, then it is necessary to return to step 1 and repeat the optimization procedure step by step. If, after step 4, it is not possible to define a set of feasible solutions, or in step 5, it is not possible to find an optimal solution, it is recommended to consider changing the values of the boundary parameters (step 6.2) and repeat the procedure from step 1. If such a change of parameters is not possible, the problem is unsolved, and the procedure should be stopped.

3. Results of Computational Experiments

3.1. Assumptions, Input Data, and Computational Experiments

The computations verifying the methodology proposed and presented in Section 2 were carried out on a MacBook Pro computer with a 3.1 GHz dual-core Intel Core processor and a Windows 10 operating system embedded in a virtual machine. The model was implemented into the Excel environment, and the computations were performed using Frontline Solvers' evolutionary optimization engines.

In the experiments, 10 sample instances were analyzed, i.e., sets consisting of 8–20 PLUs, as per customer order, containing only the units to be potentially expanded into SPLUs, see subset $A2$ on Figure 2. The complete data sets of analysed instances are attached to this article, see Tables A1–A3 in Appendix A. SPLUs were then built from the elementary PLUs based on the mathematical model presented in Section 2, Equations (1)–(11). The results obtained are shown below in Table 4. The analysis included such parameters as the following:

- Minimum and maximum values: weight, height, and number of PLUs of *top* and *base* type;
- The structure of storage carriers, including PLU *top* and *base* types;
- Boundary parameters of height and weight of built SPLUs.

The results were analyzed taking into account the following:

- The number of SPLUs constructed; the value of the decision variable y_j;
- The degree of filling the available height and weight of SPLUs;
- Computing time to solve the problem.

Table 4. Results of computational experiments.

Instance	Parameters [1]					Results [2]			
	N° PLU	min w_i; h_i	max w_i; h_i	t_i [3]	f_i [4]	N° SPLU	% H^{max}	% W^{max}	CT
1	9	186; 311	256; 416	1; 8	1; 1; 7; 0; 0	3	91.0–95.3	78.1–82.2	0.7
2	9	175; 295	259; 421	2; 7	5; 3; 1; 0; 0	3	83.0–92.9	70.4–79.6	0.01
3	8	236; 405	361; 530	3; 5	1; 4; 3; 0; 0	4	77.6–83.0	69.8–77.4	0.8
4	8	245; 414	349; 518	2; 6	2; 3; 3; 0; 0	4	72.8–82.0	63.0–76.0	1.8
5	20	130; 300	390; 520	7; 13	2; 5; 4; 2; 7	9	40.9–99.2	41.9–86.6	51.2
6	20	182; 353	246; 433	5; 15	1; 7; 8; 4; 0	7	69.6–100.0	55.0–77.6	2.6
7	20	175; 344	341; 533	5; 15	6; 3; 2; 6; 3	8	80.7–99.8	70.1–78.7	1.6
8	17	148; 359	273; 442	5; 12	7; 10; 0; 0; 0	6	70.8–99.6	55.5–73.0	2.7
9	15	129; 347	266; 438	2; 13	6; 6; 3; 0; 0	5	90.6–99.6	53.7–72.9	1.6
10	19	225; 444	267; 571	7; 12	8; 3; 0; 3; 5	10	47.5–90.3	31.4–60.6	1.6

[1] Parameters are expressed in: N° PJŁ (units), w (kg), h (mm), t_i (-), and f (-). [2] Results are expressed in: N° SPLU (units), % H^{max} (%), % W^{max} (%), and CT (s). [3] Number of PLUs of *top*; *base* status. [4] Number of PLUs of fragility class 1; 2; 3; 4; 5.

3.2. Results and Discussion

The results of the experiments presented in Table 4 allow for the following conclusions and observations:

- The application of the proposed mathematical model allows for a significant reduction of the space required to locate PLUs in the transportation and/or warehousing process, thus creating fewer required pallet positions. In the analyzed instances, 3–10 SPLUs were created from 8–20 PLUs, which means a two- to three-times reduction in the demand for pallet positions in logistics processes. A smaller number of PLUs translates directly into lower transportation costs in logistics networks, as well as less demand for rack slots in the process of storing PLUs.
- The smaller the number of *top* status PLUs in the structure of stacked units, the better the chance of building a smaller number of SPLUs while maintaining all the constraints. In instances 1, 2, and 9 with at most two PLUs *top* type, the number of SPLUs is three times less than the number of PLUs; in the other instances, the higher number of *top* type PLUs results in at most 2–2.5 times reduction of the number of SPLUs compared to PLUs.
- The higher the values of weight and height of individual PLUs subjected to stacking, the fewer multi-layered SPLUs can be built. The structure of the SPLUs built depends on the specificity of the items located on the individual PLUs. In all instances (1–10), a higher degree of achievement of the max height parameter (82.0–100%) is obtained compared to the max weight parameter (60.6–86.6%). This means that the analyzed instances constitute a set of relatively light but hight products.
- The computing time to solve the problem of optimal stacking pallet load units is similar for all analyzed instances. It is 0.7–2.6 s. for 8–20 PLUs per instance, except 51.2 s. for 20 PLUs, see instance 5.

4. Conclusions

4.1. Summary of the Results Obtained

The research carried out on the construction of stacked palletized load units SPLU resulted in the development of a mathematical model for this decision problem. While the phenomenon of the construction of palletized load units itself is a problem that has been studied for over 50 years, the authors of the publication have recently extended the existing state of knowledge in the direction of off-line type algorithms and the consideration of constraints strongly related to the specifics of the logistics industry, where these problems occur in daily operations. This article deals with the latter trend. Indeed, the authors focused on the phenomenon of PLU stacking in a situation where the process of picking pallet load units for the customer, and more broadly for the ordering party, produces multi-product units that significantly deviate from the permissible height and weight parameters of pallet units. In addition, there is a high probability that some of the PLUs built in this way will be characterized by a lack of susceptibility to stacking in the process of SPLU construction. This results directly from the fact that there is not enough *top*-type PLU layer support area for the construction of subsequent layers, i.e., subsequent PLUs.

As indicated in the section on the state-of-the-art review, the phenomenon of PLU stacking itself is not a new issue. However, it concerns the situation of parallel PLUs construction planning and vehicle loading planning. The authors of this article addressed the issue of PLU stacking planning in a broader context and as a stand-alone decision problem. Indeed, based on practical experience in the logistics industry, it is noticeable that the need to build SPLUs is not only related to the stacking of load carriers in the vehicle cargo space but also applies to the construction of SPLUs that are created well in advance of the loading process, and therefore, must be subject to periodic storage in racking slots or other spaces with more stringent height parameters compared to the vehicle cargo space. For this reason, the authors decided to develop a mathematical model to meet the requirements of the wider logistics process.

The summary of the research presented in this article, i.e., a comparison of methodological similarities and differences between the classical bin packing problem (BPP) and the proposed methodology of stacking pallet load units problem (SPLUP), is shown in Table 5. The main conclusion is that the SPLUP considered by the authors is different from typical BPP, even combined problem of bin packing and vehicle loading. The mathematical model of SPLUP includes typical BPP constraints and additional restrictions such as filling the top layer, fragility, and height homogeneity. Due to the limited number of similar works, a comparison of different approaches with the perspective of computational experiments, which would indicate the effectiveness of the proposed approach in relation to other studies, is not included. However, in principle, this was not the aim of this work at this stage of research. Finally, the verification of the proposed methodology for the SPLUP solution is limited to the experimental computations of analyzed real-world cases.

The computational experiments illustrated the correctness of the developed model as they show the potential for reducing the space occupied in logistic processes, including the necessary space on the vehicle load bed or in the racking slot of the distribution warehouse. The research focused neither on a large number of instances analyzed nor on orders in which a large number of PLUs created that need to be stacked. This will constitute the subsequent step of the research. The computation time for the analyzed instances and using evolutionary engines was no more than several seconds, except one instance with computation time of around 1 min., which constitutes the entire procedure promising.

Table 5. Comparison of methodological similarities and differences between classical bin packing problem and stacking pallet load units problem.

Comparative Parameter		Literature [2]		Presented Research [2]
The subject of decision problem	Packaging of products	All [1] exc [3,4,16]	+	-
	Packaging of pallet load units	-	—	+
	Packaging of products and vehicle loading	[3,4,16]	+	-
Type of decision problem	NP-hard type	All [1]	+	+
Dimensions of analysis	1D	[15]	+	-
	2D	[2,7]	+	+
	3D	[1–3,6,12–14]	+	-
Item shape regularity	Regular shape	All[1] exc [13]	+	+
	Irregular shape	[13]	+	-

Table 5. *Cont.*

Comparative Parameter		Literature [2]		Presented Research [2]
Spectrum of applied constraints	Weight of items	All [1]	+	+
	Height of items	All [1]	+	+
	Vertical support/Filling the layer	[11]	+	+
	Load bearing/Fragility	[11,12]	+	+
	Stability	[12]	+	-
	Height homogeneity	[12]	+	+
	Compatibility of items	[17]	+	+
	Sequencing	[11]	+	+
Dependence on time	On-line procedure	[6,10]	+	-
	Off-line procedure	All [1] exc [6,10]	+	+
Comparison of different approaches/methods	Methodological comparison	[2,6,9]	+	-
	Comparison of CPU time	[4,11,15,16]	+	-
Number of items per instance	Hard case (200−1000)	[11,15,17]	+	-
	Soft case (up to 200)	All [1] exc [11,15,16]	+	+

[1] all the reference articles, see references; [2] (+) means the comparative parameter is present, (−) otherwise.

4.2. Directions for Further Research

Based on the results obtained and the research assumptions made, the authors of this article plan to develop the research thread undertaken. The following studies are proposed:

- Conducting analytical tests for a wider range of test sets, both regarding the number of PLUs to be stacked within a single order, as well as the number of instances included in the experiments. The orders to be analyzed will also be selected so as to examine a bigger number of *top* and *base* PLUs as well as a different weight and height structure of PLUs, i.e., relatively high and light PLUs vs. low and heavy PLUs.
- Conducting a series of experiments assuming different ranges of parameters of acceptable weight and height of SPLUs built.
- Developing a procedure to support purchasing decisi ons, in which, based on the analysis of the result of the PLU stacking planning process, a possible slight modification of the order size will be indicated. The goal is to determine the benefits of eliminating or significantly reducing the number of *top* type PLUs in the planning process and thus assess the possibility of further reducing the number of SPLUs

Author Contributions: Conceptualization, P.S. and H.S.; methodology, P.S.; software, P.S.; validation, H.S.; formal analysis, H.S.; investigation, P.S.; resources, P.S.; data curation, P.S.; writing—original draft preparation, P.S. and H.S.; writing—review and editing, P.S. and H.S.; visualization, P.S.; supervision, H.S.; project administration, P.S.; funding acquisition, P.S. and H.S. All authors have read and agreed to the published version of the manuscript.

Funding: This research was funded by the Ministry of Science and Higher Education, Republic of Poland, and was performed at Poznan University of Technology, Faculty of Civil and Transport Engineering, grant number 0416/SBAD/0004.

Institutional Review Board Statement: Not applicable.

Informed Consent Statement: Not applicable.

Data Availability Statement: The data suplemental to this research is attached to the article as Appendix A, see Tables A1–A3.

Acknowledgments: The authors of the article are grateful to Frontline Systems Inc. for providing the Analytic Solver software and Solver Engines for research purposes.

Conflicts of Interest: The authors declare no conflict of interest.

Appendix A

This section contain data supplemental to the research, see Tables A1–A3, necessary to reproduce all the experiments that results are presented in Section 3 of this article.

Table A1. The data applied to the experimental part of the research on SPLUP, instances 1–5.

Instance	i	w_i	h_i	f_i	t_i	Instance	i	w_i	h_i	f_i	t_i
1	1	227	372	2	0	4	1	245	414	1	0
	2	240	392	2	0		2	347	516	3	1
	3	233	381	2	0		3	260	429	1	0
	4	240	392	2	0		4	329	498	3	0
	5	256	416	3	1		5	276	445	2	0
	6	186	311	1	0		6	291	460	2	0
	7	214	353	2	0		7	349	518	3	1
	8	219	360	2	0		8	297	466	2	0
	9	222	365	2	0	5	1	390	520	5	1
	-	-	-	-	-		2	197	357	2	0
2	1	199	331	1	0		3	345	481	5	1
	2	248	405	3	1		4	151	317	1	0
	3	197	329	1	0		5	224	379	2	0
	4	202	336	1	0		6	356	491	5	1
	5	259	421	3	1		7	297	441	4	0
	6	233	383	3	0		8	130	300	1	0
	7	196	327	1	0		9	221	377	2	0
	8	175	295	1	0		10	240	393	3	0
	9	221	364	2	0		11	244	396	3	0
	-	-	-	-	-		12	207	365	2	0
3	1	360	529	3	1		13	234	388	3	0
	2	348	517	3	1		14	355	490	5	1
	3	361	530	3	1		15	359	494	5	1
	4	282	451	2	0		16	271	420	3	0
	5	303	472	2	0		17	303	446	4	0
	6	236	405	1	0		18	363	497	5	1
	7	290	459	2	0		19	227	382	2	0
	8	297	466	2	0		20	364	498	5	1

[1] Parameters are expressed in: i (-), w_i (kg), h_i (mm), f_i (-), t_i (-).

Table A2. The data applied to the experimental part of the research on SPLUP, instance 6.

Instance	i	w_i	h_i	f_i	t_i	Instance	i	w_i	h_i	f_i	t_i
6	1	238	422	4	1	6	11	204	380	2	0
	2	244	430	4	1		12	201	377	2	0
	3	219	399	3	0		13	201	377	2	0
	4	182	353	1	0		14	211	389	2	0
	5	230	412	3	0		15	229	411	3	0
	6	208	386	2	0		16	229	411	3	0
	7	239	424	4	1		17	196	370	2	0
	8	226	408	3	0		18	234	418	3	1
	9	224	405	3	0		19	246	433	4	1
	10	222	402	3	0		20	214	393	2	0

Table A3. The data applied to the experimental part of the research on SPLUP, instances 7–10.

Instance	Parameters [1]					Instance	Parameters [1]				
	i	w_i	h_i	f_i	t_i		i	w_i	h_i	f_i	t_i
7	1	331	523	5	1	9	1	266	438	3	1
	2	197	369	1	0		2	146	358	1	0
	3	325	516	5	1		3	228	413	3	0
	4	341	533	5	1		4	129	347	1	0
	5	308	496	4	1		5	219	407	2	0
	6	178	347	1	0		6	257	432	3	1
	7	283	468	4	0		7	175	377	2	0
	8	232	410	2	0		8	211	401	2	0
	9	292	477	4	0		9	159	367	1	0
	10	190	361	1	0		10	169	373	1	0
	11	175	344	1	0		11	179	380	2	0
	12	297	483	4	0		12	138	352	1	0
	13	267	449	3	0		13	195	391	2	0
	14	237	415	2	0		14	181	381	2	0
	15	194	365	1	0		15	130	347	1	0
	16	228	405	2	0		-	-	-	-	-
	17	247	427	3	0		-	-	-	-	-
	18	300	486	4	1		-	-	-	-	-
	19	209	383	1	0	10	1	261	552	4	1
	20	288	473	4	0		2	231	463	1	0
8	1	273	442	2	1		3	225	444	1	0
	2	222	408	2	0		4	238	483	2	0
	3	152	359	1	0		5	232	467	1	0
	4	154	363	1	0		6	228	455	1	0
	5	189	386	1	0		7	235	474	2	0
	6	148	361	1	0		8	227	450	1	0
	7	216	404	2	0		9	248	513	2	0
	8	234	416	2	0		10	267	571	5	1
	9	214	403	2	0		11	226	447	1	0
	10	166	371	1	0		12	255	534	4	0
	11	166	371	1	0		13	228	454	1	0
	12	155	364	1	0		14	226	447	1	0
	13	155	364	2	0		15	267	570	5	1
	14	252	428	2	1		16	260	550	4	1
	15	212	402	2	1		17	265	565	5	1
	16	237	418	2	1		18	266	568	5	1
	17	256	431	2	1		19	265	564	5	1

References

1. Sawicki, P.; Sawicka, H. Optimisation of the Two-Tier Distribution System in Omni-Channel Environment. *Energies* **2021**, *4*, 7700. [CrossRef]
2. Dell'Amico, M.; Magnani, M. Solving a Real-Life Distributor's Pallet Loading Problem. *Math. Comput. Appl.* **2021**, *26*, 53. [CrossRef]
3. Morabito, R.; Morales, S.R.; Widmer, J.A. Loading optimization of palletized products on trucks. *Transp. Res. Part E Logist. Transp. Rev.* **2000**, *36*, 285–296. [CrossRef]
4. Alonso, M.T.; Alvarez-Valdes, R.; Parreño, F.; Tamarit, J.M. Algorithms for Pallet Building and Truck Loading in an Interdepot Transportation Problem. *Math. Probl. Eng.* **2016**, *2016*, 3264214. [CrossRef]
5. Lim, A.; Zhang, X. The Container Loading Problem. In Proceedings of the 2005 ACM Symposium on Applied Computing—SAC'05, Santa FE, NM, USA, 13–17 March 2005; ACM Press: New York, NY, USA, 2005; p. 913. [CrossRef]
6. Ali, S.; Ramos, A.G.; Carravilla, M.A.; Oliveira, J.F. On-line three-dimensional packing problems: A review of off-line and on-line solution approaches. *Comput. Ind. Eng.* **2022**, *168*, 108122. [CrossRef]
7. Young-Gun, G.; Kang, M.K. A fast algorithm for two-dimensional pallet loading problems of large size. *Eur. J. Oper. Res.* **2001**, *134*, 193–202. [CrossRef]
8. Akkaya, S.B.; Gül, A.; Coşkun, Z.; Karaman, C.; Öztop, H.; Mullaoğlu, G. The Distributor's Pallet Loading Problem: A Case Study. In *Proceedings of the International Symposium for Production Research 2018*; Durakbasa, N.M., Gencyilmaz, M.G., Eds.; Springer International Publishing: Berlin/Heidelberg, Germany, 2019; pp. 937–948. [CrossRef]
9. Silva, E.; Oliveira, J.F.; Wäscher, G. The pallet loading problem: A review of solution methods and computational experiments: The pallet loading problem: A review of solution methods and computational experiments. *Int. Trans. Oper. Res.* **2016**, *23*, 147–172. [CrossRef]

10. Lin, B.; Li, J.; Bai, R.; Qu, R.; Cui, T.; Jin, H. Identify Patterns in Online Bin Packing Problem: An Adaptive Pattern-Based Algorithm. *Symmetry* **2022**, *14*, 1301. [CrossRef]
11. Gzara, F.; Elhedhli, S.; Yildiz, B.C. The Pallet Loading Problem: Three-dimensional Bin Packing with Practical Constraints. *Eur. J. Oper. Res.* **2020**, *287*, 1062–1074. [CrossRef]
12. Tresca, G.; Cavone, G.; Carli, R.; Cerviotti, A.; Dotoli, M. Automating Bin Packing: A Layer Building Matheuristics for Cost Effective Logistics. *IEEE Trans. Autom. Sci. Eng.* **2022**, *19*, 1599–1613. [CrossRef]
13. Zuo, Q.; Liu, X.; Chan, W.K.V. A Constructive Heuristic Algorithm for 3D Bin Packing of Irregular Shaped Items. *Comp. Sci. Computl. Geom.* **2022**. [CrossRef]
14. Elhedhli, S.; Gzara, F.; Yildiz, B. Three-Dimensional Bin Packing and Mixed-Case Palletization. *Inf. J. Optim.* **2019**, *1*, 323–352. [CrossRef]
15. El-Ashmawi, W.H.; Elminaam, D.S.A. A modified squirrel search algorithm absed on improved best fit heuristic and operator strategy for bin packing problem. *Appl. Soft. Comp.* **2019**, *82*, 105565. [CrossRef]
16. Moura, A.; Bortfeldt, A. A two-stage packing problem procedure. *Int. Trans. Oper. Res.* **2017**, *24*, 43–58. [CrossRef]
17. Moura Santos, L.F.O.; Iwayama, R.S.; Cavalcanti, L.B.; Turi, L.M.; de Souza Morais, F.E.; Mormilho, G.; Cunha, C.B. A variable neighborhood search algorithm for the bin packing problem with compatible categories. *Expert Syst. Appl.* **2019**, *124*, 209–225. [CrossRef]

Article

Forecasting Steel Production in the World—Assessments Based on Shallow and Deep Neural Networks

Balduíno César Mateus [1,2,*], **Mateus Mendes** [3,4], **José Torres Farinha** [4,5], **António J. Marques Cardoso** [2], **Rui Assis** [1] **and Lucélio M. da Costa** [6]

[1] EIGeS—Research Centre in Industrial Engineering, Management and Sustainability, Lusófona University, Campo Grande, 376, 1749-024 Lisboa, Portugal
[2] CISE—Electromechatronic Systems Research Centre, University of Beira Interior, Calçada Fonte do Lameiro, 6201-001 Covilhã, Portugal
[3] Instituto Superior de Engenharia de Coimbra, Polytechnic of Coimbra, 3045-093 Coimbra, Portugal
[4] Institute of Systems and Robotics, University of Coimbra, 3004-531 Coimbra, Portugal
[5] Centre for Mechanical Engineering, Materials and Processes—CEMMPRE, University of Coimbra, 3030-788 Coimbra, Portugal
[6] Department of Electrical Engineering, University of Coimbra, 3030-788 Coimbra, Portugal
* Correspondence: balduino.mateus@ubi.pt

Abstract: Forecasting algorithms have been used to support decision making in companies, and it is necessary to apply approaches that facilitate a good forecasting result. The present paper describes assessments based on a combination of different neural network models, tested to forecast steel production in the world. The main goal is to find the best machine learning model that fits the steel production data in the world to make a forecast for a nine-year period. The study is important for understanding the behavior of the models and sensitivity to hyperparameters of convolutional LSTM and GRU recurrent neural networks. The results show that for long-term prediction, the GRU model is easier to train and provides better results. The article contributes to the validation of the use of other variables that are correlated with the steel production variable, thus increasing forecast accuracy.

Keywords: steel production; time series; neural networks; CNN; MLP; LSTM; GRU

Citation: Mateus, B.C.; Mendes, M.; Farinha, J.T.; Cardoso, A.J.M.; Assis, R.; da Costa, L.M. Forecasting Steel Production in the World—Assessments Based on Shallow and Deep Neural Networks. *Appl. Sci.* **2023**, *13*, 178. https://doi.org/10.3390/app13010178

Academic Editor: Panagiotis Tsarouhas

Received: 20 November 2022
Revised: 12 December 2022
Accepted: 16 December 2022
Published: 23 December 2022

1. Introduction

Globalization increasingly demonstrates the importance of predicting different phenomena that may occur and have huge economic and social implications. As a result, companies are increasingly more interested in technological solutions, including forecasting algorithms, which allow them to anticipate scenarios and support decisions.

There is a relationship between energy intensity and production costs in the steel industry [1]. The impact of investments on the energy intensity of the steel industry is extended over time.

The steel industry is a strategic sector for industrialized economies. Because it is demanding in terms of capital and energy, companies have consistently emphasized technological advances in the production process to increase productivity [2]. The steel industry is a key industrial sector in the modern world. It consists of the economical agents that perform the processes for obtaining steel-based products [3,4]. Those products must have characteristics of safety and quality according to the standards defined for the consumer market [5].

Good decision making helps the companies produce high-quality products, innovate, fulfill customers' needs, and grow.

As the companies grow, the countries' economies also grow [6], and more social needs are satisfied. As a result, market players and countries have been concerned with long-term policies that lead to sustainability in the long run. Governments have proposed solutions that help companies that follow the best practices, including activities such as recycling and energy management, in a perspective of a circular economy. That is increasingly more important, and contrasts with obsolete managing practices of massive exploitation of natural resources [7–9]. Hence, in order to implement the best managing practices, the best management tools are required. That includes tools which provide information about the past, present, and expected future values of key indicators and variables. Statistical and machine learning forecasting models have been applied with success to fulfill the need to anticipate future scenarios, based on data of the present and past [10–12].

The present work uses a public dataset of steel production in the world. The data are used to create and train different predictive models and forecast steel production in the coming years. The models evaluated are machine learning methods based on shallow and deep neural networks.

Global world data were chosen because nowadays the world is more and more integrated and the world steel production gives a strong indication of how the economy evolves [13]. The data necessary for the analysis are publicly available.

The objectives and contributions of this paper are as follows:

- to identify the models' behavior and performance, comparing the different methods with a number of different parameters;
- to forecast steel production in the world in the short and long term using neural networks;
- to evaluate the prediction performance of recurrent neural networks and convolutional neural networks;
- to evaluate the robustness of the preceding approaches to forecast nonstationary time series; and
- to contribute to validate the robustness of the model by using other variables correlated with the steel production ' main variables.

To analyze the use of forecasting methods, a review of bibliographic databases was performed by using the research databases Web of Science, Scopus, and Google Scholar. The keywords used were "steel production", "time series", and "neural networks".

Figure 1 shows the result after searching the topic under study, which shows that the databases returned four articles in total. They focus on the following issues: hybrid static-sensory data modeling for prediction tasks in the basic oxygen furnace process [14]; temperature prediction for a reheating furnace by the closed recursive unit approach [15]; detecting and locating patterns in time series using machine learning [16]; and deep learning for blast furnaces [17].

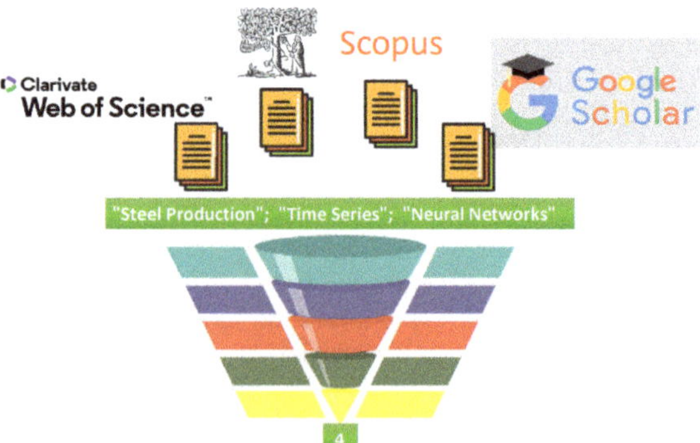

Figure 1. Results of the searches in the scientific articles databases.

Other studies found focus on steel forecasting in China [18], in Japan [19], and in Poland [20]. By using other methodologies, there is also a long-term scenario forecast focusing on energy consumption and emissions for the Chinese steel industry [21].

The paper is structured as follows. In Section 2, a global overview of the steel production around the world is given. In Section 3, the concepts of artificial intelligence methods used are presented. In Section 4, a study to evaluate and validate the forecasting models, is presented. In Section 5 the results are discussed. Finally, in Section 6, some conclusions are drawn and future work is highlighted.

2. Steel Production around the World

There are many different types of steel. Each type has specific characteristics, such as chemical composition, heat treatments, and mechanical properties. The metals are produced according to the market needs and demands. That means supply must be adjusted according to the requirements of demand. Specific applications have appeared in the market and they require specific types of steel [22].

It is possible to identify more than 3500 different types of steels. About 75% of them have been developed in the last 20 years. This shows the great evolution that the sector has experienced [23]. Figure 2 shows that most of the steel produced worldwide is consumed by the civil construction sector, which uses 52% of the material. This is due to the demand for bigger and better buildings, which use increasingly more steel. As countries evolve, more and more buildings and infrastructure are needed, and steel construction is also increasingly more popular for houses and warehouses, due to the small amount of labour required. Next to this sector, more than 16% of the steel produced worldwide is used in mechanical equipment, and 12% is used in the automotive industry.

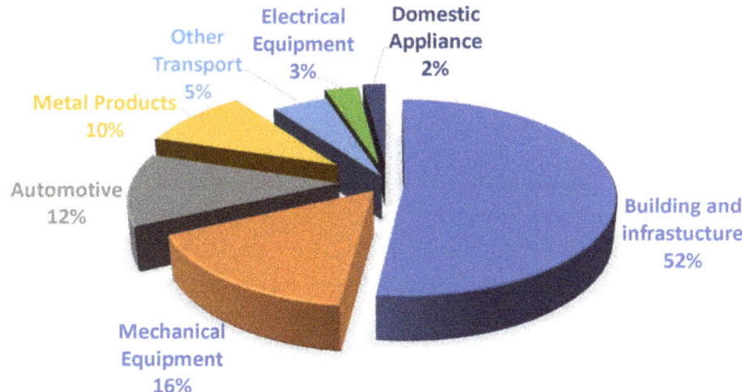

Figure 2. Annual crude steel production, per industrial consumer sector; Graph created from data taken from source [24].

The World Steel Association report states that Asia produced 1374.9 Mt of crude steel in 2020, an increase of 1.5% compared to 2019. China, in 2020, reached 1053.0 Mt, an increase of 5.2% compared to 2019. Figure 3 shows China's share of global crude steel production increased from 53.3% in 2019 to 56.5% in 2020. In 2019, China's apparent consumption of crude steel was about 940 million tons. India's production was 99.6 million tons in 2020, down 10.6% from 2019 [25].

Japan produced 83.2 Mt of crude steel in 2020, down 16.2% in 2019. South Korea produced 67.1 Mt, down 6.0% in 2019. The EU produced 138.8 Mt of crude steel in 2020, a reduction of 11.8% compared to 2019.

Germany produced 35.7 Mt of crude steel in 2020, a decrease of 10.0% compared to 2019. In short, global crude steel production reached 1864.0 million tons (Mt) in the year 2020, a decrease of 0.9% compared to 2019.

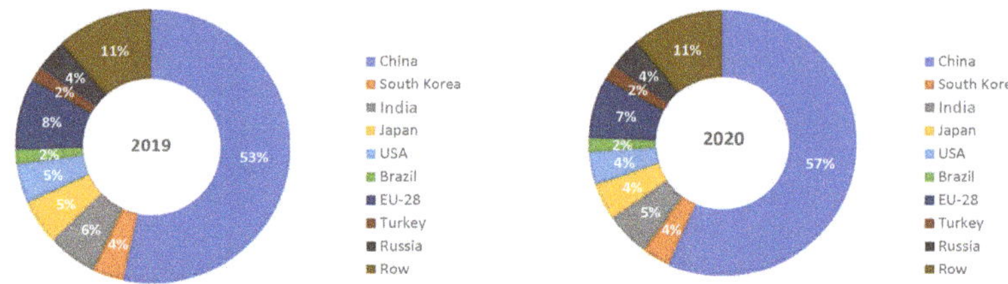

Figure 3. Annual crude steel production for 2019 and 2020; adapted from [25].

The 2009 economic crisis led to a market recession in industrial activity and the corresponding demand for steel, which remains 27% below precrisis levels. As a result, several production sites have been closed or their production has decreased, with consequeces in unemployment at the European level in recent years. Approximately 40,000 jobs were lost. Therefore, the pressure on this industry to restructure and reduce production capacity will continue to be one of the main challenges for the European industry in the short and midterm [26].

In the time interval from 1970 to 2012, the results of the same study indicate that the consumption of steel per capita in Europe fell down up to 50%, due to the development of new materials and the corresponding replacement in some sectors. An important example is the automotive industry, wherein many parts of automobiles were replaced by lighter and cheaper parts made of plastic, aluminium, or other synthetic materials [27].

According to studies by [28], the steel demand in most industries will peak before 2025. The total steel demand has increased from 600 Mt in 2010 to 702 Mt in 2015 and will increase to 753 Mt in 2025. From then on, gradually, it is expected to decrease to approximately 510 Mt in 2050.

Total steel demand will only decrease by 8% in 2030 when the average useful life of buildings increases by 30%. However, this influence becomes very obvious after 2030, because a 23% reduction in steel demand is expected to happen by 2050 [28].

Because of the need to plan production in advance, many different methods were proposed with the objective of forecasting future demand and production [29–31].

3. Artificial Intelligence Predictive Models

Generally speaking, artificial intelligence (AI) is the knowledge field that is concerned with the development of techniques that allow computers to act in a way that looks like an intelligent organism, the most important model being a human brain [32]. According to [33], AI can be defined as the computer simulation of the human thought process. AI techniques include expert systems, reasoning based on fuzzy logic (FL) and artificial neural networks (ANN), among several other tools.

As technology evolves, more computation power becomes available to use algorithms and tools that have not been deployed in the past due to the lack of resources [34]. Nowadays, industries take advantage of some technological tools that provide them with significant advantages, such as state of the art non-linear machine learning (ML) methods, including evolutionary algorithms, neural networks, and other artificial intelligence (AI) techniques. Those methods have been considered fundamental to achieve informed and automated decision making based on big data and AI algorithms [35–37]. Big data analysis, AI, and ML, applied to the Internet of things (IoT), allow for real-time predictions of manufacturing equipment, making it possible to predict many equipment faults before they happen. Therefore, it is possible to launch a work order before the fault occurs, effectively preventing it from happening at all. This allows one to plan the maintenance procedures and resources, like the technicians and spares [38–42], some time in advance, which facilitates optimization of human and material resources.

Neural networks are one of the most popular AI techniques for performing tasks such as classification, object detection and prediction. They have been successfully applied to solve many maintenance and quality-control problems, such as detection of defects in power systems, as is mentioned by [43,44]. Lippmann [45] has used ANN for pattern recognition due to its ability to generalize and respond to unexpected patterns.

3.1. Multilayer Perceptron

Multilayer perceptron neural networks (MLP) are a type of feedforward neural network. They are generally used due to their fast operation, ease of implementation, and training set requirements [46–48]. MLP's have been successfully applied to a wide range of problems, such as complex and nonlinear modeling, classification, prediction, and pattern recognition.

In the MLP, each artificial neuron, receives data from the network input or from other neurons, and produces an output which is a function of the input values and the neuron's internal training values.

An MLP model with an insufficient or excessive number of neurons in the hidden layer, for the type of problem being solved, is likely to have problems learning. Convergence is difficult and that leads to time-consuming adjustments. Too few neurons cannot retain enough data and there is underfitting. Too many neurons learn the problem instead of generalizing, and there is overfitting of the data. As yet, there is no analytical method by which to determine the right number of neurons in the hidden layer, only practical recommendations based on experience. In general, a rule is to put a hidden layer with a number of neurons that is the average between the input and the output layer. This number is then optimized through trial-and-error experiments [49,50].

Usually, the selection of the training algorithm is dependent on factors such as the organization of the network, the number of hidden layers, weights and biases, learning error rate, etc. Error gradient-based algorithms are common. They work by adjusting the weights in function of the gradient of the error between the output desired and the output obtained. Slow convergence and high dependence on initial parameters, and the tendency to get stuck in local minima are the limitations of gradient-based algorithms [51].

For a given neuron j, each neural input signal (x_k) is multiplied by their respective corresponding weight values (w_{kj}), and the resulting products are added to generate a total weight in the form of $w_{j1}x_1 + w_{j2}x_2 + \cdots + w_{jm}x_m$. The sum of the weighted inputs and the bias Equation (1) forms the input for the activation function, φ. The activation function φ processes this sum and provides the output, U_j neuron output, according to Equation (2) and Figure 4 [52,53]. We have

$$S_j = \sum_{k=1}^{m_j} w_{jk} \times x_k j + b_j \tag{1}$$

$$U_j = \varphi(S_j) = \varphi(\sum_{k=1}^{m_j} w_{jk} \times x_k j + b_j). \tag{2}$$

The symbol w_{ji}^H denotes the synaptic weight between the neuron leaving the hidden layer and the neuron entering the output layer: the symbol b_j^H denotes the neuron bias in the hidden layer; the superscript O is the output layer. In the figure, the green circles indicate biases, which are constants corresponding to the intersection in the conventional regression model [54].

This type of network can be trained by using many different algorithms. Backpropagation is possibly the most common. Other approaches include resilient backpropagation (RPROP) with or without weight backtracking, described in [55], or the globally convergent version (GRPROP) [56].

MLP neural networks are supported by many popular machine learning frameworks. In the Python sklearn library, they are implemented by MLPClassifier and MLPRegressor functions [57].

In the deep learning framework Tensorflow, the function neuralnet implements a sequential model with dense layers [58].

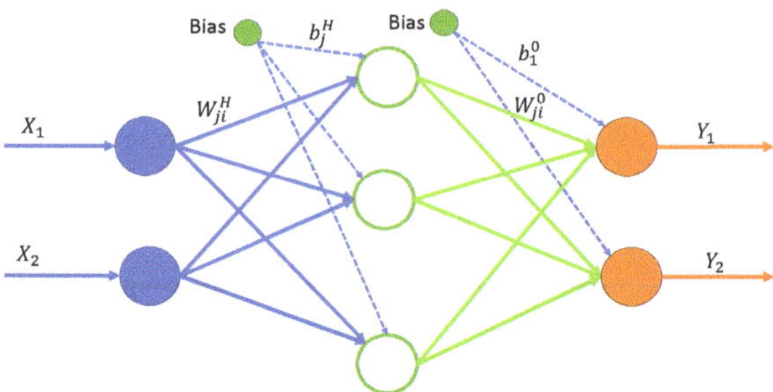

Figure 4. Representation of a multilayer perceptron, with input X, one hidden layer, and output Y.

3.2. Convolutional Neural Networks

This type of neural network is specially adequate to process images. They have outstanding results in object detection. Recently, they have also been used in other fields, especially in combination with other neural networks [59–62]. The first aplication of a CNN dates back from 1998, by Yann LeCunn. Since then, there have been many contributions in order to make those networks more powerful and faster [63]. The operation of a CNN has the principle of filtering features such as lines, curves, and edges in an image. With each added layer, the filtered features form a more complex image of selected details [64]. This neural network has the ability to process features and objects with good accuracy, depending on the type of architecture applied [65–69]. Therefore, it may also be used to process other types of data, in addition to images.

3.3. Recorrent Neural Networks

Recurrent, or feedback, networks have their outputs fed back as input signals. This feedback works as a kind of state memory, thus making the networks very suitable to use to process time-varying systems. The signal travels over the network in two directions, having dynamic memory and the ability to represent states in dynamic systems. The present paper particularly focuses on recurrent neural networks known as long short-term memory (LSTM), whose representation is shown in the Figure 5 and gated recurrent unit (GRU).

Hochreiter and Schmidhuber [70] proposed the LSTM cell, which is a popular recurrent model. The LSTM has a capacity to remember, and is controlled by introducing a "gate" into the cell. Since this pioneering work, LSTMs were made popular and used by many researchers [71,72]. According to [73], the internal calculation formula of the LSTM cell is defined as follows:

$$f_t = \sigma(x_t W_f + h_{t-1} U_f + b_f) \tag{3}$$

$$i_t = \sigma(x_t W_i + h_{t-1} U_i + b_i) \tag{4}$$

$$o_t = \sigma(x_t W_o + h_{t-1} U_o + b_o) \tag{5}$$

$$\tilde{C}_t = tan[(x_t W_C + h_{t-1} U_c + b_c] \tag{6}$$

$$C_t = \sigma(f_t \times C_{t-1} + i_t \times \tilde{C}_t) \tag{7}$$

$$h_t = tanh(C_t) \times o_t, \tag{8}$$

where t is timestep, x_t is input to the current t, W_f is weight associated with the input, h_{t-1} is the hidden state of the previous timestamp, U_f is the weight matrix associated with hidden state, W_i is weight matrix of input, U_i is weight matrix of input associated with hidden state, C_t is memory from current block, and h_t output of current block. The f_t is the forget gate, i_t is input gate, o_t is the output gate, C_t is new cell state, $\tilde{C}_t$ is values generated by tanh with all the possible values between -1 and 1, and b_f, b_i, b_o, b_c are the bias terms.

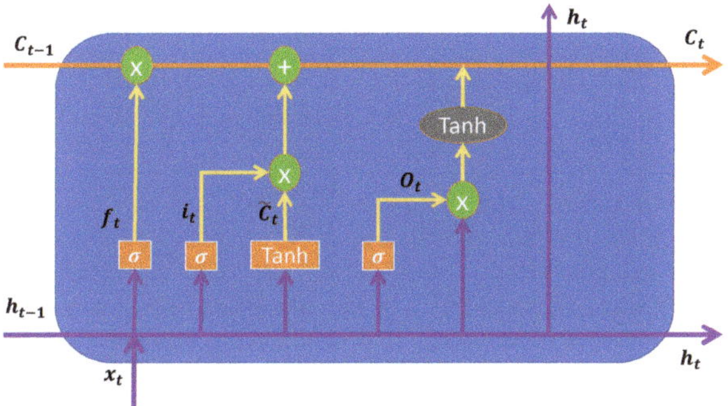

Figure 5. Architecture of an LSTM cell; adapted from [73–75].

The learning capacity of the LSTM cell is superior to that of the standard recurrent cell. However, the additional parameters add to the computational load. Cho et al. [76] introduced the gated recurring unit (GRU) model, which is a type of modified version of the LSTM.

Figure 6 shows the internal architecture of a GRU unit cell. These are the mathematical functions used to control the locking mechanism in the GRU cell:

$$z_t = \sigma(x_t W^z + h_{t-1} U^z + b_z) \tag{9}$$

$$r_t = \sigma(x_t W^r + h_{t-1} U^r + b_r) \tag{10}$$

$$\tilde{h}_t = tan(x_t W_h + (r_t \times h_{t-1}) U_h + b)_h \tag{11}$$

$$h_t = (1 - z_t) \times \tilde{h}_t + z_t \times h_{t-1}, \tag{12}$$

where x_t is input vector, h_t is output vector, $\hat{h}_t$ is candidate activation vector, z_t is update gate vector, r_t is reset gate vector and, W^z, W^r, W^h denote the weight matrices for the corresponding connected input vector. U^z, U^r, U represent the weight matrices of the previous time step, and b_r, b_z, b_h are biases.

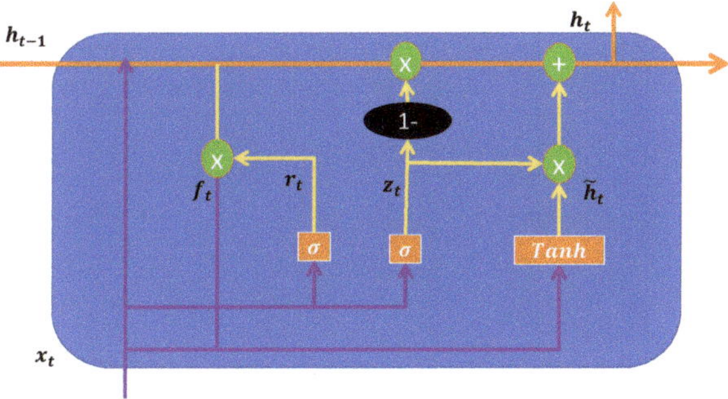

Figure 6. Architecture of a GRU unit; adapted from [76,77].

3.4. Model Evaluation

For evaluation of the performance of the models, the metrics used are the root mean square error (RMSE) and the mean absolute percentage error (MAPE). RMSE is calculated as in Formula (13) and MAPE is calculated as in Formula (14).

$$\text{RMSE} = \sqrt{\frac{1}{n}\sum_{t=1}^{n}(Y_t - \hat{Y})^2}. \tag{13}$$

Y_t is the actual value, and $\hat{Y}$ is the predicted value obtained from the models. n is the number of samples in the set being evaluated. The RMSE is a type of absolute error, whereas MAPE is an error which is relative to the variable being predicted. The MAPE can only be calculated if the zeros are removed from the dataset. We have

$$\text{MAPE} = \frac{1}{n}\sum_{t=1}^{n}\frac{|Y_t - \hat{Y}_t|}{|Y_t|}. \tag{14}$$

4. Empirical Examination

4.1. Data Analysis

The present study was carried out by using a dataset consisting of samples available on the World Steel website (https://www.worldsteel.org/steel-by-topic/statistics/steel-statistical-yearbook.html (accessed on 7 June 2022)).

Those samples represent the steel production, per year, in millions of metric tons, in the world, from 1900 to 2021. The values are plotted in Figure 7. Table 1 shows a summary of statistical parameters of the variable. The values show growth over time, as mentioned above, reflecting growth of some sectors such as transport and civil construction. As the chart shows, the production has been small and with a small growth rate in the beginning of the century. After 1945, the growth accelerates, until 1970. Then there is a a period of no growth, followed by a decline and another period of stagnation. After 1995 there is a sharp increase, where the growth accelerated very quickly. The variable is particularly difficult to predict, because the periods described above are clearly distinct.

Table 1. Characteristics of the variable annual crude steel production in the world. std is the standard deviation, Q_1 is the 25% percentile, Q_2 is the 50% percentile, Q_3 is the 75% percentile. Max is the maximum value, Min the minimum value and Mean the average value.

Mean	Std	Min.	Max	Q_1	Q_2	Q_3
4.999142×10^5	5.081208×10^5	2.830000×10^4	1.953304×10^6	9.382500×10^4	3.488500×10^5	6.755030×10^5

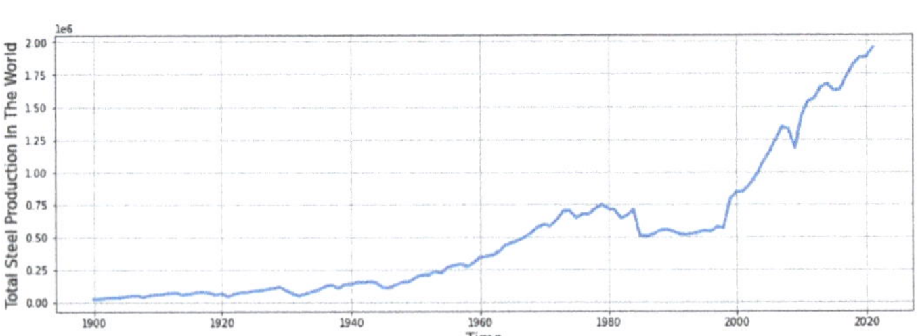

Figure 7. Annual crude steel production in the world, in millions of metric tons.

4.2. Experiments

The experiments performed consist of testing different model architectures and hyperparameters, in order to find the best predictive models. All expriments used the same sequence of actions present in Figure 8.

The data has a sampling rate per year; for all three variables the sample set has a size of 72 samples. In this phase the data were adjusted in order to have the same sample size in the same time period.

The dataset was divided in two parts, one for training and the other for testing. For the first test, the train subset is 70% of the data, the test subset is the remaining 30%. The second test where use the architecture encoder–decoder was used in the train subset is 80% and the test subset is 20%. The training process was allowed to proceed for up to 2000 epochs.

The experiments were performed in order to understand the impact of varying the size of the window of past samples, as well as the the number of years in advance. Naturally, the larger the number of years ahead that is being predicted (gap), the larger the error expected. The larger the size of the historical window used, the smaller the error expected, up to a certain size. As mentioned above, the evolution of the variable has distinct patterns over the years. Therefore, too many samples from the past may carry patterns that are not applicable to the future.

Figures 9 and 10 show a summary of the best results obtained for different neural network models, namely MLP, CNN-MLP, LSTM, CNN-LSTM, GRU, and CNN-GRU. The window size varied between 4 and 16 years. The gap was varied between 1 and 9 years. Table 2 summarizes the architectures that presented the best performance.

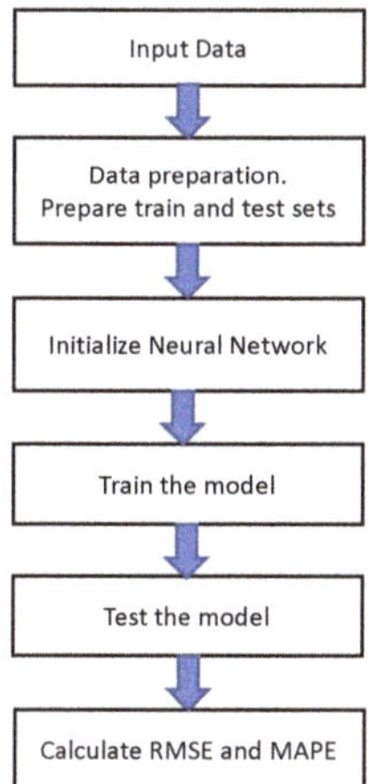

Figure 8. Procedure used to train and test the models to predict steel production in the world.

Table 2. Predictive models with the best results.

Model		
Architecture	**Unit**	**Activation Function**
CNNLSTM	CNN = 64, LSTM = 50	Relu
MLP	MLP = 50	Relu
GRU encoder decoder	GRU = 200	Relu

Figures 9 and 10 shows the RMSE and MAPE errors for different window sizes and gaps. As the Figure 11 show, the CNN LSTM and the MLP offers lower errors for wider gaps, even with small windows.

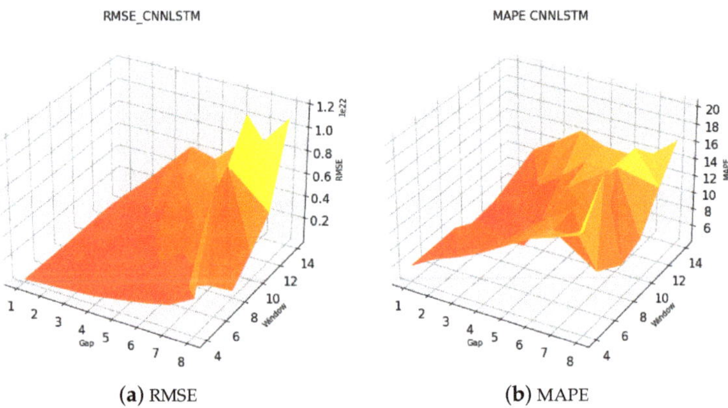

(**a**) RMSE

(**b**) MAPE

Figure 9. RMSE and MAPE for different historical window sizes (number of past samples) and gaps (years ahead), from CNNLSTM model.

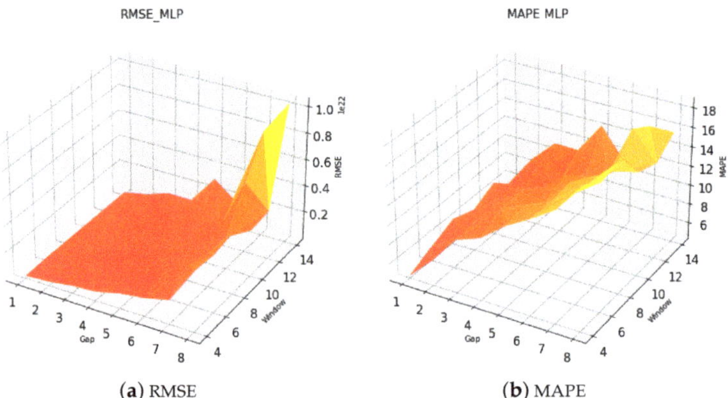

(**a**) RMSE

(**b**) MAPE

Figure 10. RMSE and MAPE for different windows (number of past samples) and gaps (number of years ahead) from the MLP model.

Table 3 shows the description of the variables, as they present different magnitudes. Because we do not intend to have this type of problem, we performed the normalization of all variables as shown in Figure 12.

Table 3. Description of the variables steel production, producer price index by commodity, and world population.

	Steel Production	**Iron and Steel**	**World Population**
count	7.200000	72.000000	7.200000
mean	7.863977×10^5	107.355084	5.008493×10^9
std	4.856869×10^5	77.016810	1.655973×10^9
min	1.916000×10^5	19.058333	2.499322×10^9
25%	5.038660×10^5	29.931250	3.528970×10^9
50%	6.130825×10^5	104.683333	4.905897×10^9
75%	9.953132×10^5	137.208333	6.414362×10^9
max	1.953304×10^6	356.707750	7.909295×10^9

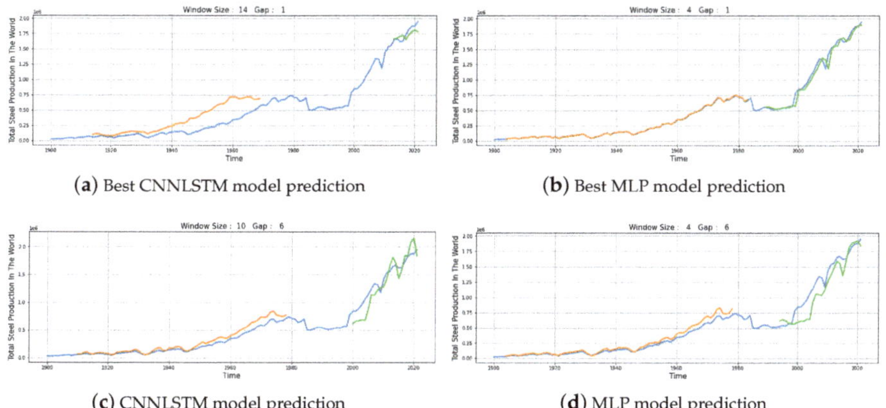

(**a**) Best CNNLSTM model prediction

(**b**) Best MLP model prediction

(**c**) CNNLSTM model prediction

(**d**) MLP model prediction

Figure 11. Predictions of the best forecast models based on CNNLSTM and MLP models, with and without the convolutional layer. The blue line is the actual value. The orange line is the prediction based on the training set, and the gray line is the prediction based on the test set. The predictions were obtained with window width 4 and gap 1.

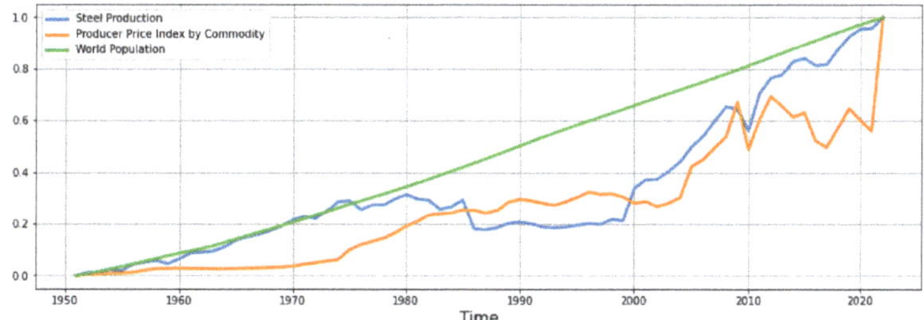

Figure 12. Steel production, producer price index by commodity, and world population in format time series.

The models performed well for both short- and long-term forecasts. Models with encoder–decoder architecture were also tested in order to improve the steel production forecast results by using the market variables and their derivatives that affect the steel production variable (producer price index by commodity https://fred.stlouisfed.org/series/PPIACO (accessed on 14 June 2022), world population https://www.macrotrends.net/countries/WLD/world/population (accessed on 14 June 2022)) as shown in Figure 12.

As can be seen from Figure 13, the variables have a good correlation. In the case of the derivatives, the correlations are weak, and the derivative that has a reasonable correlation in relation to steel production is the world population.

By using the same architecture with GRU units, we conducted two tests, the first consisting of adding one variable at a time to the model until reaching the six variables with their derivatives. With this, the results of Figure 14 achieved a good prediction of the variable under study with the input of the three variables as is shown in the graph.

The second test had the goal of testing the neural network, having as input two variables in the network. One of them is fixed (steel production) and the other is variable. As can be seen in Figure 15, the variables steel production and world population, or steel production and world population derivative, are the ones that present the best results.

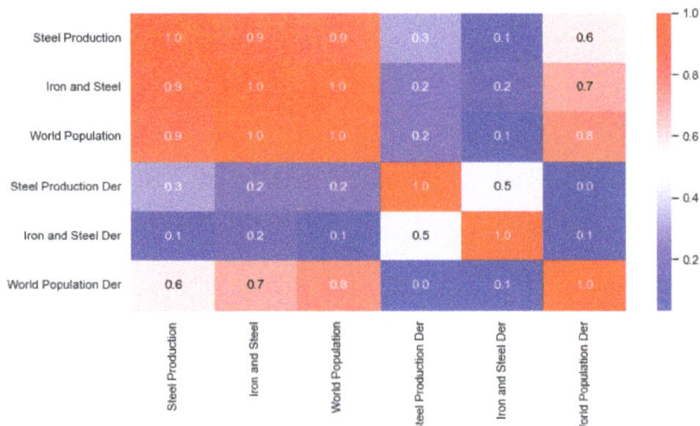

Figure 13. Correlation of the variables steel production, producer price index by commodity, world population, and their first derivates in time series.

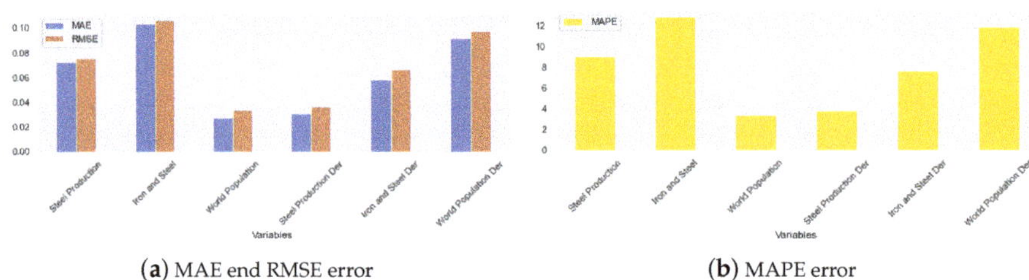

(**a**) MAE end RMSE error (**b**) MAPE error

Figure 14. The test error one, used the same architecture with GRU unit.

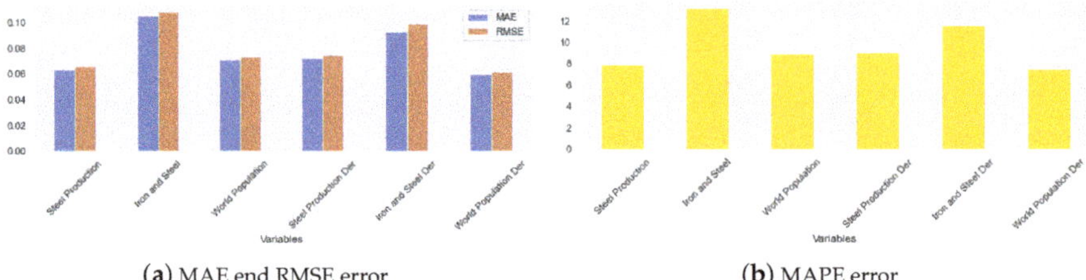

(**a**) MAE end RMSE error (**b**) MAPE error

Figure 15. The test error two, used the same architecture with GRU unit.

5. Discussion

Artificial intelligence is the field of knowledge concerned with developing techniques that allow computers to act in a way that looks like an intelligent organism. Table 4 shows a summary of MAPE errors for the models tested, for different gaps (1 to 9 years ahead) and for different historical windows (4–14 samples). The table shows that the models, especially the ones with higher sliding windows and a convolutional layer, present a better MAPE. The convolutional layer contributes to improve convergence during learning.

The table also shows that for all models, the greater the value of the gap, the larger the errors, which is an expected result. In general it is possible to obtain errors of about 5% for the next year, but for an advance of three years and beyond, the errors less than 10% are very rare.

Table 4. MAPE errors of tests with different historical windows (4–14 samples) and gaps (years ahead) ranging from 1 to 8.

Window	Gap	MLP	LSTM	GRU	CNNMLP	CNNLSTM	CNNGRU
	1	**4.54**	5.18	4.67	4.75	5.9	5.59
	2	6.93	7.5	7.99	7.29	7.57	6.93
	3	9.84	10.12	9.66	9.0	8.67	9.67
	4	10.77	11.91	11.55	10.82	10.06	10.49
4	5	12.7	13.97	12.76	13.06	11.74	12.09
	6	14.15	15.24	13.44	13.05	13.36	12.84
	7	15.77	16.76	16.24	15.09	14.44	15.74
	8	18.78	18.08	18.23	17.35	15.96	19.54
	1	5.65	4.86	4.85	**4.73**	5.56	6.89
	2	9.42	8.94	7.71	7.87	9.16	9.12
	3	10.04	10.38	9.17	9.68	10.76	11.18
	4	11.97	11.46	10.49	10.55	9.43	11.49
6	5	12.97	12.5	12.1	10.84	11.27	11.13
	6	14.03	14.17	14.21	14.52	12.23	10.55
	7	15.71	15.82	16.0	14.01	12.42	11.81
	8	17.46	16.81	18.23	16.24	20.43	15.6
	1	5.85	5.11	5.23	**4.59**	4.85	5.52
	2	8.62	9.54	7.18	6.82	6.21	7.16
	3	10.03	10.29	9.71	10.32	7.3	7.58
	4	11.89	11.06	10.63	11.71	6.71	7.63
8	5	11.62	14.9	12.2	11.26	10.79	8.71
	6	14.37	14.39	13.84	12.03	12.55	15.48
	7	15.46	14.63	15.3	18.41	6.22	9.3
	8	17.88	15.46	17.13	15.17	7.88	6.35
	1	5.61	**5.02**	5.09	5.46	5.06	5.41
	2	9.74	14.0	7.73	7.03	6.59	7.16
	3	9.66	10.88	9.44	9.96	7.35	8.11
	4	10.67	11.78	12.35	10.19	7.99	11.23
10	5	12.6	14.67	13.81	10.96	12.42	12.07
	6	13.11	13.78	13.58	10.96	9.72	16.83
	7	14.43	15.57	14.47	13.0	16.32	11.52
	8	15.05	16.14	17.25	16.15	8.81	**4.82**
	1	**5.28**	7.33	5.74	5.59	5.42	5.66
	2	7.84	10.03	8.77	8.48	5.61	9.4
	3	10.2	11.14	10.36	10.66	9.13	8.96
	4	9.56	13.64	12.02	9.71	11.11	11.14
12	5	9.92	13.13	12.01	11.98	13.51	14.32
	6	11.69	13.66	12.17	10.89	13.35	18.05
	7	17.06	17.13	15.76	12.62	16.6	27.02
	8	14.25	14.22	16.95	18.39	12.83	23.63
	1	5.69	5.71	5.5	6.13	**4.31**	6.31
	2	8.89	12.5	8.65	7.99	10.84	10.35
	3	11.3	11.88	8.86	9.06	12.64	14.12
	4	11.25	12.83	11.89	9.26	14.58	13.7
14	5	14.51	10.34	12.58	10.59	13.47	15.48
	6	11.35	14.93	9.6	10.02	13.37	13.88
	7	15.73	14.47	12.37	12.4	14.09	15.97
	8	15.77	11.62	13.67	13.47	16.4	30.38

By inputting other variables that have a significant correlation with the steel production variables, it was found that these and their derivatives can have an impact on the predicted values.

The results also show that the size of the historical sliding window is important, but larger windows are not always better than smaller windows. A window of 10 samples offers some of the best results when the convolution layer is used, for the LSTM and GRU models. The convolutional layer seems to offer some stability in the LSTM and GRU models for midterm predictions, when historical windows of 10–12 samples are used.

6. Conclusions

Production management is the process or activity by which resources flow within a defined system, and are grouped and transformed in a controlled manner to add value, according to the policies defined by a company.

As presented in the literature review, steel plays an active role in human activities. Figure 7 shows the amount of steel produced annually around the world. Because steel is a finite and important natural resource, monitoring it is a key strategic issue for companies and others; hence the importance of forecasting so that decisions can be made in advance.

A reliable forecasting method can be a key asset to anticipate good and bad periods, giving the management opportunity to react and take measures in time and possibly avoid losses.

The present paper demonstrates the behavior of different predictive models, particularly the combination of convolutional layers for the MLP, LSTM, and GRU. The results show that each model presents a particularity regarding the delay windows and the advance windows.

Table 4 shows that for the delay window equal to 4 and 12 the MLP model shows better accuracy. For the delay window equal to 6 and 8, it is the CNNMLP model that shows better accuracy. For the delay window equal to 10, it is the LSTM model that shows better accuracy, and for the delay window equal to 14, it is the CNNLSTM model that shows better accuracy.

The results of the encoder–decoder model presented the best result, although it turned out that these results had a great influence on the input variables of the model. The results also show that it is possible to improve the model if information is added that correlates in some way with the variable being predicted. It should also be noted that much of this information is included in the variables' derivatives, so differencing can greatly reduce prediction errors.

Future work includes experiments to combine other dependent variables, such as GDP growth, to improve predictions, as well as apply the same models to predict the production of other commodities.

Author Contributions: Conceptualization, J.T.F., A.J.M.C. and M.M.; methodology, J.T.F. and M.M.; software, B.C.M. and M.M.; validation, B.C.M. and M.M.; formal analysis, J.T.F. and M.M.; investigation, B.C.M. and M.M.; resources, J.T.F. and A.J.M.C.; writing—original draft preparation, B.C.M.; writing—review and editing, J.T.F., M.M., R.A. and L.M.d.C.; project administration, J.T.F. and A.J.M.C.; funding acquisition, J.T.F. and A.J.M.C. All authors have read and agreed to the published version of the manuscript.

Funding: This work was supported by the COFAC/EIGeS.

Data Availability Statement: The data that support the findings of this study are openly available in [World Steel website] at [https://www.worldsteel.org/steel-by-topic/statistics/steel-statistical-yearbook.html].

Acknowledgments: The authors thanks to COFAC by the funding and all support to the development of this research.

Nomenclature/Notation

The following abbreviations are used in this manuscript:

AI	Artificial Intelligence
ANN	Artificial Neural Networks
CNN	Convolution Neural Networks
GRU	Gated Recurrent Units

IoT	Internet of Things
LSTM	Long Short-Term Memory
MLP	Multilayer Perceptron Neural Networks
MAPE	Mean Absolute Percentage Error
RMSE	Root Mean Square Error
RNN	Recurrent Neural Networks

References

1. Gajdzik, B.; Sroka, W.; Vveinhardt, J. Energy Intensity of Steel Manufactured Utilising EAF Technology as a Function of Investments Made: The Case of the Steel Industry in Poland. *Energies* **2021**, *14*, 5152. [CrossRef]
2. Tang, L.; Liu, J.; Rong, A.; Yang, Z. A review of planning and scheduling systems and methods for integrated steel production. *Eur. J. Oper. Res.* **2018**, *133*, 1–20. [CrossRef]
3. Pei, M.; Petäjäniemi, M.; Regnell, A.; Wijk, O. Toward a Fossil Free Future with HYBRIT: Development of Iron and Steelmaking Technology in Sweden and Finland. *Metals* **2020**, *10*, 972. [CrossRef]
4. Liu, W.; Zuo, H.; Wang, J.; Xue, Q.; Ren, B.; Yang, F. The production and application of hydrogen in steel industry. *Int. J. Hydrogen Energy* **2021**, *46*, 10548–10569. [CrossRef]
5. Eisenhardt, K.M. Strategy as strategic decision making. *MIT Sloan Manag. Rev.* **1999**, *40*, 65.
6. Anderson, K. The Political Market for Government Assistance to Australian Manufacturing Industries. In *World Scientific Reference on Asia-Pacific Trade Policies*; World Scientific: Singapore, 2018; pp. 545–567. [CrossRef]
7. Redclift, M. *Sustainable Development: Exploring the Contradictions*; Routledge: London, UK, 1987. [CrossRef]
8. Holmberg, J.; Sandbrook, R. Sustainable Development: What Is to Be Done? In *Policies for a Small Planet*; Routledge: London, UK, 1992; 20p.
9. Colla, V.; Branca, T.A. Sustainable Steel Industry: Energy and Resource Efficiency, Low-Emissions and Carbon-Lean Production. *Metals* **2021**, *11*, 1469. [CrossRef]
10. Rodrigues, J.; Cost, I.; Farinha, J.T.; Mendes, M.; Margalho, L. Predicting motor oil condition using artificial neural networks and principal component analysis. *Eksploat. Niezawodn.-Maint. Reliab.* **2020**, *22*, 440–448. [CrossRef]
11. Iruela, J.R.S.; Ruiz, L.G.B.; Capel, M.I.; Pegalajar, M.C. A TensorFlow Approach to Data Analysis for Time Series Forecasting in the Energy-Efficiency Realm. *Energies* **2021**, *14*, 4038.
12. Ramos, D.; Faria, P.; Vale, Z.; Mourinho, J.; Correia, R. Industrial Facility Electricity Consumption Forecast Using Artificial Neural Networks and Incremental Learning. *Energies* **2020**, *13*, 4774. [CrossRef]
13. Coccia, M. Steel market and global trends of leading geo-economic players. *Int. J. Trade Glob. Mark.* **2014**, *7*, 36. [CrossRef]
14. Sala, D.A.; Yperen-De Deyne, V.; Mannens, E.; Jalalvand, A. Hybrid static-sensory data modeling for prediction tasks in basic oxygen furnace process. *Appl. Intell.* **2022**. [CrossRef]
15. Chen, C.J.; Chou, F.I.; Chou, J.H. Temperature Prediction for Reheating Furnace by Gated Recurrent Unit Approach. *IEEE Access* **2022**, *10*, 33362–33369. [CrossRef]
16. Janka, D.; Lenders, F.; Wang, S.; Cohen, A.; Li, N. Detecting and locating patterns in time series using machine learning. *Control Eng. Pract.* **2019**, *93*, 104169. [CrossRef]
17. Kim, K.; Seo, B.; Rhee, S.H.; Lee, S.; Woo, S.S. Deep Learning for Blast Furnaces: Skip-Dense Layers Deep Learning Model to Predict the Remaining Time to Close Tap-Holes for Blast Furnaces. In Proceedings of the 28th ACM International Conference on Information and Knowledge Management, CIKM '19, Beijing, China, 3–7 November 2019; Association for Computing Machinery: New York, NY, USA, 2019; pp. 2733–2741. [CrossRef]
18. Xuan, Y.; Yue, Q. Forecast of steel demand and the availability of depreciated steel scrap in China. *Resour. Conserv. Recycl.* **2016**, *109*, 1–12. [CrossRef]
19. Crompton, P. Future trends in Japanese steel consumption. *Resour. Policy* **2000**, *26*, 103–114. [CrossRef]
20. Gajdzik, B. Steel production in Poland with pessimistic forecasts in COVID-19 crisis. *Metalurgija* **2021**, *60*, 169–172.
21. Wang, P.; Jiang, Z.-Y.; Zhang, X.-X.; Geng, X.-Y.; Hao, S.-Y. Long-term scenario forecast of production routes, energy consumption and emissions for Chinese steel industry. *Chin. J. Eng.* **2014**, *36*, 1683–1693. [CrossRef]
22. Sharma, A.; Kumar, S.; Duriagina, Z. *Engineering Steels and High Entropy-Alloys*, 1st ed.; BoD—Books on Demand: London, UK, 2020.
23. Javaid, A.; Essadiqi, E. Final report on scrap management, sorting and classification of steel. *Gov. Can.* **2003**, *23*, 1–22.
24. Carilier, M. Distribution of Steel END-Usage Worldwide in 2019, by Sector. 2021. Available online: https://www.statista.com/statistics/1107721/steel-usage-global-segment/ (accessed on 2 February 2022).
25. Worldsteel I Global Crude Steel Output Decreases by 0.9% in 2020. 2021. Available online: http://www.worldsteel.org/media-centre/press-releases/2021/Global-crude-steel-output-decreases-by-0.9--in-2020.html (accessed on 2 February 2022).
26. Moya, J.A.; Pardo, N. The potential for improvements in energy efficiency and CO_2 emissions in the EU27 iron and steel industry under different payback periods. *J. Clean. Prod.* **2013**, *52*, 71–83. [CrossRef]
27. Crompton, P. Explaining variation in steel consumption in the OECD. *Resour. Policy* **2015**, *45*, 239–246. [CrossRef]
28. Yin, X.; Chen, W. Trends and development of steel demand in China: A bottom–up analysis. *Resour. Policy* **2013**, *38*, 407–415. [CrossRef]

29. Mohr, S.H.; Evans, G.M. Forecasting coal production until 2100. *Fuel* **2009**, *88*, 2059–2067. [CrossRef]

30. Berk, I.; Ediger, V.S. Forecasting the coal production: Hubbert curve application on Turkey's lignite fields. *Resour. Policy* **2016**, *50*, 193–203. [CrossRef]

31. Mengshu, S.; Yuansheng, H.; Xiaofeng, X.; Dunnan, L. China's coal consumption forecasting using adaptive differential evolution algorithm and support vector machine. *Resour. Policy* **2021**, *74*, 102287. [CrossRef]

32. Raynor, W.J. *The International Dictionary of Artificial Intelligence*, Library ed.; Glenlake Pub. Co.; Fitzroy Dearborn Pub: Chicago, IL, USA, 1999; Volume 1, OCLC: Ocm43433564.

33. Cirstea, M.; Dinu, A.; McCormick, M.; Khor, J.G. *Neural and Fuzzy Logic Control of Drives and Power Systems*; Google-Books-ID: pXVgBWRMdgQC; Elsevier: Amsterdam, The Netherlands, 2002.

34. Mateus, B.C.; Mendes, M.; Farinha, J.T.; Cardoso, A.J.M. Anticipating Future Behavior of an Industrial Press Using LSTM Networks. *Appl. Sci.* **2021**, *11*, 6101. [CrossRef]

35. Gubbi, J.; Buyya, R.; Marusic, S.; Palaniswami, M. Internet of Things ([IoT]): A vision, architectural elements, and future directions. *Future Gener. Comput. Syst.* **2013**, *29*, 1645–1660. [CrossRef]

36. Vaio, A.D.; Hassan, R.; Alavoine, C. Data intelligence and analytics: A bibliometric analysis of human–Artificial intelligence in public sector decision-making effectiveness. *Technol. Forecast. Soc. Chang.* **2022**, *174*, 121201. [CrossRef]

37. Niu, Y.; Ying, L.; Yang, J.; Bao, M.; Sivaparthipan, C.B. Organizational business intelligence and decision making using big data analytics. *Inf. Process. Manag.* **2021**, *58*, 102725. [CrossRef]

38. Chen, J.; Gusikhin, O.; Finkenstaedt, W.; Liu, Y.N. Maintenance, Repair, and Operations Parts Inventory Management in the Era of Industry 4.0. *IFAC-PapersOnLine* **2019**, *52*, 171–176. [CrossRef]

39. Farinha, J.M.T. *Asset Maintenance Engineering Methodologies*, 1st ed.; Taylor & Francis Ltd.: Abingdon, UK, 2018.

40. Asri, H. Big Data and IoT for real-time miscarriage prediction A clustering comparative study. *Procedia Comput. Sci.* **2021**, *191*, 200–206. [CrossRef]

41. Soltanali, H.; Khojastehpour, M.; Farinha, J.T.; Pais, J.E.d.A.e. An Integrated Fuzzy Fault Tree Model with Bayesian Network-Based Maintenance Optimization of Complex Equipment in Automotive Manufacturing. *Energies* **2021**, *14*, 7758. [CrossRef]

42. Martins, A.; Fonseca, I.; Farinha, J.T.; Reis, J.; Cardoso, A.J.M. Maintenance Prediction through Sensing Using Hidden Markov Models—A Case Study. *Appl. Sci.* **2021**, *11*, 7685. [CrossRef]

43. Sidhu, T.S.; Singh, H.; Sachdev, M.S. Design, implementation and testing of an artificial neural network based fault direction discriminator for protecting transmission lines. *IEEE Trans. Power Deliv.* **1995**, *10*, 697–706. [CrossRef]

44. Das, R.; Kunsman, S.A. A novel approach for ground fault detection. In Proceedings of the 57th Annual Conference for Protective Relay Engineers, College Station, TX, USA, 1 April 2004 ; pp. 97–109. [CrossRef]

45. Lippmann, R. An introduction to computing with neural nets. *Expert Syst. Appl.* **1987**, *4*, 4–22. [CrossRef]

46. Orhan, U.; Hekim, M.; Ozer, M. EEG signals classification using the K-means clustering and a multilayer perceptron neural network model. *Expert Syst. Appl.* **2011**, *38*, 13475–13481. [CrossRef]

47. Lee, S.; Choeh, J.Y. Predicting the helpfulness of online reviews using multilayer perceptron neural networks. *Expert Syst. Appl.* **2014**, *41*, 3041–3046. [CrossRef]

48. Ferasso, M.; Alnoor, A. Artificial Neural Network and Structural Equation Modeling in the Future. In *Artificial Neural Networks and Structural Equation Modeling*; Springer: Berlin/Heidelberg, Germany, 2022; pp. 327–341. [CrossRef]

49. Hazarika, N.; Chen, J.Z.; Tsoi, A.; Sergejew, A. Classification of EEG signals using the wavelet transform. *Signal Process* **1997**. [CrossRef]

50. Oğulata, S.N.; Şahin, C.; Erol, R. Neural network-based computer-aided diagnosis in classification of primary generalized epilepsy by EEG signals. *J. Med. Syst.* **2009**, *33*, 107–112. [CrossRef]

51. Faris, H.; Aljarah, I.; Al-Madi, N.; Mirjalili, S. Optimizing the Learning Process of Feedforward Neural Networks Using Lightning Search Algorithm. *Int. J. Artif. Intell. Tools* **2016**, *25*, 1650033. [CrossRef]

52. Bonilla Cardona, D.A.; Nedjah, N.; Mourelle, L.M. Online phoneme recognition using multi-layer perceptron networks combined with recurrent non-linear autoregressive neural networks with exogenous inputs. *Neurocomputing* **2017**, *265*, 78–90. [CrossRef]

53. Ding, H.; Dong, W. Chaotic feature analysis and forecasting of Liujiang River runoff. *Soft Comput.* **2016**, *20*, 2595–2609. [CrossRef]

54. Zhang, Z. Neural networks: Further insights into error function, generalized weights and others. *Ann. Transl. Med.* **2016**, *4*, 300. [CrossRef] [PubMed]

55. Riedmiller, M.; Braun, H. A direct adaptive method for faster backpropagation learning: The RPROP algorithm. In Proceedings of the IEEE International Conference on Neural Networks, San Francisco, CA, USA, 28 March–1 April 1993; Volume 1, pp. 586–591. [CrossRef]

56. Anastasiadis, A.D.; Magoulas, G.D.; Vrahatis, M.N. New globally convergent training scheme based on the resilient propagation algorithm. *Neurocomputing* **2005**, *64*, 253–270. [CrossRef]

57. Pedregosa, F.; Varoquaux, G.; Gramfort, A.; Michel, V.; Thirion, B.; Grisel, O.; Blondel, M.; Prettenhofer, P.; Weiss, R.; Dubourg, V.; et al. Scikit-learn: Machine Learning in Python. *J. Mach. Learn. Res.* **2011**, *12*, 2825–2830.

58. Abadi, M.; Agarwal, A.; Barham, P.; Brevdo, E.; Chen, Z.; Citro, C.; Corrado, G.S.; Davis, A.; Dean, J.; Devin, M.; et al. TensorFlow: Large-Scale Machine Learning on Heterogeneous Systems. 2015. Available online: tensorflow.org (accessed on 4 May 2022).

59. Badawi, A.A.; Chao, J.; Lin, J.; Mun, C.F.; Sim, J.J.; Tan, B.H.M.; Nan, X.; Aung, K.M.M.; Chandrasekhar, V.R. Towards the AlexNet Moment for Homomorphic Encryption: HCNN, theFirst Homomorphic CNN on Encrypted Data with GPUs. *arXiv* **2020**, arXiv:1811.00778.
60. Arya, S.; Singh, R. A Comparative Study of CNN and AlexNet for Detection of Disease in Potato and Mango leaf. In Proceedings of the 2019 International Conference on Issues and Challenges in Intelligent Computing Techniques (ICICT), Ghaziabad, India, 27–28 September 2019; Volume 1, pp. 1–6. [CrossRef]
61. Zou, M.; Zhu, S.; Gu, J.; Korunovic, L.M.; Djokic, S.Z. Heating and Lighting Load Disaggregation Using Frequency Components and Convolutional Bidirectional Long Short-Term Memory Method. *Energies* **2021**, *14*, 4831. [CrossRef]
62. Seo, D.; Huh, T.; Kim, M.; Hwang, J.; Jung, D. Prediction of Air Pressure Change Inside the Chamber of an Oscillating Water Column–Wave Energy Converter Using Machine-Learning in Big Data Platform. *Energies* **2021**, *14*, 2982. [CrossRef]
63. Zhu, F.; Ye, F.; Fu, Y.; Liu, Q.; Shen, B. Electrocardiogram generation with a bidirectional LSTM-CNN generative adversarial network. *Sci. Rep.* **2019**, *9*, 6734. [CrossRef]
64. Bouzerdoum, M.; Mellit, A.; Massi Pavan, A. A hybrid model (SARIMA–SVM) for short-term power forecasting of a small-scale grid-connected photovoltaic plant. *Sol. Energy* **2013**, *98*, 226–235. [CrossRef]
65. Zheng, L.; Xue, W.; Chen, F.; Guo, P.; Chen, J.; Chen, B.; Gao, H. A Fault Prediction Of Equipment Based On CNN-LSTM Network. In Proceedings of the 2019 IEEE International Conference on Energy Internet (ICEI), Nanjing, China, 27–31 May 2019. [CrossRef]
66. Pan, S.; Wang, J.; Zhou, W. *Prediction on Production of Oil Well with Attention-CNN-LSTM*; IOP Publishing: Bristol, UK, 2021; Volume 2030, p. 012038. [CrossRef]
67. Li, X.; Yi, X.; Liu, Z.; Liu, H.; Chen, T.; Niu, G.; Yan, B.; Chen, C.; Huang, M.; Ying, G. Application of novel hybrid deep leaning model for cleaner production in a paper industrial wastewater treatment system. *J. Clean. Prod.* **2021**, *294*, 126343. [CrossRef]
68. He, Y.; Liu, Y.; Shao, S.; Zhao, X.; Liu, G.; Kong, X.; Liu, L. Application of CNN-LSTM in Gradual Changing Fault Diagnosis of Rod Pumping System. *Math. Probl. Eng.* **2019**, *2019*, 4203821. [CrossRef]
69. Ren, L.; Dong, J.; Wang, X.; Meng, Z.; Zhao, L.; Deen, M.J. A Data-Driven Auto-CNN-LSTM Prediction Model for Lithium-Ion Battery Remaining Useful Life. *IEEE Trans. Ind. Inform.* **2021**, *17*, 3478–3487. [CrossRef]
70. Hochreiter, S.; Schmidhuber, J. Long Short-Term Memory. *Neural Comput.* **1997**, *9*, 1735–1780. [CrossRef] [PubMed]
71. Rodrigues, J.; Farinha, J.; Mendes, M.; Mateus, R.; Cardoso, A. Comparison of Different Features and Neural Networks for Predicting Industrial Paper Press Condition. *Energies* **2022**, *15*, 6308. [CrossRef]
72. Gers, F.A.; Schmidhuber, E. LSTM recurrent networks learn simple context-free and context-sensitive languages. *IEEE Trans. Neural Netw.* **2001**, *12*, 1333–1340. [CrossRef] [PubMed]
73. Yu, Y.; Si, X.; Hu, C.; Zhang, J. A review of recurrent neural networks: LSTM cells and network architectures. *Neural Comput.* **2019**, *31*, 1235–1270. Available online: https://direct.mit.edu/neco/article-pdf/31/7/1235/1053200/neco_a_01199.pdf (accessed on 4 May 2022). [CrossRef] [PubMed]
74. Mateus, B.C.; Mendes, M.; Farinha, J.T.; Assis, R.; Cardoso, A.J.M. Comparing LSTM and GRU Models to Predict the Condition of a Pulp Paper Press. *Energies* **2021**, *14*, 6958. [CrossRef]
75. Costa Silva, D.F.; Galvão Filho, A.R.; Carvalho, R.V.; de Souza, L.; Ribeiro, F.; Coelho, C.J. Water Flow Forecasting Based on River Tributaries Using Long Short-Term Memory Ensemble Model. *Energies* **2021**, *14*, 7707. [CrossRef]
76. Cho, K.; van Merrienboer, B.; Bahdanau, D.; Bengio, Y. On the Properties of Neural Machine Translation: Encoder-Decoder Approaches. *arXiv* **2014**, arXiv:1409.1259.
77. Li, P.; Luo, A.; Liu, J.; Wang, Y.; Zhu, J.; Deng, Y.; Zhang, J. Bidirectional Gated Recurrent Unit Neural Network for Chinese Address Element Segmentation. *ISPRS Int. J. Geo-Inf.* **2020**, *9*, 635. [CrossRef]

Article

A Stochastic Queueing Model for the Pricing of Time-Sensitive Services in the Demand-Sharing Alliance

Jianpei Wen

Department of Industrial Engineering and Management, Peking University, Beijing 100871, China; wenjianpei@pku.edu.cn

Abstract: The medical alliance has developed rapidly in recent years. This kind of alliance established by multiple hospitals can alleviate the imbalance of medical resources. We investigate the benefit of demand sharing between a hospital with large demand (HD) and another hospital with large supply (HS). Two hospitals are modeled as queueing systems with finite service rates. Both hospitals set prices to maximize the revenues by serving their time-sensitive patients. We adopt a cooperative game theoretic framework to determine when demand sharing is beneficial. We also propose an optimal allocation of this benefit through a commission fee, which makes the alliance stable. We find that demand sharing may not be beneficial even if HS has a low capacity utilization. Demand sharing becomes beneficial for both hospitals only when the idle service capacity of HS exceeds a threshold, which depends on the potential demand rate of the HS and the unit waiting cost of hospitals. Furthermore, if the idle service capacity of HS is smaller than another threshold, which depends on the potential demand of the two hospitals and the service capacity of HD, then the benefit of demand sharing will be independent of the service capacity and potential demand of HD. We also examine the effect of system parameters on revenue gains due to demand sharing.

Keywords: queueing; pricing; demand sharing; cooperative game theory

Citation: Wen, J. A Stochastic Queueing Model for the Pricing of Time-Sensitive Services in the Demand-Sharing Alliance. *Appl. Sci.* **2022**, *12*, 12121. https://doi.org/10.3390/app122312121

Academic Editor: Panagiotis Tsarouhas

Received: 24 October 2022
Accepted: 23 November 2022
Published: 27 November 2022

Publisher's Note: MDPI stays neutral with regard to jurisdictional claims in published maps and institutional affiliations.

1. Introduction

The mismatch between supply and demand has been a prevalent problem in the healthcare system. For example, in some Chinese remote rural areas or new peri-urban areas, the medical resources for some kinds of illness is poor [1]. In this case, the local medical service providers are faced with a potential demand that is larger than what it can handle. Meanwhile, the medical resources in metropolis are plentiful. Due to stiff competition, these service providers in big city may not attract enough demand.

The imbalance of medical resources makes patients spend too much time waiting for medical services. Studies have shown that waiting time is one of the main indicators for evaluating patient satisfaction, which has a great impact on overall satisfaction [2,3]. Song et al. (2019) [4] analyze patients' willingness to make their first visit to primary care institutions and shows that the convenience of consultation is the key factor which patients choose primary care institutions for the first diagnosis. In a hospital survey in Nanjing, China, waiting time accounted for a higher proportion of the total waiting time for medical treatment, which affected patient utility [5].

In order to reduce this mismatch, in China, hospitals have combined into a medical alliance. One example is the cross-regional specialized medical alliance [6] in China. This alliance aims to integrate medical resources and use telemedicine so that patients from remote areas can enjoy medical services of a similar quality to those available in metropolis [7]. In this alliance, rural hospitals can share demand and medical resource with the hospital located in major cities. In this case, the urban hospital is considered as the under-demanded service provider. This form is developing very rapidly in China, there are many medical alliances under construction in Chinese cities (Beijing, Shanghai, Wuhan, Shenzhen, etc.) [8].

Forming an alliance is a reasonable strategy for hospitals to solve the imbalance between supply and demand. Such a strategy would be feasible if the HD shares a part of its demand with the HS who pays an appropriate commission fee. It is called as demand sharing strategy, and both HD and HS can benefit from this strategy. Evidently, the HD can generate extra revenue by collecting the commission fee, and HS can utilize its idle capacity to serve more patients. It seems that the demand sharing strategy is a win-win thing for the two hospitals. However, as shown in this study, this intuitive win-win situation can only occur under some conditions. These conditions are very important for practitioners considering this demand sharing strategy by forming an alliance of hospitals facing demand and supply mismatch.

In the cross-regional specialized medical alliance, since each member of the alliance is decentralized managed, a reasonable method of profit distribution is one of the key factors to achieve cooperation. Otherwise, the HD may prefer to expand its facilities and serve more patients, the HS may tends to reduce its investment, and the patients are more likely to go to the HD because they do not believe in the diagnosis and treatment ability of HS. The mismatch problem becomes more serious, and enter a virtuous circle. Therefore, without a reasonable method of profit distribution, it is difficult to achieve cooperation.

This problem also shows in many other service markets due to asymmetric information or other reasons. Owing to some barriers to market entry (such as exclusive licenses, anti-competitive subsidization, and tariff protection), a service provider (monopoly) with limited service capacity in the market can be faced with a potential demand that is larger than what it can handle.

In this paper, we try to study the alliance, where some participants are often in short supply, while others are facing an oversupply situation. For example, the demand of hospitals in remote areas exceeds supply, or the demand of core hospitals of the medical alliance exceeds supply. HD will transfer part of their demand to HS, which can improve the efficiency of HS. There are several main research questions to answer; they are as follows: (1) When does demand sharing benefit both hospitals? (2) With demand sharing, what are the optimal pricing decisions for both hospitals? (3) How to divide the revenue gains from the demand sharing alliance between two hospitals?

Two hospitals serve their own stream of delay-sensitive patients in two different regions. Each service provider is modeled as an $M/M/1$ queueing system with different potential Poisson arrival rate and exponential service rate. If the HS participates in this alliance, then it should serve two classes of patients from two different regions. These two classes differ in their arrival rates and delay sensitivities. Based on the price and delay information, each arriving patient in HD is faced with three options—joining the line of HD, switching to the line of HS, and balking. Patients in Region-2 (or HS region) decide whether to join the line of HS or balk. The term 'balk' means that the patient will not join the system and leave without service. Patients do not observe the actual queue length at their arrivals. However, they are informed of the average delay (long-term statistics). We also assume that the patient type (HS region or HD region) is known to HS upon patients' arrival, and different prices can be set for different patient types. All patients are served on a first-come first-served (FCFS) basis. Both hospitals try to maximize their revenues. The two hospitals use the pricing mechanisms to control the demand of each stream, and allocate the revenue gain by the commission fee. If this operation benefit both hospitals, then the HD and HS will collaborate, and a demand sharing alliance will be established. This study adopts a cooperative game theoretic framework to model the cooperation between HD and HS.

The major contributions of this study are as follows:

1. we obtain the threshold condition for HS's idle capacity under which the demand sharing works.
2. we provide the optimal (or stable) commission fee charged by HD that makes demand sharing alliance work.

The rest of this paper is organized as follows. Section 2 summarizes related work on capacity sharing and co-sourcing and compares them with our demand sharing model.

Section 3 applies a rational queueing framework to model the service process of HD and HS. Section 4 identifies when demand sharing is beneficial and presents the optimal allocation of the benefit through commission fees if demand sharing is beneficial. Finally, Section 5 concludes with a summary.

2. Related Literature

Medical alliance is an effective way to alleviate the imbalance of medical resources. We will first review the relevant literature of the Medical Alliance. Demand sharing is our main strategy for building a medical alliance, which is the key to improving the uneven supply and demand. A stream of operation management literature related to our work studies the advantage of the inter-firm collaboration strategy in dealing with demand fluctuation or the economies of scope. Therefore, we then review the work on resource pooling, capacity sharing, co-sourcing, and the on-demand service platform. Another related study area is service pricing in queueing systems with self-interested customers.

The medical alliance or referral system have been studied in several articles. Li and Zhang (2015) [9] studied different kinds of contract used in the medical alliance, and address the problem that how to design a contract to to motivate the service provider to do their best to serve patients with mild diseases. Chen et al. (2015) [10] used rational queuing theory to study the capacity planning problem in the referral system. In this paper, we also study the medical alliance, but we focus on the analysis of the benefit allocation mechanism of the demand sharing alliance.

Several papers studied resource pooling and cost sharing among queueing systems. Anily and Haviv (2010) [11] use a cooperative game to study the cost allocation mechanism in the case, wherein a number of servers pool their capacity and customer streams into a single $M/M/1$ system, and show that the game possesses non-empty cores. Yu et al. (2015) [12] adopted a queueing model and cooperative game to identify settings wherein several firms investing and sharing with one facility can improve the service level and reduce the cost. Zeng et al. (2017) [13] formulate a cooperative game to study the capacity transfer among several $M/M/1$ systems or $M/M/s$ systems, and propose cost-sharing rules that are at the core of the corresponding game. Anily and Haviv (2017) [14] study the line-balancing in a parallel $M/M/1$ system and $M/M/1/1$ system with a cooperative game and show that the core is non-empty. These aforementioned studies assume that the demand is exogenous and can be routed or pooled among all lines without considering customers' interest. Contrarily, our study assumes that the demand is endogenous—customers are treated as players in the game.

Concerning co-sourcing, it occurs when a firm outsources part of its service to another firm for strategic (customer segmentation) or operational (demand uncertainty) reasons [15]. Aksin et al. (2008) [16] analyzed the optimal capacity and pricing decisions in call center settings under each contract type when a firm adopted outsourcing. Lee et al. (2012) [17] studied how to outsource one level of a two-level service process, wherein the first level diagnoses the request's complexity. Their studies focused on the problem of coordinating the service providers serving heterogeneous customers. It must be noted that the firm that outsources has complete control over the product price (or owns the demand). Contrarily, in a demand sharing alliance, the HS can set its price and decide how much demand it needs to serve.

To deal with demand fluctuation, a stream of literature studies the capacity sharing between firms who compete on price. Li and Zhang (2015) [9] studied the benefit of capacity sharing in shipping industries. They compared the capacity reservation model, wherein shipping forwarders are allowed to reserve shipping capacity before the demand is realized and have an option to trade capacity after the demand is realized, with the passive capacity sharing model, wherein shipping forwarders only have an option to trade capacity after the demand is realized. They found that capacity reservation model helps the carrier firm to squeeze more profits out of the shipping forwarders. Guo and Wu (2016) [18] studied capacity sharing between two firms that engage in price competition under ex-ante and ex-post capacity sharing price schemes. They found that the equilibrium outcome under ex-ante contracting was more sensitive to variations in market parameters than ex-post contracting. Cetinkaya et al. (2012) [19] studied the capacity collaboration under the scenario wherein two firms build capacity before the demand is realized and make production decisions after they receive a demand signal. These works used a traditional demand function capturing the relationship between demand and price without considering the customer delay cost. Customer delay cost is a fundamental factor in modeling service demand. Contrarily, our study examines demand sharing in service industries where demand depends on both price and delay cost.

Our study is also related to some studies focusing on the on-demand service platforms, such as Uber and DiDi. Tang et al. (2016) and Taylor (2016) [20,21] studied how to set the wage for part-time employees and the price for delay sensitive customers to match the demand with the supply. In our study, the HS is an independent firm with its own demand rather than being the firm that plays the role of part-time employees of an HD. The HS makes its decision based on its own demand and capacity.

Finally, the current study is related to a research stream on pricing for queueing systems with self-interested customers, which is started by [22,23]. See [24–26] for an excellent review. Note that we focus on the static pricing policy in this paper since the dynamic price changes may not allowed in many industries. Chen and Frank (2004) [27] studied monopoly pricing an unobservable queue with homogeneous customers with joining and balking options. Chen and Wan (2003) [28] studied simultaneous price competition between two firms in a market with homogeneous customers. Our model is closely related to that of the model by [29], which investigated the service provider's optimal pricing service for two types of customers who cannot observe the queue length. The pricing policy prescribes different prices for different types of customers. Suk and Wang (2020) [30] consider a tandem queueing system with price sensitive but delay insensitive heterogeneous customers, and study static pricing policy and dynamic pricing policy. In this study, we study the pricing problem under the cooperation between HD and HS, which form a parallel-server queueing system.

3. Model Formulation

3.1. Two Hospitals with Unbalanced Congestion

First, we describe the mismatch between supply and demand of two hospitals based on a rational queueing framework. We consider two independent hospitals; each hospital has a M/M/1 queueing system with processing rate $\mu_i, i = \{1,2\}$. Their potential patient arrival rate is Λ_i. Patients must wait when the doctor is busy. Patient's delay cost is proportional to the system delay. The cost of stay per unit time is denoted by c_i. Each arriving patient must decide whether to request for the medical service (joining) or not (balking). Patients have a reserved value for each service, which is denoted by V_i. The utility function of patients is formulated as $U_i = V_i - c_i w_i - p_i$, where w_i is the average sojourn time and p_i is the price. We assume that customers can make decision based on the long-term average sojourn time and system parameters such as stay cost per unit time. In several practical situations, the average stay time is available public information. For instance, in Canada [31] or Hong Kong [32], the average stay time for some non-emergent medical service in public hospitals is posted on the website. A patient purchases the service if the patient's net surplus is positive, that is, $U_i \geq 0$. The effective arrival rate, which represents the patient arrivals with purchase, is denoted by λ_i. The revenue function of the two independent hospitals can be written as $\pi_i^0 = p_i \lambda_i$. We assume that the service capacity is fixed and its investment is sunk cost. Previous literature [27] provides the optimal pricing strategy of a monopoly firm via queueing modeling as follows,

1. if $\Lambda_i \geq \mu_i - \sqrt{c_i \mu_i / V_i}$, then the optimal price is $p_i^* = V_i - \sqrt{c_i V_i / \mu_i}$, the optimal effective arrival rate is $\lambda_i^* = \mu_i - \sqrt{c_i \mu_i / V_i}$, and the corresponding optimal revenue is $\pi_i^0 = (\sqrt{V_1 \mu_1} - \sqrt{c_1})^2$.

2. if $\Lambda_i < \mu_i - \sqrt{c_i\mu_i/V_i}$, then the optimal price is $p_i^* = V_i - c_i/(\mu_i - \Lambda_i)$, the optimal effective arrival rate is $\lambda_i^* = \Lambda_i$, and the corresponding optimal revenue is $\pi_i^0 = \Lambda_2[V_2 - c_2/(\mu_2 - \Lambda_2)]$.

This optimal price need to balance the trade off between the revenue from a patient and the externality caused by this patient.

The hospital HD, indexed by $i = 1$, which is a large general hospital, attracts a demand which is too large to be met in China because of the medical habits of Chinese patients, the distrust of patients in the treatment capacity of small hospitals, and poor doctors in small hospitals, etc. We define this large demand scenario as the case where the potential patient arrival rate is larger than the effective arrival rate under the optimal pricing strategy, i.e., $\mu_1 - \sqrt{c_1\mu_1/V_1} < \Lambda_1$. Accordingly, we make the following assumption for the HD.

Assumption 1. *HD is characterized by a limited service rate that satisfies* $\mu_1^0 < \mu_1 < \mu_1^1$, *where* μ_1^0 *and* μ_1^1 *are defined as* $\mu_1^0 = \frac{c_1}{V_1}$; *and* $\mu_1^1 = \Lambda_1 + \frac{c_1}{2V_1} + \sqrt{\frac{c_1^2}{4V_1^2} + \frac{c_1\Lambda_1}{V_1}}$.

In Assumption 1, the condition $\mu_1 < \mu_1^1$ is equivalent to $\mu_1 - \sqrt{c_1\mu_1/V_1} < \Lambda_1$. The other condition $\mu_1^0 < \mu_1$ is equivalent to $V_1 - c_1/\mu_1 > 0$, which ensures that the HD can attract at least one patient. Accordingly, we can get the optimal revenue of HD when it operates independently, that is $\pi_1^0 = (\sqrt{V_1\mu_1} - \sqrt{c_1})^2$.

The hospital HS, indexed by $i = 2$, which is a community hospital, cannot attract patients due to its lack of medicine and poor doctors. However, it can provide alternative services for common diseases as that of the HD. Given the market segmentation, since the coverage of medical insurance is only local or the distance between regions is relatively long, the HS cannot directly serve patients in the HD's region (region-1). One case that satisfies the above two types of hospitals is the medical alliance constructed between hospitals in remote areas and specialized hospitals in big cities, such as the "Wu Jieping Urological Medical Center". In this kind of alliance, hospitals in remote areas are hospitals HD, since the medical resources in their areas are poor, which makes it difficult to treat some intractable diseases. On the other hand, specialized hospitals in big cities are hospital HS due to the sufficient medical resources in the large cities and the fiercer competition among hospitals. Moreover, due to geographical restrictions, the two hospitals can be considered to be in two separate markets.

The hospital HS is faced with a small potential demand due to its lack of medicine and poor doctors, such as it is a new hospital. Similarly, we make the following assumption to define this under-utilized scenario.

Assumption 2. *HS is characterized by a large service rate that satisfies* $\mu_2 > \mu_2^0$, *where* μ_2^0 *is defined as* $\mu_2^0 = \Lambda_2 + \frac{c_2}{2V_2} + \sqrt{\frac{c_2^2}{4V_2^2} + \frac{c_2\Lambda_2}{V_2}}$.

In Assumption 2, the condition $\mu_2 > \mu_2^0$ is equivalent to $\mu_2 - \sqrt{c_2\mu_2/V_2} > \Lambda_2$; it means that the HS is faced with the potential demand that is smaller than the demand level under the optimal policy. HS's optimal revenue, when it operates independently, is $\pi_2^0 = \Lambda_2[V_2 - c_2/(\mu_2 - \Lambda_2)]$. It is increased in the potential arrival rate Λ_2.

Note that the above two scenarios are defined based on their own region characteristics and optimal pricing strategy. We do not give assumption for the relationship between the congestion of two hospitals. Thus, the congestion of the under-utilized HS is possibly more heavy than the over-demanded HD.

3.2. Demand Sharing between Two Hospitals

We can see that both HD and HS are faced with a mismatch between supply and demand. It seems that it is beneficial for both hospitals if the HD shares a part of its patients with the HS. We consider one type of patients sharing alliance between the two hospitals, wherein the HD adds an alternative for its patients that they can opt to use HS's service, as shown in Figure 1. The price of this option p_{12} is different from the original price of HD's service p_1. If a patient opts this alternative and purchases HS's service, then the HD can get a commission fee s from this transaction.

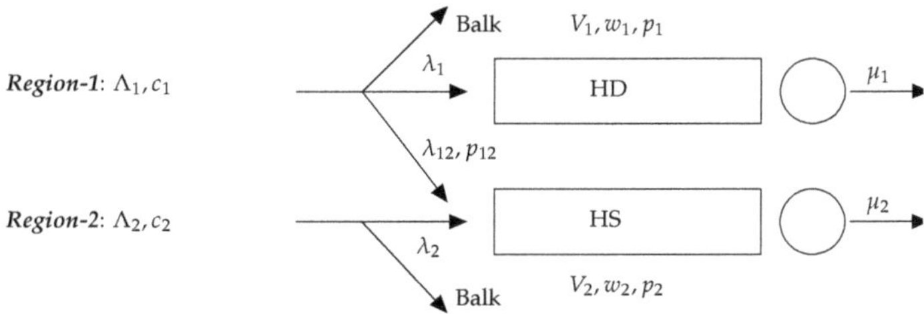

Figure 1. Graphic representation of the system configuration.

In this alliance, HS is faced with two classes of patients. Two classes are completely differentiated and observed by the HS upon their arrivals. HS can control the arrival rate of the two streams by pricing. We assume that all patients are served on a FCFS basis. Thus, the expected waiting times postulated by the two classes are the same. We also assume that the patients from HD are more impatient than those from Region-2, that is, $c_1 \geq c_2$. We make this assumption due to two main reasons. First, if $c_1 \geq c_2$, then it will be more profitable for HS to serve patients from Region-2 under the same waiting time. As a result, the balking patient must belong to Market-1, if it exists, which is more appropriate for most scenarios. Second, if we make an opposite assumption, that is, $c_1 < c_2$, then the balking patient must belong to Region-2, if it exists. This would produce similar results because the HS's trade-off between revenue gain from higher utilization and revenue loss due to the externality of new patients would still hold.

We now describe the interaction between hospitals and patients in this alliance. Two hospitals need to make pricing decisions. HD decides the price of its service p_1 and the commission fee s. HS decides its price for Region-2 p_2 and the price for patients switching from HD p_{12}.

Given the prices, each arriving patient in HD is faced with three options—joining the queue of HD, switching to the queue of HS, and balking. Patients must choose one option with the largest net surplus. Concerning patients in Region-2, they join the queue of HS if the patient's net surplus is positive. In equilibrium, a patient should be indifferent between either alternative. Hence, two classes of patient's decentralized decision behavior for a given pricing strategy results in the following relations:

$$\begin{cases} V_1 - p_1 - \frac{c_1}{\mu_1 - \lambda_1} = V_2 - p_{12} - \frac{c_1}{\mu_2 - \lambda_{12} - \lambda_2} \geq 0, \\ V_2 - p_2 - \frac{c_2}{\mu_2 - \lambda_{12} - \lambda_2} \geq 0, \\ \lambda_1 + \lambda_{12} \leq \Lambda_1, \end{cases} \tag{1}$$

wherein λ_{12} represents the arrival rate of patients that switch from HD to HS.

Both HD and HS attempt to maximize their revenues by pricing their services. The HD utilizes the price p_1 and commission fee s to control the demand served by itself, λ_1, and the part shared with the HS, λ_{12}. Hence, the HD's optimization problem is as follows:

$$\max_{p_1, s} \pi_1(p_1, s) = p_1 \lambda_1 + s \lambda_{12}. \tag{2}$$

Given that each patient type is known upon arrival, the HS can set different prices for different classes of patients. There exists a trade-off between making extra revenue and increasing the delay of patients by serving more patients that switch from the HD. The

HD employs two prices—p_2 and p_{12}—to balance the two types of demands. The HS's optimization problem is as follows:

$$\max_{p_2, p_{12}^0} \pi_2(p_2, p_{12}^0) = p_2 \lambda_2 + p_{12}^0 \lambda_{12}, \tag{3}$$

where $p_{12}^0 = p_{12} - s$.

4. Analysis

A medical alliance can be reached by the pricing mechanisms that provides adequate incentives. First, we must identify the setting under which collaboration is essential. We solve the optimal pricing problem for the medical alliance. Subsequently, we compare the optimal revenue of the medical alliance with the total optimal revenues of the two independently operating hospitals. Collaboration becomes beneficial only when the former is greater than the latter. If demand sharing is feasible, then we can formulate a two-players cooperative game to characterize the bargaining processes for HD and HS and to determine the optimal allocation of the extra benefit due to demand sharing through commission fees.

4.1. Total Benefit of Demand Sharing

We first analyze the medical alliance that generates the maximum benefit possible from demand sharing. The objective is to maximize the total revenue with three optimal prices (i.e., p_1, p_2, p_{12}) as follows.

$$\pi_c(p_1, p_2, p_{12}) = p_1 \lambda_1 + p_{12} \lambda_{12} + p_2 \lambda_2. \tag{4}$$

In (4), we can see that the direct benefit of demand sharing is $p_{12} \lambda_{12}$, since, otherwise, this part of patients cannot get service. However, the waiting time will increase if HS would serve more patients; this may drive the HS to set a lower price to compensate its own patients (from Region-2) for the extra delay. This may cause a revenue reduction (lower p_2) in the Market-2. Hence, whether demand sharing can raise total revenue depends on HS's market characteristics. Proposition 1 specifies a lower threshold for the HS service rate, which provides the condition for revenue gain via demand sharing. Proofs of propositions and corollaries are in the Appendixes A and B.

Proposition 1. *There exists a threshold*

$$\underline{\mu_2} = \Lambda_2 + \frac{c_1}{2V_2} + \sqrt{\frac{c_1^2}{4V_2^2} + \frac{c_2 \Lambda_2}{V_2}}, \tag{5}$$

such that if and only if $\mu_2 > \underline{\mu_2}$, then the medical alliance via demand sharing will result in a revenue gain compared to the independent hospitals without demand sharing.

Proposition 1 shows that even if the HS has some idle capacity, the demand sharing may not be necessarily beneficial. For the demand sharing to be feasible, the HS's idle capacity(service rate) must exceed a threshold, that is, $\mu_2 > \underline{\mu_2}$. The threshold $\underline{\mu_2}$ implies a condition as per which the HS's revenue loss, as a result of lowering the price p_2 for its region, must be equal to its revenue gain, as a result of serving more patients from the HD's region.

Given $\mu_2 > \underline{\mu_2}$, wherein the demand sharing is beneficial, we derive the optimal pricing strategy with the medical alliance. Intuitively, with more idle capacity, the HS will serve more patients who make the switch to increase total revenue. HD can at least share patient flow with an arrival rate of $\Lambda_{12}^0 = \Lambda_1 - \lambda_1^0$, which does not affect its optimal pricing strategy in Region-1. If HD shares a demand that is larger than Λ_{12}^0, then the demand sharing may result in a new mismatch between demand and supply. Hence, whether to share a demand that is larger than Λ_{12}^0 would depend on the trade-off between the revenue loss caused by the new mismatch and the revenue gain from HS's higher utilization. An extreme case would be if HS has infinite service rate, then it may be optimal for HD to share all of its demand. The following proposition provides an upper threshold for the HS service rate that can characterize the optimal strategies if the two hospitals cooperate and make an alliance.

Proposition 2. *There exists a threshold*

$$\overline{\mu_2} = \Lambda - \mu_1 + \sqrt{\frac{c_1\mu_1}{V_1}} + \frac{c_1}{2V_2} + \sqrt{\frac{c_1^2}{4V_2^2} + \frac{c_1\Lambda_1 + c_2\Lambda_2 - c_1\mu_1 + c_1\sqrt{c_1\mu_1/V_1}}{V_2}}. \tag{6}$$

If $\mu_2 < \mu_2 \leq \overline{\mu_2}$, then the HD would utilize its own optimal pricing strategy and share its remaining patients Λ_{12}^0 with the HS; additionally, the HS would serves all the patients from Region-2, that is, $\lambda_2^* = \Lambda_2$ and accept a part of the demand switching from the HD; which is given as:

$$\lambda_{12}^* = \mu_2 - \Lambda_2 - \sqrt{\frac{c_1\mu_2 - c_1\Lambda_2 + c_2\Lambda_2}{V_2}}. \tag{7}$$

The corresponding total optimal revenue is as follows:

$$\pi_c^* = V_1\mu_1 + V_2\mu_2 + 2c_1 - 2\sqrt{c_1\mu_1 V_1} - 2\sqrt{V_2(c_1\mu_2 - c_1\Lambda_2 + c_2\Lambda_2)}.$$

Otherwise, that is, $\overline{\mu_2} < \mu_2$, under the medical alliance, the HD will share at least Λ_{12}^0 with the HS and may deviate from its own optimal strategy; the HS will accept all patients shared by HD. Hence, all patients from both Region-1 and Region-2 will be served by the two hospitals.

Proposition 2 shows the condition, that is, $\mu_2 < \mu_2 \leq \overline{\mu_2}$, that even with demand sharing and medical alliance, the HD will still hold its optimal strategy and the two hospitals will be unable to serve all the patients in both the regions. On the other hand, if $\overline{\mu_2} < \mu_2$, then the HS can serve more patients than HD's remaining demand under its optimal strategy. In this case, the HD will not implement its own optimal strategy due to share more patients with the HS. Compared to the independent operations, HD will earn less revenue by serving its own region. However, this revenue loss will be lesser than the revenue gain as a result of HS's higher utilization. In this case, the two hospitals can serve all patients from the two regions.

The following lemma characterizes the market shares of the two hospitals under the centralized operation when the HS capacity is sufficiently large.

Lemma 1. *Suppose that $\overline{\mu_2} < \mu_2$, if there exists an equilibrium for the meidical alliance, then the corresponding equilibrium effective arrival rate of HD λ_1^* should be the solution to the equation*

$$(V_1 - V_2)(\mu_1 - \lambda_1^*)^2(\mu_2 - \Lambda + \lambda_1^*)^2 + [c_1\mu_2 - c_1\Lambda_2 + c_2\Lambda_2](\mu_1 - \lambda_1^*)^2$$
$$-c_1\mu_1(\mu_2 - \Lambda + \lambda_1^*)^2 = 0, \tag{8}$$

where $\Lambda = \Lambda_1 + \Lambda_2$.

If the two hospitals' service quality is equal, that is, $V_1 = V_2$, then there will be a unique equilibrium, and the demand shared by HD will be given by

$$\lambda_{12}^* = \Lambda_1 - \mu_1 - (\mu_1 + \mu_2 - \Lambda)\frac{c_1\mu_1 - \sqrt{c_1\mu_1(c_2\Lambda_2 + c_1\mu_2 - c_1\Lambda_2)}}{c_2\Lambda_2 + c_1\mu_2 - c_1\Lambda_2 - c_1\mu_1}. \tag{9}$$

Equation (8) is the necessary condition for equilibrium, which is the first order condition of (4). When the service quality of the two hospitals become equal, they become different in waiting time and pricing for the patients in Region-1. Lemma 1 gives the optimal demand sharing strategy for the medical alliance when $V_1 = V_2$. Here, we focus on the case of $V_1 = V_2$, and discuss the case of different service quality later.

Let us use $\nabla \pi_c$ to denote the total revenue gain of the medical alliance. It is defined as the difference between the total revenue of the medical alliance and the sum of revenues of two independent hospitals, that is, $\nabla \pi_c = \pi_c - \pi_1^0 - \pi_2^0$. We conduct a numerical study to illustrate the impact of the service rate of the HS, μ_2, on the total revenue gain, when $V_1 = V_2$, as shown in Figure 2. We observe that the revenue gain ratio curve can be divided into three segments. When $\mu_2 < \underline{\mu_2}$, there is no revenue gain. When $\underline{\mu_2} < \mu_2 \leq \overline{\mu_2}$, the revenue gain is increasing quickly in the service capacity. In this case, we can prove that the revenue gain is convex increasing in μ_2. When $\overline{\mu_2} < \mu_2$, the revenue gain is increasing at a slower rate in μ_2.

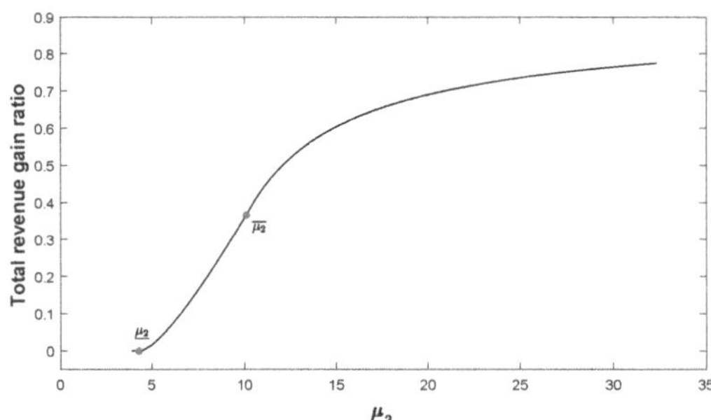

Figure 2. Total revenue gain ratio is increasing in μ_2. ($V_1 = V_2 = 2.5, c_1 = 2, \mu_1 = 10, \Lambda_1 = 12,$ $c_2 = 0.5, \Lambda_2 = 3, \alpha = \beta = 0.5$). Total revenue gain ratio: $\nabla \pi_c / (\pi_1^0 + \pi_2^0)$.

Such a behavior can be explained as follows: when $\mu_2 \leq \overline{\mu_2}$, two hospitals still cannot serve the total patients in both the regions. The total revenue gain, as μ_2 increases, is mainly due to serving more patients from the HD region. Therefore, the gain ratio will increase at an accelerated pace and the demand sharing will not change the HD's optimal pricing strategy in its market. However, when $\overline{\mu_2} < \mu_2$, the two hospitals can serve all the patients. An increase in the HS's capacity will not result in an increase in the demand. Thus, the total revenue gain is mainly attributed to the improvement of workload balance between the two hospitals in medical alliance. This result is consistent with the monopoly case by [27], if we consider the two hospitals in medical alliance with demand sharing.

4.2. Revenue Allocation in the Medical Alliance

Although demand sharing can significantly increase the overall revenue of both hospitals, in the case of $\mu_2 > \overline{\mu_2}$, the collaboration can occur only when each hospital can earn more revenue individually when compared to the independent operation without demand sharing. Therefore, how to allocate the revenue gain between two hospitals in such a way that both hospitals have the incentive to collaborate is an important issue. This section shows that an optimal allocation can be performed by determining an appropriate commission fee.

We now use a two-player bargaining game to characterize the collaboration between HD and HS. Bargaining games have been extensively utilized to model the negotiation processes on prices between sellers and buyers, firm mergers, and acquisitions of small firms. This research area has been surveyed by [33]. Given that the negotiation over the commission fee for each switching patients involves HD and HS, we solve the bargaining game between the HD and the HS by utilizing the cooperative-game solution (also called the generalized Nash bargaining solution (NBS)) [34]. Our analysis framework can be applied to other types of bargaining solutions, such as the Kalai–Smorodinsky bargaining solution [35,36]. The cooperative-game solution can be obtained by solving the following problem:

$$\max_{p_1, p_2, p_{12}^0, s} \quad [\pi_1^*(p_1, s) - \pi_1^0]^\alpha [\pi_2^*(p_2, p_{12}^0) - \pi_2^0]^\beta$$
$$\text{s.t.} \quad \pi_1^*(p_1, s) - \pi_1^0 > 0; \tag{10}$$
$$\pi_2^*(p_2, p_{12}^0) - \pi_2^0 > 0.$$

The disagreement payoffs for the HD and the HS are their revenues earned when they operate independently. Let $\pi_1^*(s)$ and $\pi_2^*(p_{12}^0, s)$ denote the revenues of HD and HS, respectively, when they cooperate. The parameters $\alpha, \beta > 0$, $(\alpha + \beta = 1)$ represent the bargaining power of the HD and the HS, respectively. The two constraints require that both hospitals can make more revenue with demand sharing than without it (or independent operations). The next lemma gives a method for determining the commission fee that facilitates demand sharing.

Lemma 2. *In the equilibrium, the two hospitals set their optimal prices as a centralized hospital. The optimal commission fee s^* can be derived by solving*

$$\max_s \quad [\pi_1^*(p_1^*, s) - \pi_1^0]^\alpha [\pi_2^*(p_2^*, p_{12}^* - s) - \pi_2^0]^\beta$$
$$\text{s.t.} \quad \pi_1^*(p_1^*, s) - \pi_1^0 > 0; \tag{11}$$
$$\pi_2^*(p_2^*, p_{12}^* - s) - \pi_2^0 > 0,$$

where p_1^*, p_2^*, p_{12}^* are given in the proofs of the propositions and lemma in Section 4.1.

Subsequently, the revenue π_1^*, π_2^* and the optimal commission fee s^* can be derived according to Proposition 2 and Lemma 1. For a certain range of the HS's capacity, the optimal commission fee can be expressed explicitly as stated in the following proposition.

Proposition 3. *For $\underline{\mu_2} < \mu_2 \leq \overline{\mu_2}$, in equilibrium, the negotiated commission fee s^* for each switching patient is independent of HD's service capacity μ_1 and potential demand Λ_1, which is given as*

$$s^* = \frac{\alpha V_2}{\alpha + \beta} \left[1 - \sqrt{\frac{c_1}{V_2(\mu_2 - \Lambda_2)} + \frac{c_2 \Lambda_2}{V_2(\mu_2 - \Lambda_2)^2}}\right]. \tag{12}$$

Additionally, the equilibrium commission fee s^* decreases in c_1, c_2, Λ_2, but increases in V_2, μ_2.

Under the condition of $\underline{\mu_2} < \mu_2 \leq \overline{\mu_2}$, the HD shares a part of patients with the HS, without experiencing any impact on its optimal pricing strategy for Region-1. Since the HS now accepts some extra patients from the HD, the expected waiting time becomes longer than the expected waiting time without demand sharing. This means that the HS must reduce its price for his original patients in Region-2. Hence, to compensate for this revenue loss of HS, the commission fee is less than a proportion of the co-payment of the switching patients, that is, $s^* < \alpha p_{12}^*/(\alpha + \beta)$. In addition, we find that this policy of revenue allocation mainly depends on HS's region characteristics, its service rate, and switching patients' delay sensitivity. Although the result of additional idle HS capacity leading to higher total revenue gain and commission fee is intuitive, quantifying the HS capacity range and the optimal commission fee offers medical personnel valuable information.

We present a numerical example in Figure 3 to illustrate the impact of the parameters of HS, c_2, Λ_2, μ_2, on the revenue gain due to demand sharing. Figure 3 shows the following: (1) the revenue gain can be significant (e.g., more than 20%) depending on the parameters (i.e., Λ_2 and μ_2); (2) the revenue gain is monotonous with respect to c_2, Λ_2, μ_2, it is more sensitive to the demand level at the HS market.

Unlike the case of $\underline{\mu_2} < \mu_2 \leq \overline{\mu_2}$, wherein the HD can share its demand without affecting its optimal strategy, the case of $\overline{\mu_2} < \mu_2$ is more complicated. We can still get the equilibrium commission fee for this case with $V_1 = V_2$. The non-equal service value case will be discussed later.

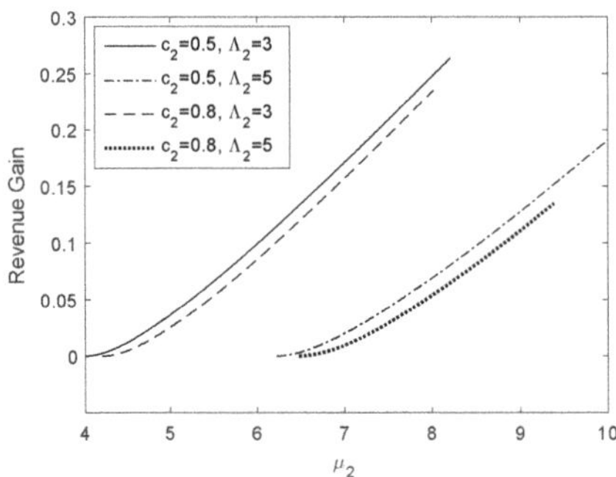

Figure 3. Revenue gain of the collaboration when $\mu_2 < \mu_2 \leq \overline{\mu_2}$. ($V = 2.5, c_1 = 1, \mu_1 = 10$, $\Lambda_1 = 12, \alpha = \beta = 0.5$, Revenue gain ratio: $\nabla \pi = (\nabla \pi_1 + \nabla \pi_2)/(\pi_1^0 + \pi_2^0)$)

Lemma 3. *For $V_1 = V_2$ and $\overline{\mu_2} < \mu_2$, in equilibrium, the negotiated commission fee s^* for each switching patient is*

$$s = \frac{\alpha}{\alpha + \beta}[V_2 + \frac{(\sqrt{V_1 \mu_1} - \sqrt{c_1})^2 - V_2 \Lambda_2}{\lambda_{12}^*} + \frac{c_2 \Lambda_2}{(\mu_2 - \Lambda_2)\lambda_{12}^*}$$
$$- \frac{c_1}{\mu_2 - \Lambda_2 - \lambda_{12}^*} + \frac{c_1 \Lambda_1 - c_2 \Lambda_2 - c_1 \lambda_{12}^*}{(\mu_2 - \Lambda_2 - \lambda_{12}^*)\lambda_{12}^*}],$$

where λ_{12}^* is given in Lemma 1.

Unfortunately, although the closed-form of the equilibrium commission fee is available for the $\overline{\mu_2} < \mu_2$ case with the condition $V_1 = V_2$, we cannot get the properties of Proposition 3 analytically. Thus, we conduct a numerical study to investigate the impact of the two regions' parameters (i.e., $c_1, c_2, \Lambda_1, \Lambda_2, \mu_1, \mu_2$) on the equilibrium commission fee and revenue gain. The total revenue gain is equal to $(V_1 - c_1/(\mu_1 - \Lambda_1 + \lambda_{12}^*))(\Lambda_1 - \lambda_{12}^*) - (\sqrt{V_1 \mu_1} - \sqrt{c_1})^2 + s^* \lambda_{12}^*$, which is actually the revenue from commission $s^* \lambda_{12}^*$ minus the revenue loss in Region-1 $\pi_1^0 - (V_1 - c_1/(\mu_1 - \Lambda_1 + \lambda_{12}^*))(\Lambda_1 - \lambda_{12}^*)$. The revenue loss is attributed to the fact that the HD shares more demand than its own optimal pricing strategy.

The impacts of all parameters, except for c_1, are similar to the case $\mu_2 < \mu_2 \leq \overline{\mu_2}$. The total revenue gain, i.e., $(V_1 - c_1/(\mu_1 - \Lambda_1 + \lambda_{12}^*))(\Lambda_1 - \lambda_{12}^*) - (\sqrt{V_1 \mu_1} - \sqrt{c_1})^2 + s^* \lambda_{12}^*$, is not necessarily monotonous with respect to the unit waiting cost c_1. Figure 4 shows that the revenue from commission fee, that is $s^* \lambda_{12}^*$, is decreasing in the delay sensitivity of patients in Region-1 c_1. This is mainly because, with more delay sensitive patients, there is an increase in the contribution of the idle capacity of HS toward this alliance, which reduces HD's revenue allocating from the medical alliance. However, in Region-1, the revenue loss, i.e., $\pi_1^0 - (V_1 - c_1/(\mu_1 - \Lambda_1 + \lambda_{12}^*))(\Lambda_1 - \lambda_{12}^*)$, is also decreasing and concave in c_1. The reason is that, with the medical alliance, the revenue from Region-1, i.e., $(V_1 - c_1/(\mu_1 - \Lambda_1 + \lambda_{12}^*))(\Lambda_1 - \lambda_{12}^*)$, is less sensitive to patients' unit waiting cost when compared to the case without demand sharing, i.e., π_1^0. The difference between the two revenues, that is $\pi_1^0 - (V_1 - c_1/(\mu_1 - \Lambda_1 + \lambda_{12}^*))(\Lambda_1 - \lambda_{12}^*)$, decreases as patients become more delay sensitive. Finally, the total revenue gain, which is a combination of the trend of revenue from commission and revenue loss in Region-1, first increases and subsequently decreases in the delay sensitivity of patients in the Region-1. The total revenue of the alliance, which is equal to $\pi_1^*(\alpha + \beta)/\alpha$, has the same trend.

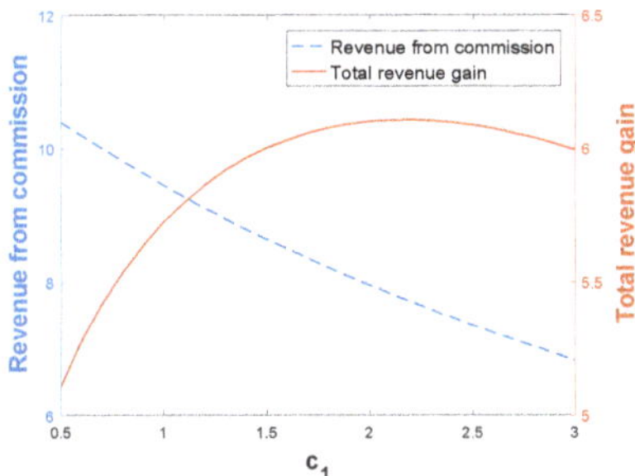

Figure 4. The impact of c_1 when $\overline{\mu_2} < \mu_2$. Total revenue gain: $\pi_1^* - \pi_1^0$. Revenue from commission: $s^*\lambda_{12}^*$. ($\alpha = \beta = 0.5, V_1 = V_2 = 2.5, c_2 = 0.5, \mu_1 = 10, \mu_2 = 15, \Lambda_2 = 3, \Lambda_1 = 12$.)

Subsequently, we consider the case wherein the service quality of the two hospitals are different, that is, $V_1 < V_2$ or $V_1 > V_2$. Equation (8) is the equilibrium condition, which is the first order condition. It must be noted that closed-form solutions cannot be obtained since they are roots of a quadratic equation. We conduct a numerical study for the case with a unique equilibrium, as shown in Figure 5. We try to investigate the impact of the service quality on the commission fee, the rate of sharing patient flow, the revenue loss in Region-1, and the total revenue gain. Figure 5 shows that the total revenue gain and the demand sharing rate λ_{12} are all decreasing in the HD's service quality V_1. This observation implies that the HD with a lower service quality has more incentives to share its demand with the HS, which means the HS can replace the HD to a greater extent. However, the commission fee is not necessarily monotonous with respect to V_1. In fact, the commission fee is increasing in V_1 when $V_1 < V_2$, and decreasing in V_1 when $V_1 > V_2$ (reaching the maximum at $V_1 = V_2$).

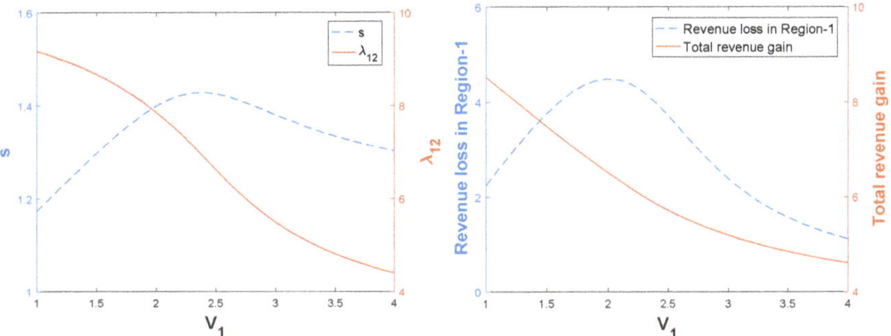

Figure 5. The impact of V_1 when $\overline{\mu_2} < \mu_2$. Total revenue gain: $\pi_1^* - \pi_1^0$. Revenue from commission: $s^*\lambda_{12}^*$. ($\alpha = \beta = 0.5, V_2 = 2.5, c_1 = 1, c_2 = 0.5, \mu_1 = 10, \mu_2 = 15, \Lambda_2 = 3, \Lambda_1 = 12$.)

5. Conclusions

In this study, we have analyzed the benefit of demand sharing between an over-demanded hospital (HD) and an under-demanded hospital (HS). Both hospitals set their prices to maximize their revenues from serving delay-sensitive patients. We adopted a cooperative game theoretic framework to characterize the situation where the demand sharing is beneficial. We also obtain the commission fee, which guarantees that the medical alliance is stable.

We find that collaboration is not always beneficial even if HS has idle capacity. Demand sharing becomes beneficial for both HD and HS only when the service rate of HS is larger than the first threshold (lower one). This threshold depends on the HS's rate of patient flow and the patients' unit waiting cost.

Furthermore, if the service rate of HS is less than the second threshold (higher one), which depends on the rate of patient flow of both markets and the HD's capacity, then the total revenue gain from demand sharing would be independent of the service rate and potential demand of the HD.

We also showed the effects of the two hospitals' parameters on the revenue gains from demand sharing. Results show that the medical alliance can be more beneficial with less delay sensitive (or more patient) patients in the moderate service rate case (the HS's service rate is less than the second threshold). However, if the HS's service rate exceeds the second (higher) threshold, then the revenue gains may not be necessarily monotonic with respect to the unit waiting costs of both regions. These insights can help service hospitals to form and manage medical alliances.

Our study assumes that patients of each region are homogeneous. However, it would also be interesting to consider the heterogeneity of the patients in each market.

Funding: This research received no external funding.

Data Availability Statement: Not applicable.

Conflicts of Interest: The authors declare no conflict of interest.

Appendix A. Proof for Proposition 1, Proposition 2 and Lemma 1

According to the Theorem 1 of [29], facing two classes of patients, which are characterized by $c_1 > c_2$, the optimal pricing and admission control policy for HS is that the lower delay cost enters first, followed by, possibly, part of the other class. It means that HS should first accept patients only from Market-2 and continue doing this until the class is captured; subsequently, if there is sufficient service capacity, then HS should accept patients from Market-1 until it captures the whole sharing demand from Market-1. HS uses different prices to control two streams of demand. However, if HS accepts two types of patients, all patients are faced with the same waiting time in the queue.

Proof. The optimal problem is to maximize the total revenues shown in Equation (4) under the equilibrium constraints of the patient behavior, which is defined in Equation (1). Before further analysis, let us discuss HS's optimal problem, which is given as follows,

$$\max_{p_2, p_{12}} \pi_2(p_2, p_{12}) = p_{12}\Lambda_{12} + p_2\lambda_2. \tag{A1}$$

According to Theorem 1 of [29], given the potential sharing demand rate Λ_{12}, the optimal solution of the problem defined in Equation (A1) satisfies the following:

1. Suppose that $\mu_2^0 < \mu_2 \leq \underline{\mu_2}$, then the optimal pricing strategy is as follows:

$$p_2^* = V_2 - \sqrt{\frac{c_2}{\mu_2 - \Lambda_2}}, \ p_{12}^* > V_2 - \sqrt{\frac{c_1}{\mu_2 - \Lambda_2}}.$$

Additionally, the corresponding actual demand rates are as follows:

$$\lambda_2^* = \Lambda_2, \ \lambda_{12}^* = 0.$$

2. Suppose that $\underline{\mu_2} < \mu_2 \leq \overline{\mu_2'} = \Lambda_2 + \Lambda_{12} + \frac{c_1}{2V_2} + \sqrt{\frac{c_1^2}{4V_2^2} + \frac{c_2\Lambda_2 + c_1\Lambda_{12}}{V_2}}$, then the optimal prices satisfy the following:

$$p_2^* = V_2 - c_2\sqrt{\frac{V_2}{c_1\mu_2 - c_1\Lambda_2 + c_2\Lambda_2}}, \ p_{12}^* = V_2 - c_1\sqrt{\frac{V_2}{c_1\mu_2 - c_1\Lambda_2 + c_2\Lambda_2}}.$$

Additionally, the corresponding actual demand rates are as follows:

$$\lambda_2^* = \Lambda_2, \ \lambda_{12}^* = \mu_2 - \Lambda_2 - \sqrt{\frac{c_1\mu_2 - c_1\Lambda_2 + c_2\Lambda_2}{V_2}}.$$

3. Suppose that $\overline{\mu_2'} < \mu_2$, then the optimal prices satisfy the following:

$$p_2^* = V_2 - \frac{c_2}{\mu_2 - \Lambda_2 - \Lambda_{12}}, \ p_{12}^* = V_2 - \frac{c_1}{\mu_2 - \Lambda_2 - \Lambda_{12}}.$$

Additionally, the corresponding actual demand rates are as follows:

$$\lambda_2^* = \Lambda_2, \ \lambda_{12}^* = \Lambda_{12}.$$

Subsequently, let us analyze HD's optimal problem, which is given as follows,

$$\max_{p_1} \pi_1(p_1) = p_1\lambda_1. \tag{A2}$$

We can obtain the optimal strategy of HD based on the monopolistic pricing results of [27], which is given as follows,

1. Suppose that $\Lambda_1 \geq \mu_1 - \sqrt{\frac{c_1\mu_1}{V_1}}$, then the optimal pricing strategy is $p_1^* = V_1 - \sqrt{\frac{c_1V_1}{\mu_1}}$.
   Additionally, the corresponding actual demand rate is $\lambda_1^* = \mu_1 - \sqrt{\frac{c_1\mu_1}{V_1}}$.

2. Suppose that $\Lambda_1 < \mu_1 - \sqrt{\frac{c_1\mu_1}{V_1}}$, then the optimal price is $p_1^* = V_1 - \frac{c_1}{\mu_1 - \Lambda_1}$. Additionally, the corresponding actual demand rate is $\lambda_1^* = \Lambda_1$.

Here, we analyze the centralized optimal problem shown in Equation (4). Since the threshold μ_2 does not depend on the HD's decision, and the optimal strategy of the HS is to capture the demand of Market-2 only when $\mu_2^0 < \mu_2 \leq \underline{\mu_2}$, the two hospitals are independent. Two hospitals independently operate as a monopolist. Hence, in this case, centralized operation will not result in a revenue gain.

However, if $\mu_2 > \underline{\mu_2}$, then HS can raise its revenue by continuing to accept a part of the sharing demand. Hence, we prove the Proposition 1.

In addition, we find that if $\mu_2 \leq \overline{\mu_2'}$, the real sharing demand accepted by HS would satisfy $\lambda_{12}^* < \Lambda_{12}$. Hence, if $\Lambda_{12} \leq \Lambda_1 - \mu_1 + \sqrt{\frac{c_1\mu_1}{V_1}}$, which results in the condition $\mu_2 \leq \overline{\mu_2}$, then the two hospitals can independently obtain their own optimal strategy. Hence, we get the first part of Proposition 2.

As the service capacity of the HS increases, that is, $\overline{\mu_2'} < \mu_2$, HS adopts the optimal strategy to capture both the streams of demand Λ_1, Λ_{12}. Hence, in the case $\mu_2 > \overline{\mu_2}$, no patient would balk from the two hospitals. The demand shared by HD will be at least $\Lambda_{12}^0 = \Lambda_1 - \mu_1 + \sqrt{\frac{c_1\mu_1}{V_1}}$. Hence, we prove the second part of Proposition 2.

Subsequently, let us prove Lemma 1. In this case, two hospitals cannot independently obtain their own optimal strategy anymore. The centralized operation must facilitate a trade-off between the marginal revenue of HD and HS. Subsequently, the optimal problem would change into

$$\max_{p_2,p_{12},p_1} \pi_c(p_1,p_2,p_{12}) = p_1\lambda_1 + p_{12}\lambda_{12} + p_2\lambda_2, \ s.t. \lambda_1 + \lambda_{12} = \Lambda_1.$$

In this scenario, if we increase the potential demand rate of the switching patients Λ_{12}, then the optimal revenue of HD would decrease, but the revenue of HS would increase. Subsequently, the optimal problem can be written as follows:

$$\max_{p_2, p_{12}, p_1} \pi_c(p_1, p_2, p_{12}) = (V_1 - \frac{c_1}{\mu_1 - \lambda_1})\lambda_1 + (V_2 - \frac{c_1}{\mu_2 - \Lambda_2 - \Lambda_{12}})(\Lambda_1 - \lambda_1)$$
$$+ (V_2 - \frac{c_2}{\mu_2 - \Lambda_2 - \Lambda_{12}})\Lambda_2.$$

According to the first order condition of this problem, we have,

$$(V_1 - V_2)(\mu_1 - \lambda_1)^2(\mu_2 - \Lambda + \lambda_1)^2 + [c_1\mu_2 - c_1\Lambda_2 + c_2\Lambda_2](\mu_1 - \lambda_1)^2$$
$$- c_1\mu_1(\mu_2 - \Lambda + \lambda_1)^2 = 0.$$

Hence, if we assume $V_1 = V_2$, then we will have

$$\lambda_{12}^* = \Lambda_1 - \mu_1 - (\mu_1 + \mu_2 - \Lambda)\frac{c_1\mu_1 - \sqrt{c_1\mu_1(c_2\Lambda_2 + c_1\mu_2 - c_1\Lambda_2)}}{c_2\Lambda_2 + c_1\mu_2 - c_1\Lambda_2 - c_1\mu_1}.$$

Hence, we have the optimal strategy defined in Lemma 1. □

Appendix B. Proof for Lemma 2, Proposition 3 and Lemma 3

Proof. To determine the NBS, we solve the optimization problem defined in Equation (10). We first determine the optimal commission fee, s^*, for the given pricing p_1^*, p_2^*, p_{12}^*. Subsequently, we solve for the optimal pricing. For the given p_1^*, p_2^*, p_{12}^*, it can be shown that Equation (10) is strictly concave in s, and hence the optimal commission fee s^* is unique. To solve for s^*, we first write the KKT conditions. Let ν_1 and ν_2 be the Lagrangian multipliers. Subsequently, the KKT conditions are as follows:

$$\lambda_{12}[p_2\lambda_2 + p_{12}\lambda_{12} - p_1\lambda_1 - 2s\lambda_{12} + (\sqrt{V\mu_1} - \sqrt{c_1})^2$$
$$- \Lambda_2(V - \frac{c_2}{\mu_2 - \Lambda_2}) + \nu_1 - \nu_2] = 0;$$
$$\nu_1[\pi_1^*(p_1, s) - (\sqrt{V\mu_1} - \sqrt{c_1})^2] = 0;$$
$$\nu_2[\pi_2^*(p_2, p_{12}^0) - \Lambda_2(V - \frac{c_2}{\mu_2 - \Lambda_2})] = 0;$$
$$\nu_1, \nu_2 \geq 0.$$

From the KKT condition, we obtain

$$s^* = \frac{p_2\lambda_2 + p_{12}\lambda_{12} - p_1\lambda_1 + (\sqrt{V\mu_1} - \sqrt{c_1})^2 - \Lambda_2(V - \frac{c_2}{\mu_2 - \Lambda_2})}{2\lambda_{12}}. \tag{A3}$$

To obtain the optimal pricing, we rewrite the Equation (10) by utilizing Equation (A3):

$$\max_{p_1, p_2, p_{12}} [\frac{p_1\lambda_1 + p_2\lambda_2 + p_{12}\lambda_{12} - (\sqrt{V\mu_1} - \sqrt{c_1})^2 - \Lambda_2(V - \frac{c_2}{\mu_2 - \Lambda_2})}{2}]^2, \tag{A4}$$

which is equivalent to the optimization problem defined in Equation (4). Hence, we have Lemma 2.

Suppose that $\underline{\mu_2} < \mu_2 \leq \overline{\mu_2}$, according to Proposition 2, the optimal problem defined in Equation (10) can be rewritten as follows:

$$\max_s (s\lambda_{12}^*)^\alpha [p_2^*\Lambda_2 + (p_{12}^* - s)\lambda_{12}^* - \Lambda_2(V_2 - \frac{c_2}{\mu_2 - \Lambda_2})]^\beta \tag{A5}$$

Given the first order condition, we have the following:

$$s = \frac{\alpha}{\alpha + \beta} p_{12}^* + \frac{\alpha[p_2^* \Lambda_2 - V_2 \Lambda_2 + c_2 \Lambda_2/(\mu_2 - \Lambda_2)]}{(\alpha + \beta)\lambda_{12}^*}$$

$$= \frac{\alpha}{\alpha + \beta}[V_2 - c_1 w^* - \frac{c_2 \Lambda_2 w^*}{\mu_2 - \Lambda_2}].$$

where the optimal waiting time is given as follows:

$$w^* = \frac{1}{\mu_2 - \Lambda_2 - \lambda_{12}^*} = \sqrt{\frac{V_2}{c_1(\mu_2 - \Lambda_2) + c_2 \Lambda_2}}$$

Hence, in this case, the equilibrium commission fee is

$$s^* = \frac{\alpha V_2}{\alpha + \beta}[1 - \sqrt{\frac{c_1}{V_2(\mu_2 - \Lambda_2)} + \frac{c_2 \Lambda_2}{V_2(\mu_2 - \Lambda_2)^2}}].$$

With the above commission given and optimal prices given in Proposition 2, the corresponding gains of optimal revenues from demand sharing can be calculated.

Suppose that $\overline{\mu_2} < \mu_2$, the optimal problem defined in Equation (10) can be rewritten as follows:

$$\max_{s} [p_1^* \lambda_1^* + s\lambda_{12}^* - (\sqrt{V_2 \mu_1} - \sqrt{c_1})^2]^\alpha [p_2^* \Lambda_2 + (p_{12}^* - s)\lambda_{12}^* - \Lambda_2(V_2 - \frac{c_2}{\mu_2 - \Lambda_2})]^\beta$$

Given the first order condition, we have the following:

$$s = \frac{\alpha}{\alpha + \beta} p_{12}^* + \frac{\alpha[p_2^* \Lambda_2 - p_1^* \lambda_1^* + (\sqrt{V_1 \mu_1} - \sqrt{c_1})^2 - V_2 \Lambda_2 + c_2 \Lambda_2/(\mu_2 - \Lambda_2)]}{(\alpha + \beta)\lambda_{12}^*},$$

which is equivalent to

$$s = \frac{\alpha}{\alpha + \beta}[V_2 + \frac{(\sqrt{V_1 \mu_1} - \sqrt{c_1})^2 - V_2 \Lambda_2 + c_2 \Lambda_2/(\mu_2 - \Lambda_2)}{\lambda_{12}^*}$$

$$- \frac{c_1}{\mu_2 - \Lambda_2 - \lambda_{12}^*} + \frac{c_1 \Lambda_1 - c_2 \Lambda_2 - c_1 \lambda_{12}^*}{(\mu_2 - \Lambda_2 - \lambda_{12}^*)\lambda_{12}^*}].$$

$\square$

References

1. Allen, D. Telemedicine Expanded to Rural China: Across the Divide. Available online: https://emag.medicalexpo.com/telemedicine-expanded-to-rural-china-across-the-divide/ (accessed on 6 October 2022).
2. Adi, L.; Yuval, W.; Carroll, J.S.; Paul, B.; dayan Yaron, B. Waiting time is a major predictor of patient satisfaction in a primary military clinic. *Mil. Med.* **2002**, *167*, 842.
3. Grossman, M. On the Concept of Health Capital and the Demand for Health. *J. Political Econ.* **1972**, *80*, 223–255 [CrossRef]
4. Song, H.; Zuo, X.; Cui, C.; Meng, K. The willingness of patients to make the first visit to primary care institutions and its influencing factors in Beijing medical alliances: A comparative study of Beijing's medical resource-rich and scarce regions. *BMC Health Serv. Res.* **2019**, *19*, 361. [CrossRef] [PubMed]
5. Wang, M.; Lu, Y.; Huang, X. Comparative Study on Medical Cost of Local and Nonlocal Patients in 4 Third Grade First Class Hospitals in Nanjing City. *Med. Soc.* **2019**, *32*, 56–59.
6. Xie, F.; Wang, Y.; Zhang, Q.; Chen, Z.; Gu, S.; Guan, W.; Li, C.; Li, T.; Li, X.; Luo, L.; et al. Development of mental health alliances in China (2017 Edition). *J. Hosp. Manag. Health Policy* **2018**, *2*. [CrossRef]
7. Wang, X. Medical Alliance Launched to Aid Rural Patients. Available online: http://www.chinadaily.com.cn/china/2016-07/27/content_26240288.htm (accessed on 27 July 2016)
8. Yang, F.; Yang, Y.; Liao, Z. Evaluation and analysis for Chinese Medical Alliance's governance structure modes based on Preker-Harding Model. *Int. J. Integr. Care* **2020**, *20*, 14. [CrossRef] [PubMed]

9.	Li, L.; Zhang, R.Q. Cooperation through capacity sharing between competing forwarders. *Transp. Res. Part E Logist. Transp. Rev.* **2015**, *75*, 115–131. [CrossRef]

10.	Chen, Y.; Zhou, W.; Hua, Z.; Shan, M. Pricing and capacity planning of the referral system with delay-sensitive patients. *J. Manag. Sci. China* **2015**, *18*, 73–83.

11.	Anily, S.; Haviv, M. Cooperation in Service Systems. *Oper. Res.* **2010**, *58*, 660–673. [CrossRef]

12.	Yu, Y.; Benjaafar, S.; Gerchak, Y. Capacity Sharing and Cost Allocation among Independent Firms with Congestion. *Prod. Oper. Manag.* **2015**, *24*, 1285–1310. [CrossRef]

13.	Zeng, Y.; Zhang, L.; Cai, X.; Li, J. Cost Sharing for Capacity Transfer in Cooperating Queueing Systems. *Prod. Oper. Manag.* **2018**, *27*, 644–662. [CrossRef]

14.	Anily, S.; Haviv, M. Line Balancing in Parallel M/M/1 Lines and Loss Systems as Cooperative Games. *Prod. Oper. Manag.* **2017**, *26*, 1568–1584. [CrossRef]

15.	Zhou, Y.P.; Ren, Z.J.; Cochran, J.J.; Cox, L.A.; Keskinocak, P.; Kharoufeh, J.P.; Smith, J.C. Service Outsourcing. In *Wiley Encyclopedia of Operations Research and Management Science*; John Wiley & Sons, Inc.: Hoboken, NJ, USA, 2010.

16.	Aksin, O.Z.; de Vericourt, F.; Karaesmen, F. Call Center Outsourcing Contract Analysis and Choice. *Manag. Sci.* **2008**, *54*, 354–368. [CrossRef]

17.	Lee, H.H.; Pinker, E.J.; Shumsky, R.A. Outsourcing a Two-Level Service Process. *Manag. Sci.* **2012**, *58*, 1569–1584. [CrossRef]

18.	Guo, L.; Wu, X. Capacity Sharing between Competitors. *Manag. Sci.* **2018**, *64*, 3554–3573. [CrossRef]

19.	Cetinkaya, E.; Ahn, H.S.; Duenyas, I. Benefits of Collaboration in Capacity Investment and Allocation. Available online: http://dx.doi.org/10.2139/ssrn.2169490 (accessed on 16 September 2012).

20.	Tang, C.S.; Bai, J.; So, K.C.; Chen, X.M.; Wang, H. Coordinating Supply and Demand on an on-Demand Platform: Price, Wage, and Payout Ratio. Available online: https://ssrn.com/abstract=2831794 (accessed on 20 December 2017).

21.	Taylor, T. On-Demand Service Platforms. *Manuf. Serv. Oper. Manag.* **2018**, *20*, 704–720. [CrossRef]

22.	Naor, P. The Regulation of Queue Size by Levying Tolls. *Econometrica* **1969**, *37*, 15–24. [CrossRef]

23.	Levhari, D.; Luski, I. Duopoly pricing and waiting lines. *Eur. Econ. Rev.* **1978**, *11*, 17–35. [CrossRef]

24.	Hassin, R. *Rational Queueing*; CRC Press: Boca Raton, FL, USA, 2016.

25.	Hassin, R.; Haviv, M. *To Queue or Not to Queue: Equilibrium Behavior in Queueing Systems*; Springer Science & Business Media: Berlin/Heidelberg, Germany, 2003; Volume 59.

26.	Ghosh, S.; Hassin, R. Inefficiency in stochastic queueing systems with strategic customers. *Eur. J. Oper. Res.* **2021**, *295*, 1–11. [CrossRef]

27.	Chen, H.; Frank, M. Monopoly pricing when customers queue. *IIE Trans.* **2004**, *36*, 569–581. [CrossRef]

28.	Chen, H.; Wan, Y.W. Price Competition of Make-to-Order Firms. *IIE Trans.* **2003**, *35*, 817–832. [CrossRef]

29.	Printezis, A.; Burnetas, A. The effect of discounts on optimal pricing under limited capacity. *Int. J. Oper. Res.* **2011**, *10*, 160. [CrossRef]

30.	Suk, T.; Wang, X. Optimal pricing policies for tandem queues: Asymptotic optimality. *IISE Trans.* **2020**, *53*, 199–220. [CrossRef]

31.	Canadian Institute for Health Information. Wait Times for Priority Procedures in Canada. Available online: https://www.cihi.ca/en/wait-times-for-priority-procedures-in-canada (accessed on 25 November 2022)

32.	Hospital Authority. Elective Cataract Surgery. Available online: https://www.ha.org.hk/visitor/ha_visitor_text_index.asp?Parent_ID=214172&Content_ID=214184 (accessed on 30 September 2022)

33.	Nagarajan, M.; Sošić, G. Game-theoretic analysis of cooperation among supply chain agents: Review and extensions. *Eur. J. Oper. Res.* **2008**, *187*, 719–745. [CrossRef]

34.	Muthoo, A. *Bargaining Theory with Applications*; Cambridge University Press: Cambridg, UK, 1999.

35.	Myerson, R.B. *Game Theory: Analysis of Conflict*; Harvard University Press: Cambridge, MA, USA, 1991.

36.	Roth, A.E. *Axiomatic Models of Bargaining*; Springer Science & Business Media: Berlin/Heidelberg, Germany, 2012; Volume 170.

applied sciences

Article

Data Matrix Based Low Cost Autonomous Detection of Medicine Packages

José Lima [1,2,3,*], Cláudia Rocha [1], Luísa Rocha [4] and Paulo Costa [1,5]

[1] INESC Technology and Science, 4200-465 Porto, Portugal
[2] Research Centre of Digitalization and Intelligent Robotics, Instituto Politécnico de Bragança, 5300-253 Bragança, Portugal
[3] Laboratório para a Sustentabilidade e Tecnologia em Regiões de Montanha (SusTEC), Instituto Politécnico de Bragança, 5300-253 Bragança, Portugal
[4] Hospital Pharmacy, Centro Hospitalar de Vila Nova de Gaia/Espinho, E.P.E., 4434-502 Vila Nova de Gaia, Portugal; lrocha@chvng.min-saude.pt
[5] Faculty of Engineering of University of Porto, 4200-465 Porto, Portugal
* Correspondence: jllima@ipb.pt

Abstract: Counterfeit medicine is still a crucial problem for healthcare systems, having a huge impact in worldwide health and economy. Medicine packages can be traced from the moment of their production until they are delivered to the costumers through the use of Data Matrix codes, unique identifiers that can validate their authenticity. Currently, many practitioners at hospital pharmacies have to manually scan such codes one by one, a very repetitive and burdensome task. In this paper, a system which can simultaneously scan multiple Data Matrix codes and autonomously introduce them into an authentication database is proposed for the Hospital Pharmacy of the Centro Hospitalar de Vila Nova de Gaia/Espinho, E.P.E. Relevant features are its low cost and its seamless integration in their infrastructure. The results of the experiments were encouraging, and with upgrades such as real-time feedback of the code's validation and increased robustness of the hardware system, it is expected that the system can be used as a real support to the pharmacists.

Keywords: data matrix; codes scan; pharmaceutical industry; traceability

Citation: Lima, J.; Rocha, C.; Rocha, L.; Costa, P. Data Matrix Based Low Cost Autonomous Detection of Medicine Packages. *Appl. Sci.* **2022**, *12*, 9866. https://doi.org/10.3390/app12199866

Academic Editor: Oscar Reinoso García

Received: 1 August 2022
Accepted: 26 September 2022
Published: 30 September 2022

Publisher's Note: MDPI stays neutral with regard to jurisdictional claims in published maps and institutional affiliations.

1. Introduction

Healthcare systems continuously seek for innovative treatments to overcome the unmet health needs as well as the unforeseeable issues, such as COVID-19, that appear without prediction. Medicines are becoming more crucial in the global health coverage and, due to the high complexity and cost of their development and manufacturing, the pharmaceutical price is increasing, which is reflected in the life quality of the end user [1,2]. According to the World Health Organization (WHO), the counterfeit medicines are problematic, which is evidenced by their entry in the pharmaceutical supply chain, has been negatively affecting the worldwide health. In low and middle income countries, in the time period between 2007 and 2016, approximately 10.5% of the pharmaceutical products were substandard or falsified [3]. Worldwide, between 2013 and 2017, 42% of all the reported counterfeit medicines were from Africa whilst 21% were from America and 21% from Europe [4]. The accounting firm PwC states that 1–30% of drugs in circulation are fake and a high number of people die annually from toxic counterfeit pharmaceuticals [5]. Actually, in 2015, a study related with substandard drugs to fight the malaria in sub-Saharan Africa estimated more than 122,000 deaths per year of children under the age of five [6]. The profitable business associated with a large number of stakeholders belonging to the distribution network (illustrated in Figure 1) increases the difficulty of controlling the market of fraudulently mislabeled medicines [7].

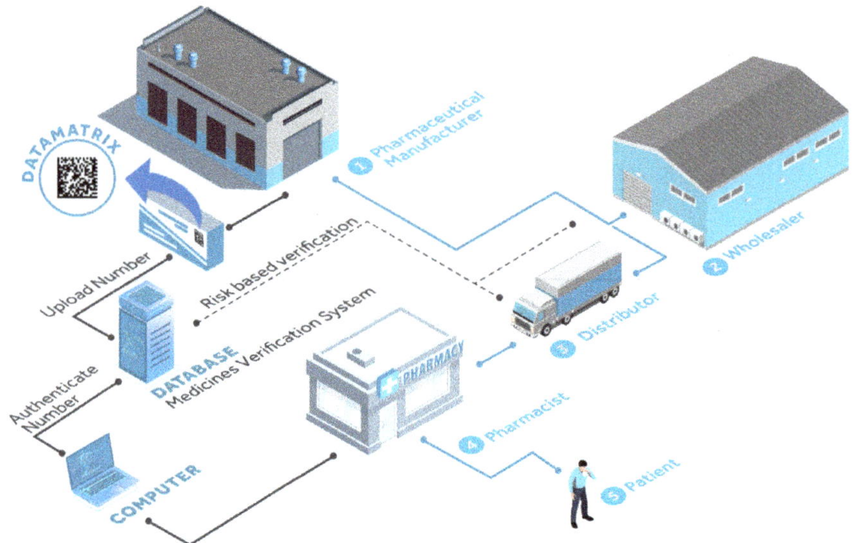

Figure 1. Product flow schematic. Adapted from [8].

The legislation applied in the Directive 2011/62/EU of the European Parliament and of the Council, and by the Commission Delegated Regulation (EU) 2016/161, looks forward to eradicate the counterfeit medicines by implementing a unique identifier which guarantees medication packaging authenticity. The regulation is set for the usage of a two-dimensional (2D) barcode—a machine-readable Data Matrix, which follows the format defined by the International Organization for Standardisation/International Electrotechnical Commission standard ('ISO/IEC') 16022:2006 [8–10]. The National Medicine Verification Organization is the entity which allows the trace of medicine from the production to the patient, ensuring its integrity [11], and holds the national repository where the data of the medicine packages circulating in the Portuguese territory resides. The national stakeholders will establish a connection with the repository to perform the verification and deactivation operations of the unique identifier of the Data Matrix. To evaluate the authenticity of the medicine, a comparison is performed between the information in the unique identifier registered by the entity responsible for uploading data to the European Medicines Verification System and the one in the medicine package [12]. This procedure is required in Portugal since the 9th of February 2019, according to the Decree-Law n.° 128/2013 [13].

Regarding the supply chain, at the pharmacy level, the medicine products are commonly packaged in three different levels: tertiary for logistics purposes (from the wholesaler to the pharmacy), secondary for inventory (performed within the pharmacy) and primary level, where the operator has direct contact with the product (e.g., vials or pills) as illustrated in Figure 2. The medicines can be traced through a bar-code which identifies the product when electronically scanned [14]. To ensure the interoperability of the pharmaceutical products verification repositories, it is mandatory to adopt a harmonized coding approach. The GS1 (Global Standards One) Data Matrix is embraced as the recommended standard, consisting in a unique identifier (product code—Global Trade Item Number (GTIN), serial number, batch number, expire date and, if required, the national registration number) and an anti-tampering device [15].

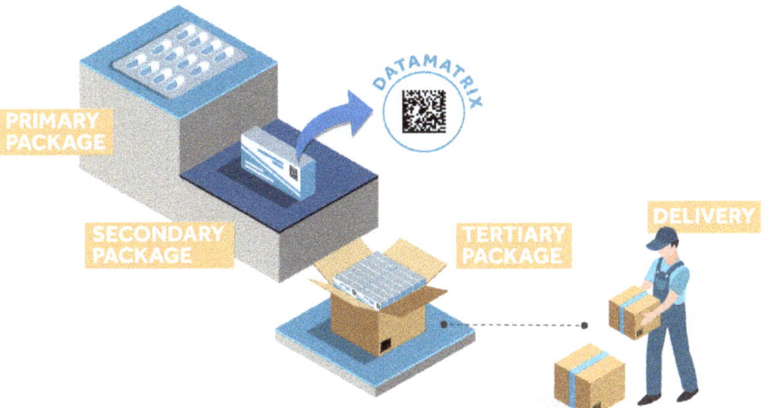

Figure 2. Packaging levels schematic. Adapted from [16].

There are actually several mathematical approaches to detect matrix-type markers, such as [17,18]. Some of the approaches are based on deep learning theory to solve the problem of barcode detection [19]. These algorithms are useful to detect the required patterns in image. The description of the markers generation can be found in [20].

Currently, some commercial solutions can achieve pharmaceutical product traceability through the Data Matrix reading. Funcode technology developed a QR Code/Data Matrix (2D Barcode) reader whose features comprise the scanning, with a Raspberry Pi, of tiny, moving and multiple barcodes using multiple cameras. The algorithm is robust to conditions such as non-uniform lighting or blurred and distorted codes. Its main applications include industrial automation, pharmaceutical products and consumer goods labels, intelligent logistics center, package tracking, automated tracking of product, among others [21]. Scandit presents a mobile barcode scanning application robust to different 1D or 2D symbologies such as Data Matrix. MatrixScan allows to locate, track and decode multiple barcodes, while displaying feedback in the device's screen as an Augmented Reality-overlay, to identify which have already been scanned. The software is able to handle barcodes in challenging conditions such as glare and metal reflections, low light and shadows, damaged, torn, and blurry barcodes, or skewed angles. Some applications in the healthcare sector emcompass shelf management, search-and-find packages, patient verification and medication tracking [22]. In the pharmaceutical/medical industry, Cognex resorts to image-based fixed-mount barcode readers (including Data Matrix symbology) to ensure the simultaneous track-and-trace of multiple medical devices on high-speed production lines, multiples sides, and even extreme angles. The algorithm is able to efficiently scan blurred, damaged or low contrast barcodes. Additional applications include the automation of inspection and distribution of primary (e.g., vials or pre-filled syringes) and secondary (e.g., vaccine boxes) packages. Cognex supplies guidance to other markets such as logistics, life sciences and aerospace [23].

To the best of the authors' knowledge, there is still not a established solution to be applied at the national hospital pharmacies. In this work, a hospital pharmacy scenario is focused, more precisely the pharmacy department of the Centro Hospitalar de Vila Nova de Gaia/Espinho, E.P.E., Portugal, where, on average, more than 27,000 medicine packages are handled per month. Due to the lack of human resources, the action of having each medicine be manually validated at the secondary level by an operator by means of a scanner is burdensome, time-consuming and not the foremost task for the operator. To prioritize some medicine packages, criteria such as their medical significance or cost might be applied. Moreover, the aforementioned action is monotonous, tedious, frequent and highly repetitive, due to the recurrent lifting and transferring of packages, and repeated hand/wrist motions to place the scan in the correct position and press the button to detect the Data Matrix code [24]. The addressed cases are common physical work-related risk factors that can cause upper limb musculoskeletal disorders to the operator [25], and therefore should be avoided. To include all the information of medicine products in a pharmaceutical system, hence enabling their traceability, as well as to contribute to the operator's ergonomy, a low cost solution for automating the simultaneous scanning of multiple medicine packages based on Data Matrix codes is proposed in this work. One important requirement was that the system should be easily integrated in the infrastructure of the Hospital Pharmacy.

The remainder of this paper is organized as follows. Section 2 presents the proposed Methodology addressing the hardware, the system architecture and the Software employed. Section 3 stresses the experimental results performed in hospital pharmacy facilities. Finally, Section 4 outlines the work with conclusions, comprising functionalities that may be interesting to incorporate in the system in the near future.

2. Proposed Methodology

Nowadays, the introduction of medicine packages into the authentication database is performed through the manual scan of their unique Data matrix, one by one by the user, becoming a repetitive and time-consuming process. To improve the users' ergonomy as well as to redirect them to other tasks, the proposed methodology addresses the simultaneous and autonomous acquisition of multiple codes. The present system is initiated by the user that presses a button and, after, the remaining process will automatically be executed. An image of the working area is acquired and the system can, autonomously, identify, extract the information of all the Data matrix codes contained in that area, and send it to the verification system database.

2.1. System Architecture

The structure of the developed system is presented in Figure 3a,b (frontal and bottom-up views, respectively), comprising a camera to acquire the images as well as a lighting system, in order to avoid being constrained by shadows or lack of brightness.

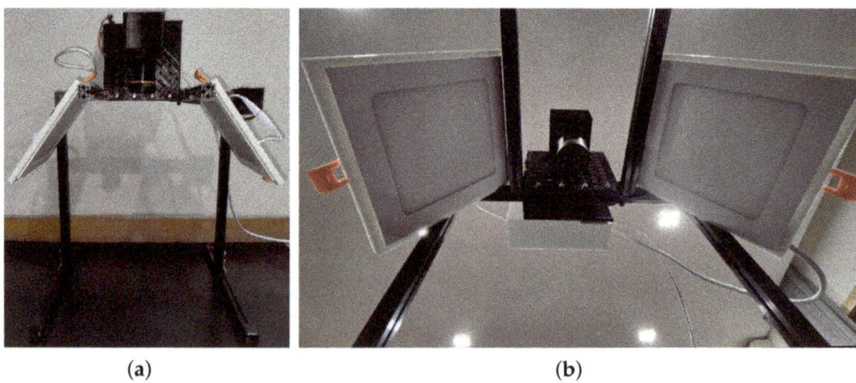

(a)   (b)

Figure 3. Developed system structure: (**a**) Frontal view; (**b**) Bottom-up view.

The acquired image can contain several Data matrix codes restricted by the size of the work area, which is 200 mm × 200 mm. The structure was assembled using MakerBeam small aluminum profiles and additive manufacturing (3D printing) parts. The prototype is a ready-to-go system since it has all the components installed and is also equipped with a power supply. It only requires to be connected by USB (Universal Serial Bus) to the computer (in this case study, from the hospital pharmacy) and the electrical grid. An illustrative image acquired by the camera is presented in Figure 4.

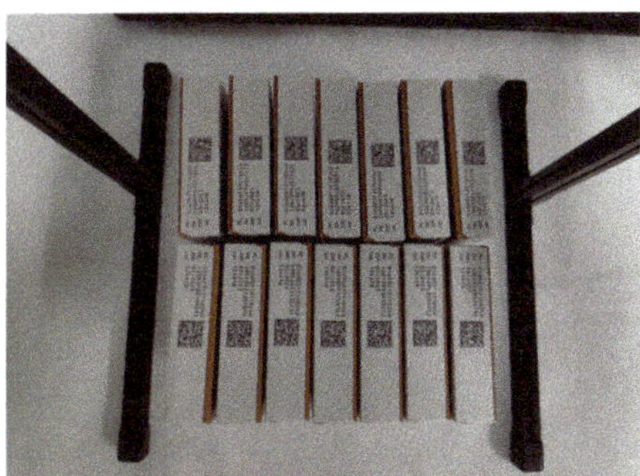

Figure 4. Acquired image from the camera according to a top-bottom perspective.

The system architecture schematic is presented in Figure 5. The proposed system is connected to the computer and it will be identified as a keyboard, the same process as the Data Matrix reader used in the current implementation of the presented case study. The Pro Micro Arduino board, composed of an ATmega32U4, is configured to be recognized as a keyboard by the MVO (Medicines Verification System) verification computer. The identified Data Matrix codes are processed by the Raspberry Pi library *libdmtx* (https: //github.com/dmtx/libdmtx, accessed on 25 September 2022) and transferred one by one through a serial port to the Arduino that further sends it to the computer (using the defined protocol). The used integrated development environment was Lazarus, into the raspbian operating system. Moreover, the Arduino board is also used to perform an Human-Machine Interface (HMI) to the user, by one RGB led, one push-button, and also to control the lighting system (with 8+8 Watt LED panels). The RGB led allows the user to infer the state of the system from the LED color, while the push-button is used to start the acquisition process of a new image containing a Data Matrix, both presented in Figure 6. Regarding the feedback from the MVO to the computer, the connection has not been implemented yet.

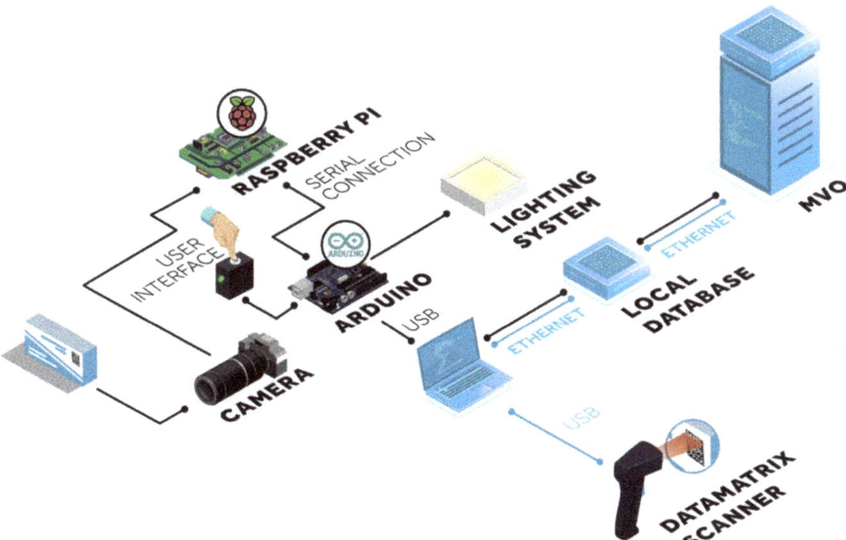

Figure 5. System architecture schematic. The current process is illustrated by the blue arrows while the proposed methodology is represented by the black ones. The connection from the MVO to the computer, in order to give visual feedback on the medicine packages, has not been implemented yet.

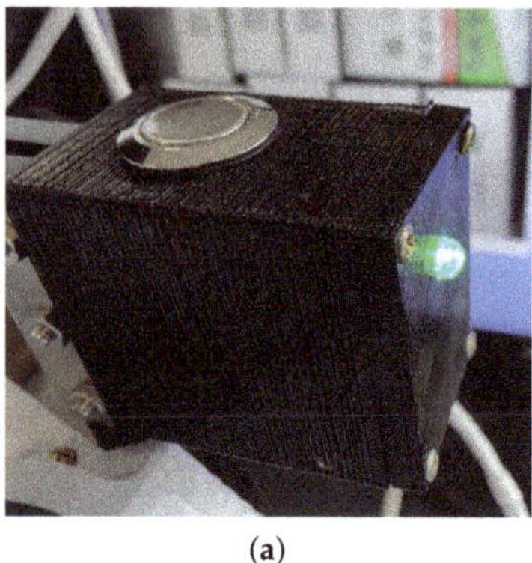
(**a**)

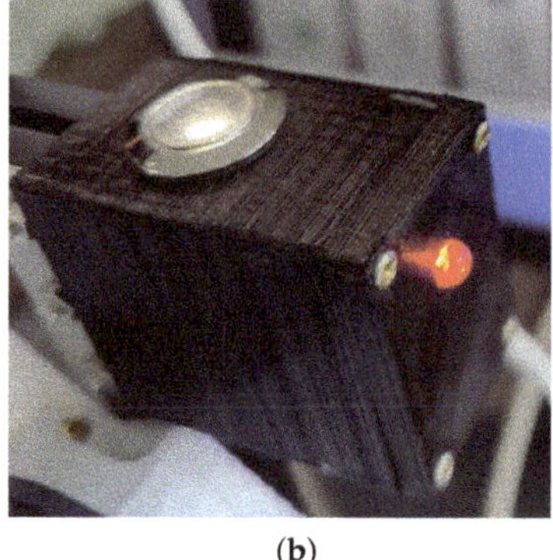
(**b**)

Figure 6. The start of the process order is made by a pushbutton: (**a**) green indicates that the system is available to start a new process; (**b**) red shows that the system is searching and decoding Data Matrix code.

The serial port communication between Arduino and Raspberry uses the channel library developed by the authors that encapsulate several data that will be decoded at the target. The image is captured with a 12 Megapixels camera connected to the Raspberry pi camera BUS attached with an 8–50 mm C-Mount Zoom Lens. The illumination system allows having a lens aperture that enables a depth of field capable of detecting Data Matrix codes in boxes at a distance of 15–25 cm. The HMI is also composed of a monitor that gives information to the user about the Data Matrix detection (Data Matrix codes are bordered by green squares) and also the total number of decoded codes, as illustrated in Figure 7.

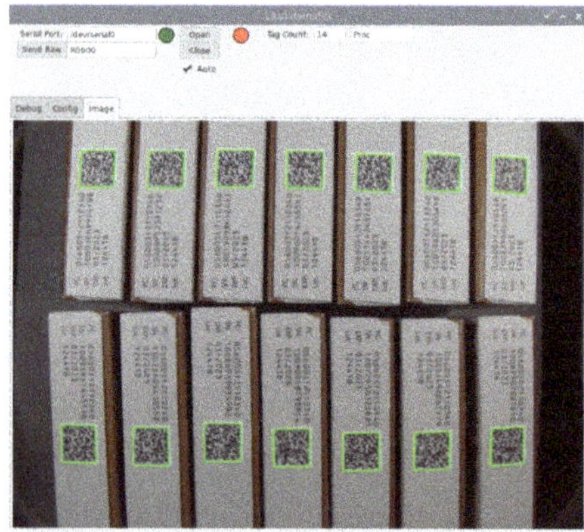

Figure 7. User interface showing the Data Matrix detection illustrated by the green squares.

2.2. Data Matrix

Data Matrix codes are two-dimensional symbols that hold a dense pattern of data with built-in error correction. Special scanners were designed to read them; however, high-resolution cameras can be used to acquire multiple Data Matrix images simultaneously. If the resolution is enough to identify each point of the Data Matrix, a library can be applied to decode its data. In the developed system, the *libdmtx* library is used to scan the Data Matrix codes in the medicines' boxes. *libdmtx* is an open-source software for reading and writing Data Matrix barcodes on Linux, Unix, OS X, Windows, and mobile devices. According to [26], several commercial scanners can be used. The software developer (for the hospital management) published the Data Formatting protocol and settings that should be used for the configuration of each matrix scanner (Datalogic Quickscan QD2403, Zebra

DS2278, Honeywell HH660, 1450G, and 7580G). Several validation tests were executed with different medicine boxes. For example, Figure 8a presents the acquired image and Figure 8b the respective *libdmtx* outputs that allowed to verify its operation (with the proper lighting system). Figure 8d exhibits the Data Matrix detected by the library, which are enclosed in green squares, and Figure 8e shows the extracted codes according to the defined protocol where each line corresponds to the decoded matrix code, as it can be read by a matrix code scanner, i.e., the output of the *libdmtx* library.

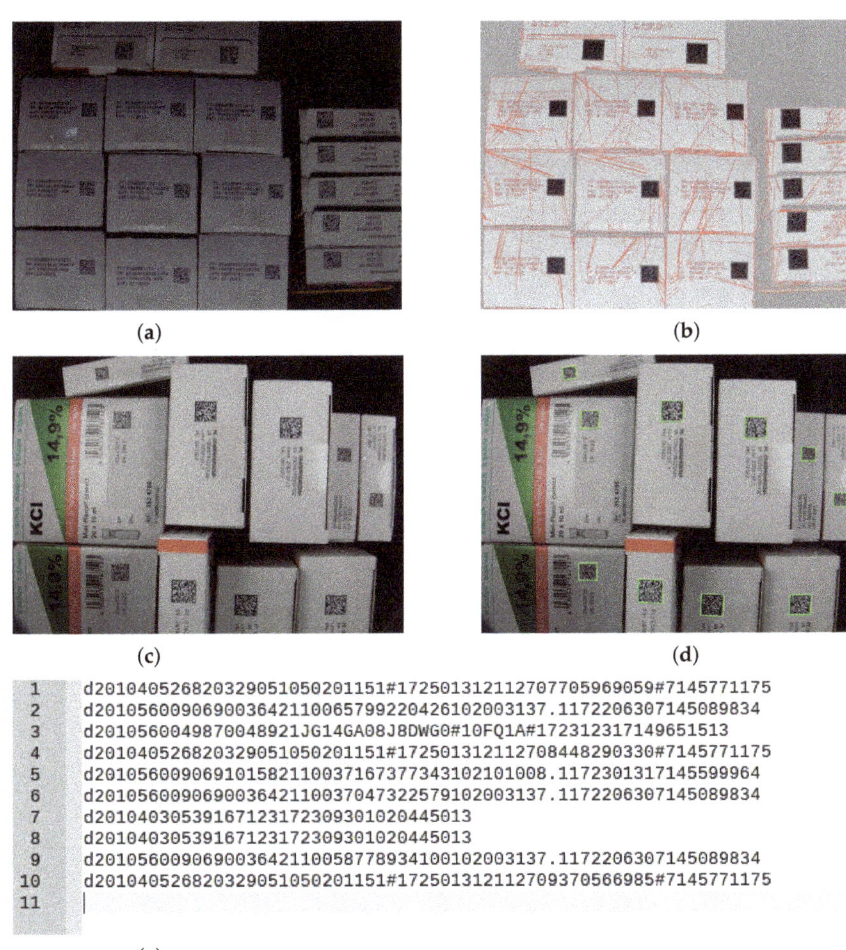

(a)

(b)

(c)

(d)

```
 1   d2010405268203290510502011151#1725013121127077059690 59#7145771175
 2   d20105600090690036421100657992204261020031 37.1172206307145089834
 3   d20105600049870048921JG14GA08J8DWG0#10FQ1A#172312317149651513
 4   d2010405268203290510502011151#1725013121127084482903 30#7145771175
 5   d20105600090691015821100371673773431021010 08.1172301317145599964
 6   d20105600090690036421100370473225791020031 37.1172206307145089834
 7   d2010403053916712317230930102044 5013
 8   d2010403053916712317230930102044 5013
 9   d20105600090690036421100587789341001020031 37.1172206307145089834
10   d2010405268203290510502011151#1725013121127093705669 85#7145771175
11   |
```

(e)

Figure 8. Scanning process: (**a**) Acquired image with 16 codes; (**b**) *libdmtx* scanning output; (**c**) Original image; (**d**) Data Matrix identified by the library; (**e**) The extracted codes according to the defined protocol (libdmtx codes output, as read by a commercial data matrix scanner).

3. Experiments and Results

The developed prototype was installed and tested at the Hospital Pharmacy of the Centro Hospitalar de Vila Nova de Gaia/Espinho, E.P.E., by technicians which perform the medicine packages registration. The proposed prototype is a stand-alone system that can be placed alongside with the conventional scanner, replacing it. A space of 300 mm × 300 mm in a table is enough to accommodate the developed reader (with a height of 400 mm). It requires to be supplied by 110 V or 230 V and it owns all the required hardware and software to introduce by USB the decodes matrix codes present on the boxes into the hospital system. The developed prototype was connected, by a USB link, to the same computer to which the original Data Matrix scanner is connected. The computer is used to send the verification codes to the central database. Instead of using the original Data Matrix scanner, the proposed system was assessed by reading the codes, decoding, and introducing them in the database by USB. Figure 9 shows the prototype installed and tested at the hospital pharmaceutical facilities. The right monitor shows the developed application and the Data Matrix identified. The left one shows the hospital computer receiving the codes to validate.

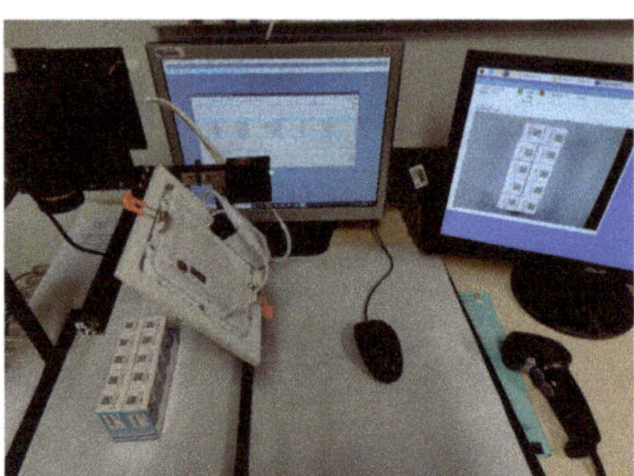

Figure 9. Prototype installed and tested at hospital pharmaceutical facilities. The right monitor presents the developed application. At left the hospital computer receiving the Data Matrix codes to validate.

The system is launched when the user presses the button and the lighting system is turned on for two seconds to stabilize the camera and capture the image. The LED turns red and the *libdmtx* is used to extract and decode each Data Matrix. Each code is introduced into the hospital computer (as a keyboard) using the same settings and protocol as the original scanner was programmed. The acquired image and its Data Matrix codes identified and delimited by green squares, allow the user to visually validate the task.

Preliminary tests were executed in order to analyze the time and the robustness of the Data Matrix code detection system. According to the commonly performed process, a technician takes between half a second and one second to scan each medicine box. In contrast, the proposed system takes about two seconds to stabilize the image and an average of one second per box. Meanwhile, while the scanning is being performed, the technician can be redirected to do another task. The proposed system has the advantage to scan multiple codes at the same time. To assess the system robustness, the scanning process was performed on boxes covered by reflector plastic as well as rounded bottles, having achieved successful results, as demonstrated by Figure 10.

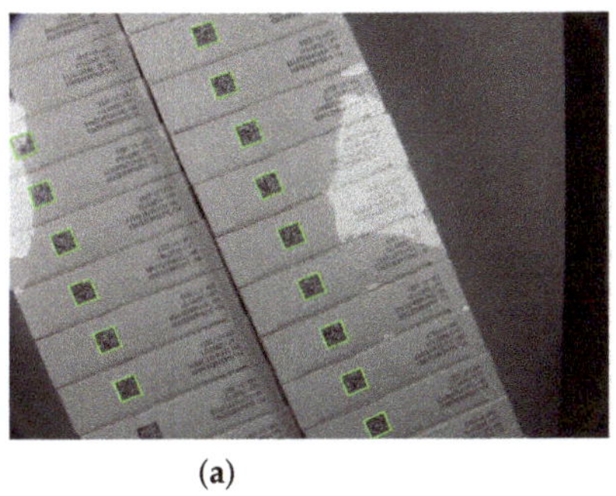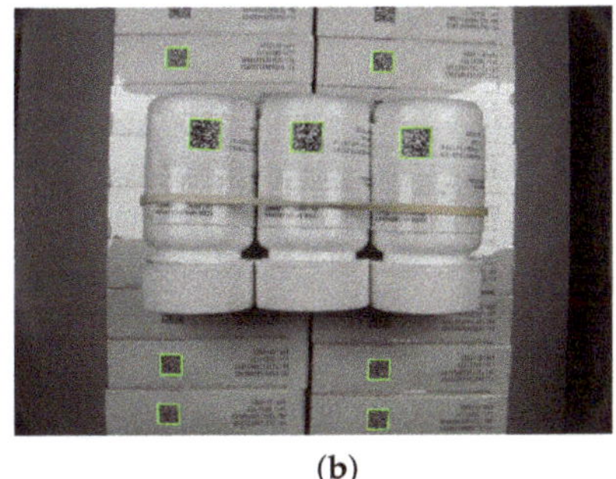

(a)   (b)

Figure 10. Medicine packages scanned: (**a**) covered by reflector plastic; and (**b**) rounded bootles.

The integration with the hospital software was tested and validated. Instead of using the original scanner to shoot Data Matrix codes one by one, the proposed system acquires one image and introduces several Data matrix codes sequentially without the intervention of the technician. The obtained feedback from four technicians (from different shifts) that used the prototype was very positive since they mentioned that the system allows them to do other tasks while introducing the codes increasing the productivity. Moreover, they pointed out some future work possibilities, such as using a conveyor belt to move the boxes in front of the camera in a FIFO (First In, First Out) cue type. Moreover, increasing the detection area was also pointed. As a statistical result, detection robustness was done with real medicine boxes. In total, about two hundred medicine boxes were placed in front of the camera (within 15 images) and 90% of Data Matrix codes were correctly identified, using the proposed system. The remaining codes had slight defects, such as being torn (as illustrated in Figure 11) or their color being too grey.

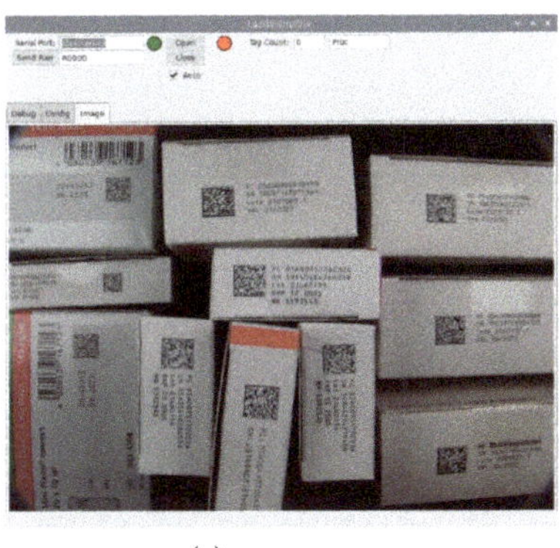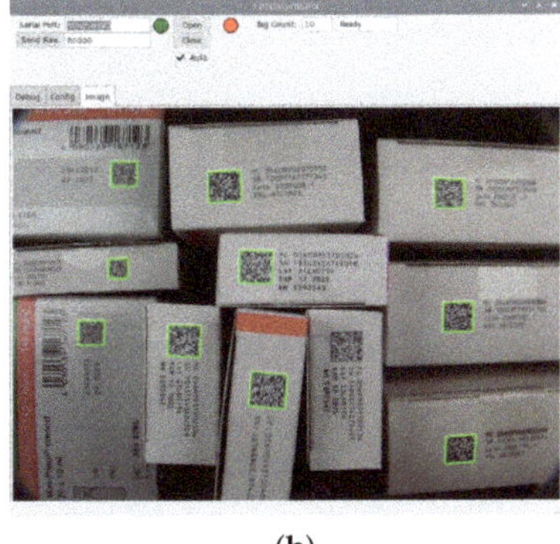

(a)   (b)

Figure 11. Medicine packages: (**a**) to be scanned; and (**b**) torn code paper is not detected by the algorithm.

In order to evaluate the illumination interference, a test was performed with different light conditions. The first test was executed turning the embedded illumination off. In this case (with a natural illuminance of 150 Lux), the extracting capacity was decreased to 18%, as presented in Figure 12 where from eleven matrix codes, only two of them were extracted.

Figure 12. Acquired image with a natural illuminance of 150 Lux and LED panels off. An extraction ratio of 18% was observed.

More tests were executed to evaluate the interference of the lighting. Table 1 summarizes the obtained ratios for data matrix extractions with different lightings. The first three tests were evaluated with both LEDS panels off for an Illuminance of 5, 90 and 150 Lux. The proposed system is very dependent on the illumination conditions. As it can be observed, one LED panel is enough but brings some reflexes that are reduced with two LED panels assembled as previously addressed. In these cases (one and two LED panels on), the ratio of detection for six image samples were 100% even when changing the external illumination from 0 to 1000 Lux.

Table 1. Extracting result ratio for different illumination conditions.

	Natural Illuminance (Lux)	Measured Illuminance (Lux)	Extracting Ratio Result (%)
LEDS off	5	5	0 (0 of 11)
LEDS off	90	90	0 (0 of 11)
LEDS off	150	150	18 (2 of 11)
LEDS on	90	1985	100 (11 of 11)
LEDS on midrule LEDS on	440 1040	2320 3200	100 (11 of 11) 100 (11 of 11)
LEDS on (1 panel)	0	980	100 (11 of 11)
LEDS on (2 panels)	0	1880	100 (11 of 11)

In order to improve the work, there are two main points that might be pointed out (to solve a limitation of the prototype), namely increasing the size of the acquired image, and a more powerful computational performance could be included to speed up the detection of matrix codes.

4. Conclusions and Future Work

The counterfeit medicine problem is still negatively affecting worldwide health. The Data Matrix code, a unique identifier that guarantees medication packaging authenticity, allows to trace a medicine from its production until its delivery to the patient. In the hospital pharmaceutical industry, at the secondary level, for the inventory task, the scanning of the medicine packages to introduce them into the management system is performed manually by the user through a Data Matrix reader. This paper proposes a low cost system to control medicine packages, acquiring a set of medicine boxes simultaneously and autonomously introducing them into the authentication database. A proof of concept was performed by installing the proposed system in a hospital pharmacy and testing it with real medicine. The achieved results were very encouraging, and it is expected that, with the addition of extra functionalities, the system can be used as a real support for the pharmacist work. The proposed prototype, works as a standalone system, that can be added to any validation system by connecting an USB port. It owns an embedded illumination system that eliminated the exterior lighting conditions. It could reach 90% of extraction ratio for real conditions and real medicines. Nevertheless, there are some limitations such as the small acquisition area and the processing time for extraction of the matrix codes.

As future work, feedback from the MVO database will be received through an ethernet port and the system will project the codes accordingly. For this purpose, a projection mapping will be used to assist the technician in the authentication operation. This task will be performed as soon as the information from the MVO is received. To improve the system flexibility and to avoid the user being forced to put the packages face with code upside, new cameras in distinct orientations will be integrated as well as the size of the work area will be increased. Additionally, including a conveyor belt in order to increase the amount of medicine packages which are autonomously scanned, and a robotic manipulator to pick the counterfeit medicine boxes, are improvements to consider. Finally, in order to improve the detection ratio, a machine learning based vision system could be added to detect matrix codes. Then, it would be possible to map the matrix codes and detect the missed ones improving their quality (e.g., rotation, histogram equalization, among others) and sent them again to the *libdmtx* library. In this way, it would be possible to improve the detection and extraction.

Author Contributions: The contributions of the authors of this work are pointed as follows: Conceptualisation, J.L. and P.C.; Methodology, All; Software, J.L. and P.C.; Validation, C.R., J.L. and L.R.; Investigation, C.R., J.L. and P.C.; Data Curation, All; Writing—Original Draft Preparation, C.R., J.L. and L.R.; Writing—Review and Editing, C.R. and J.L.; Visualisation, C.R., J.L. and P.C.; Supervision, J.L. and P.C.; Project Administration, J.L. and P.C. All authors have read and agreed to the published version of the manuscript.

Funding: This work is financed by National Funds through the Portuguese funding agency FCT - Fundação para a Ciência e Tecnologia, within the project LA/P/0063/2020.

Institutional Review Board Statement: Not applicable.

Informed Consent Statement: Not applicable.

Conflicts of Interest: The authors declare no conflict of interest.

Abbreviations

The following abbreviations are used in this manuscript:

COVID	Coronavirus Disease
EPE	Entidades Públicas Empresariais—Public Business Entities
EU	European Union
FIFO	First In, First Out
GS1	Global Standards One
GTIN	Global Trade Item Number
USB	Universal Serial Bus
WHO	World Health Organization

References

1. Aitken, M. Understanding the Pharmaceutical Value Chain. *Pharm. Policy Law* **2016**, *18*, 55–66. [CrossRef]
2. Tengler, J.; Kolarovszki, P.; Mojský, V.; Filadelfiová, L. Proposal for Drug Monitoring in the Supply Chain. In *Reliability and Statistics in Transportation and Communication*; Kabashkin, I., Yatskiv, I., Prentkovskis, O., Eds.; Springer International Publishing: Cham, switzerland, 2021; pp. 318–327.
3. WHO. *A Study on the Public Health and Socioeconomic Impact of Substandard and Falsified Medical Products*; WHO: Geneva, Switzerland, 2017.

4. WHO. *WHO Global Surveillance and Monitoring System for Substandard and Falsified Medical Products*; WHO: Geneva, Switzerland, 2017.
5. Behner, P.; Hech, M.L.; Wahl, F. *Fighting Counterfeit Pharmaceuticals. New Defenses for an Underestimated—and Growing—Menace*; PwC: London, UK, 2017.
6. Renschler, J.P.; Walters, K.M.; Newton, P.N.; Laxminarayan, R. Estimated Under-Five Deaths Associated with Poor-Quality Antimalarials in Sub-Saharan Africa. *Am. J. Trop. Med. Hyg.* **2015**, *92*, 119–126. [CrossRef] [PubMed]
7. Sweileh, W.M. Substandard and falsified medical products: Bibliometric analysis and mapping of scientific research. *Glob. Health* **2021**, *17*, 114. [CrossRef] [PubMed]
8. Introduction to the European Medicines Verification System (EMVS). Available online: https://emvo-medicines.eu/ (accessed on 27 December 2021).
9. Directive 2011/62/EU of the European Parliament and of the Council of 8 June 2011 amending Directive 2001/83/EC on the Community code relating to medicinal products for human use, as regards the prevention of the entry into the legal supply chain of falsified medicinal products. *Off. J. Eur. Union* **2011**, *54*, L 174/74–87. [CrossRef]
10. *Commission Delegated Regulation (EU) 2016/161 of 2 October 2015 Supplementing Directive 2001/83/EC of the European Parliament and of the Council by Laying Down Detailed Rules for the Safety Features Appearing on the Packaging of Medicinal Products for Human Use*; European Commission: Dublin, Ireland, 2016; Volume L 32/1, 27p.
11. MVO Portugal. A MVO Portugal. Available online: https://www.mvoportugal.pt/pt/ (accessed on 27 December 2021).
12. MVO Portugal. Sistema de Verificação de Medicamentos. 2018. Available online: https://mvoportugal.pt/pt/sistema-de-verificacao-de-medicamentos (accessed on 27 December 2021).
13. SNS Serviço Nacional de Saúde. Medicamentos de uso Humano. 2018. Available online: https://www.sns.gov.pt/noticias/2018/08/16/medicamentos-de-uso-humano/ (accessed on 27 December 2021).
14. Klein, K.; Stolk, P. Challenges and Opportunities for the Traceability of (Biological) Medicinal Products. *Drug Saf.* **2018**, *41*, 911–918. [CrossRef] [PubMed]
15. The Global Language of Business. Recommendations on a Harmonised Implementation of the EU Falsified Medicines Directive Using GS1 Standards. 2016. Available online: https://www.gs1.org/sites/default/files/docs/healthcare/position-papers/final_recommendations_harmonised_implementation_fmd_using_gs1_dec2016.pdf (accessed on 29 December 2021).
16. GS1 Brasil—Associação Brasileira de Automação. Guia de Apoio à Codificação de Medicamentos. 2017. Available online: https://www.gs1br.org/educacao-e-eventos/MateriaisTecnicos/Guia%20de%20Apoio%20%C3%A0%20Codifica%C3%A7%C3%A3o%20de%20Medicamentos.pdf (accessed on 28 December 2021).
17. Elgendy, M.; Guzsvinecz, T.; Sik-Lanyi, C. Identification of Markers in Challenging Conditions for People with Visual Impairment Using Convolutional Neural Network. *Appl. Sci.* **2019**, *9*, 5110. [CrossRef]
18. Teoh, M.K.; Teo, K.T.K.; Yoong, H.P. Numerical Computation-Based Position Estimation for QR Code Object Marker: Mathematical Model and Simulation. *Computation* **2022**, *10*, 147. [CrossRef]
19. Zhang, H.; Shi, G.; Liu, L.; Zhao, M.; Liang, Z. Detection and identification method of medical label barcode based on deep learning. In Proceedings of the 2018 Eighth International Conference on Image Processing Theory, Tools and Applications (IPTA), Xi'an, China, 7–10 November 2018; pp. 1–6. [CrossRef]
20. Kambilo, E.K.; Zghal, H.B.; Guegan, C.G.; Stankovski, V.; Kochovski, P.; Vodislav, D. A Blockchain-Based Framework for Drug Traceability: ChainDrugTrac. In Proceedings of the 37th ACM/SIGAPP Symposium on Applied Computing, Virtual Event, 25–29 April 2022; Association for Computing Machinery: New York, NY, USA, 2022; pp. 1900–1907. [CrossRef]
21. Funcode Technology. QR Code/ DataMatrix (2D Barcode) Reader for Industrial Automation. 2016. Available online: https://www.funcode-tech.com/Fun2D_QRCode_Reader_for_Industrial_Automation_AutoID40_en.html (accessed on 13 December 2021).
22. Scandit. Ways to Use Smartphone-Based Barcode Scanning in the Pharmacy. 2020. Available online: https://www.scandit.com/blog/pharmacy-apps/ (accessed on 13 December 2021).
23. Cognex. Pharmaceutical & Medical Solutions Easily and Cost-Effectively Comply with Patient Safety and Traceability Requirements. Available online: https://www.cognex.com/industries/pharmaceuticals-medical (accessed on 13 December 2021).
24. Wang, H.Y.; Feng, Y.T.; Wang, J.J.; Lim, S.W.; Ho, C.H. Incidence of low back pain and potential risk factors among pharmacists—A population-based cohort study in Taiwan. *Medicine* **2021**, *100*, e24830. [CrossRef] [PubMed]
25. Spielholz, P.; Silverstein, B.; Morgan, M.; Checkoway, H.; Kaufman, J. Comparison of self-report, video observation and direct measurement methods for upper extremity musculoskeletal disorder physical risk factors. *Ergonomics* **2001**, *44*, 588–613. [CrossRef] [PubMed]
26. MVO Portugal. LEGISLAÇÃO E DOCUMENTAÇÃO. 2018. Available online: https://mvoportugal.pt/pt/sistema-de-verificacao-de-medicamentos (accessed on 27 December 2021).

Article

Risk Treatment for Energy-Oriented Production Plans through the Selection, Classification, and Integration of Suitable Measures

Stefan Roth *, Mirjam Huber, Johannes Schilp and Gunther Reinhart

Fraunhofer Institute for Casting, Composite and Processing Technology IGCV, Am Technologiezentrum 10, 86159 Augsburg, Germany; mirjam.huber@igcv.fraunhofer.de (M.H.); johannes.schilp@igcv.fraunhofer.de (J.S.); gunther.reinhart@igcv.fraunhofer.de (G.R.)
* Correspondence: stefan.roth@igcv.fraunhofer.de

Citation: Roth, S.; Huber, M.; Schilp, J.; Reinhart, G. Risk Treatment for Energy-Oriented Production Plans through the Selection, Classification, and Integration of Suitable Measures. *Appl. Sci.* **2022**, *12*, 6410. https://doi.org/10.3390/app12136410

Academic Editor: Panagiotis Tsarouhas

Received: 19 April 2022
Accepted: 22 June 2022
Published: 23 June 2022

Publisher's Note: MDPI stays neutral with regard to jurisdictional claims in published maps and institutional affiliations.

Abstract: With rising electricity prices, industries are trying to exploit opportunities to reduce electricity costs. Adapting to fluctuating energy prices offers the possibility to save electricity costs without reducing the performance of the production system. Production planning and control play key roles in the implementation of the adjustments. By taking into account the price forecasts for the electricity markets in addition to machine utilization, work in process, and throughput time, an energy-oriented production plan is set up. The electrical energy is procured based on this plan and the associated load profile. Deviations from the forecast and the purchased amount of electricity lead to high penalties, as they can destabilize the energy system. For manufacturing companies, this means that machine failures and other unexpected events must be dealt with in a structured manner to avoid these penalty costs. This paper presents an approach to selecting, classifying, and integrating suitable measures from existing risk treatment paths into the production schedule. The selection of measures is based on a hybrid multi-criteria decision-making method in which the three relevant criteria, namely, cost, energy flexibility, and risk reduction, are weighted by applying both an analytic hierarchy process and entropy, and they are then prioritized according to multi-attribute utility theory. In the following, the subdivision into preventive and reactive measures is made in order to choose between the modification of the original plan or the creation of backup plans. With the help of mathematical optimization, the measures are integrated into the production schedule by minimizing the cost of balancing energy. The approach was implemented in MATLAB® and validated using a case study in the foundry industry.

Keywords: production planning and control; scheduling; energy flexibility; risk management; fault management; multiple-criteria decision-making

1. Introduction

Industries are facing the challenge of rising costs for electrical energy. The average electricity price for new contracts for the industry in Germany increased from €0.1207 per kWh in 2010 to €0.2138 per kWh in 2021. During this period, bulk buyers with an annual consumption of 70 million kWh or more had an increase from €0.0863 to €0.1149 per kWh with occasional drops [1]. Both small and medium-sized as well as energy-intensive industrial companies are therefore encouraged to take advantage of opportunities to reduce their electricity costs in order to maintain their competitiveness in the long term.

A promising approach to reduce these costs is adjusting the electricity demand to the variable prices on the energy markets. In the day-ahead market, the number of hours with negative electricity prices rose from 134 in 2018 and 211 in 2019 to 298 h in 2020. The mean value of the negative prices ranged in these years between −€13.70 and −€17.30 per MWh. Averaged over the respective year, the maximum daily price spread was around

€30 per MWh [2,3]. This price volatility in the electricity markets is significant for potential cost reduction.

The prerequisite to take advantage of these favorable prices and to comply with the restrictions on annual peak load and deviations is so-called energy-oriented production planning and control (PPC), which considers variable price forecasts for the electricity markets in addition to machine utilization, work in process, throughput time, and other criteria in a production system [4]. PPC is also responsible for dealing with unforeseen events such as machine failures and the corresponding adjustment of the order sequence in order to achieve the specified production targets as cost-effectively as possible [5]. With regard to electricity consumption, it is particularly important to avoid exceeding the annual peak load and deviations from the amount of electricity procured. These lead to higher grid charges or penalty costs for the imbalance caused between the available and used electrical energy in the electricity system. In order to avoid these costs, approaches within the energy-oriented PPC are required to evaluate potential risks within a production plan and to integrate suitable measures.

Roth et al. [6] introduce an approach for developing and evaluating risk treatment paths in energy-flexible production systems based on interpretive structural modeling and the calculation of conditional probabilities using Bayesian networks. Figure 1 depicts the idea of the approach in which a risk treatment path and the measures contained therein are integrated into the production plan in order to obtain a risk-treated production plan. The advantage of these paths compared to a situational reaction is that all the effects and possible interactions can be considered in advance. The approach creates a multitude of paths, as all identified risks and measures are linked based on their interactions and conditional probabilities.

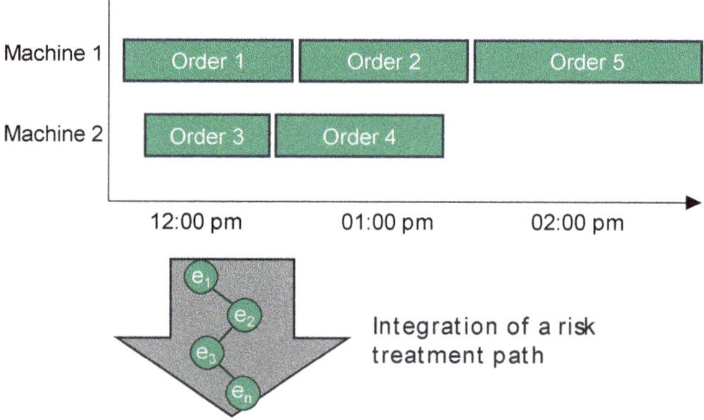

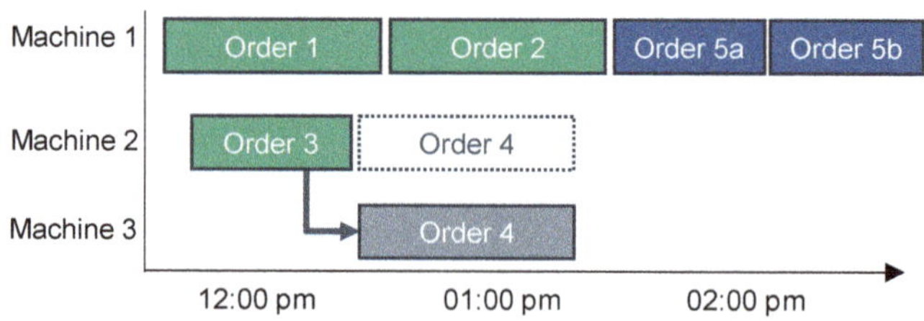

Figure 1. Integration of a risk treatment path into a production plan [6].

The approach thus leaves a need for further development with regard to the selection of a suitable path and the classification and integration of the measures it contains. In the literature, a large number of methods exist for the selection of alternatives, but there is no adaption to energy-oriented PPC. Furthermore, the state of the art offers no subdivision of energy flexibility measures into preventive and reactive, which is necessary for forward looking direct adjustments to the plan and alternative strategies if faults occur.

It is assumed that this should also follow a specific and structured approach in order to reduce the effort required for risk treatment and thus increase the acceptance of its implementation in companies. This article presents an approach that responds to these requirements. It starts with the paths shown in Figure 1 and presents how a preferred path can be selected and how the measures contained can be integrated into the production plan.

In the following, Section 2 first presents the state of the art in the relevant areas and the resulting need for action. Section 3 then introduces the scientific concept of the approach, which is described in detail in Section 4. The application based on a use case in a foundry is depicted and discussed in Section 5. The article closes with a conclusion and outlook in Section 6. The key notations are listed in Table 1.

Table 1. Notations.

Parameter	Description
A	Weighting decision matrix
a	Elements of the weighting decision matrix
$\Delta C_{tot}{}^M$	Change in measure-induced total costs
$\Delta C_{tot}{}^R$	Change in risk-induced total costs
ΔC_{tot}	Change in total costs
$cons_t$	Energy consumption in time unit t
D	Decision matrix
$duration_{ks}$	Duration of job k on station s
e	Entropy
EF	Energy flexibility
EF_{states}	Energy flexibility states
EF_t	Energy flexibility time
EF_{tech}	Energy flexibility technologies (e.g., storage facilities)
end_{ks}	Due date of job k on station s
i	Alternative
j	Criteria
k	Job
P	Probability of occurrence of risks and measures
P_{risk}	Probability of occurrence of risks
$peak_{max}$	Maximum peak load
Δpow_t	Deviation of the actual energy consumption from the forecasted load profile
$Power_{ks}$	Power consumption of job k on station s
$Power_t^{planned}$	Planned power consumption in time period t
QDC	Quantity deviation cost
r_{ij}	Rating for alternative i with criterion j
s	Station
$start_{ks}$	Start time of job k on station s
t	Time unit
u_{ij}	Marginal utility score for alternative i with criterion j
U_i	Final utility score for alternative i
$\overrightarrow{v}$	Eigenvector
w	Weighting
w_{AHP}	Subjective criteria weighting by analytic hierarchy process (AHP)
$w_{Entr,j}$	Objective criteria weighting by entropy
w_j^{final}	Final weighting for a criterion j
λ_{max}	Largest eigenvalue

2. State of the Art

This section first describes the state of the art in the field of risk management and energy-oriented PPC. Since various criteria have to be taken into account in risk management, common methods for solving complex decision problems are presented. In the following, methods for the categorization of measures are introduced. The need for action on which this contribution is based is then derived from these key areas.

2.1. Risk Management and Energy-Oriented PPC

Since failures, in general, can have far-reaching negative effects on the performance of a production system as well as on the manufacturing costs of the products, many articles offer a structure for the causes and effects of malfunctions as well as approaches for situational fault management. According to Schwartz and Voß [7], faults are events that have an effect on a process with a deviation from what was intended to occur. Greve [8] distinguishes between those faults which originate from the process and those that influence the process from outside. The causes are often used to further classify faults. A distinction is made between equipment-related, personnel-related, material-related, information-related, and order-related causes of failure by various authors [9–13]. Every cause of failure can be linked to its effects on a production system. An allocation of causes and effects, which also takes into account deviations in the electrical load profile, was worked out by Schultz [13]. In his contribution, a system is presented with which deviations in energy-flexible production systems can be counteracted with situational measures. Rösch et al. [14] introduce an approach for cost-based online scheduling to enable reactions to short-term changes and thus makes a contribution to fault management in energy-flexible production systems.

Risk management is characterized in particular by the fact that it also takes into account the probability of occurrence of the event when dealing with faults [15]. Many risk management approaches in the field of production systems are mostly based on the risk management process described in DIN ISO 31,000 [15] with risk identification, risk analysis, risk assessment, and risk treatment, whereby individual process steps are partially combined or sub-processes are supplemented. Various frameworks and approaches were developed specifically for the manufacturing sector. For example, the framework for risk management developed by Oduoza [16] enables key risk indicators in the manufacturing sector to be searched for and identified. Specific approaches can also be found in the area of PPC, such as Klöber-Koch et al. [17], who add production risks to a classic PPC system in order to make predictions about impending risk situations and to take these into account in the planning process.

If the energy consumption and energy costs in the production environment are to be considered in particular, specific work can also be found for this. Abele et al. [18] simulate disruptions in order to investigate their effects on the energy flexibility and costs of an energy-optimal production plan. The energy-optimized plan serves as an input variable for the subsequent simulation of faults. The changes in the production plan caused by disruptions lead to changed load profiles and thus also to changed electricity costs. The influence of disruptions is ultimately examined on the basis of the change in electricity costs and serves as a basis for decision-making for future investments in energy-oriented production.

Schultz et al. [19] integrate energy as a limited resource within the framework of energy-oriented production control. Exceeding the available energy resources is possible, but it is associated with disproportionately high costs. Taking into account the predominant goal in production control of adhering to delivery deadlines, the authors consider the fluctuations in energy consumption associated with the production process. The aim is to minimize the deviations from the planned load profile.

Golpîra [20] introduces smart energy-aware manufacturing plant scheduling. By proposing a new multi-objective robust optimization for a factory microgrid, the approach can be considered risk-based, as it considers the conditional value-at-risk. Coca et al. [21] illustrate the simultaneous evaluation of sustainability dimensions, containing environmental and occupational risks with an industrial case from the metal-mechanic sector. Energy flexibility in production systems is specifically considered in the work of Simon et al. [22]. The authors discuss the relevance of risk evaluation with regard to energy flexibility measures and present an approach to assess energy flexibility in production systems in terms of their risk potential on production goals, such as quality, costs, and throughput times.

2.2. Solving Complex Decision Problems

One difficulty in risk management is the different, sometimes conflicting, target values that need to be taken into account when making decisions. In the context of energy-oriented production planning, one example is that the capacity of stations should be utilized to fulfill the delivery time and that machine costs are reduced. On the other hand, the station utilization should be reduced in time windows with high electricity prices and be shifted to times with lower or negative prices.

In the literature, there are different approaches for multi-criteria decision-making (MCDM). A large number of articles in this field distinguishes between work on the development of new MCDM methods and detailed descriptions of their functioning (including Saaty [23], Ishizaka and Nemery [24], and Alinezhad and Khalili [25]), and comparative work that analyzes known MCDM methods (including Vujicic [26], Wang et al. [27], and Zanakis et al. [28]). Zavadskas et al. [29] give an initial overview of the relevant literature within the categories mentioned.

Zavadskas et al. [29] emphasize the increasing importance of hybrid MCDM methods for solving complex decision problems. An MCDM problem is generally divided into the steps of weighting the criteria and prioritizing the alternatives. Some MCDM methods are suitable for both weighting and subsequent prioritization, while other MCDM methods require a weighting of the criteria and do not offer a methodical approach for this.

For the selection of environmentally friendly technologies, Doczy and Razig [30] combine the analytic hierarchy process (AHP) and multi-attribute utility (MAUT) to give decision-makers in construction projects a method for a comprehensive assessment of sustainability without neglecting the conventional goals of construction planning. The evaluation is based on four criteria tailored to the construction sector. Şahin [31] conducts a comparative study of an MCDM problem in the context of sustainable energy generation in Turkey. The author compares the results of 42 decision problems resulting from a combination of different MCDM and weighting methods. The individual results are then summarized in an overall rating. Şahin concludes that a combination of several MCDM methods can compensate for the disadvantages of individual methods and thus enable a more accurate selection of the best alternative. Feizi et al. [32] combine a weighting based on AHP and Shannon entropy with a technique for order of preference by similarity to ideal solution (TOPSIS) to find an optimal mining location. The enormous number of 260,400 alternatives should be emphasized here. Ren et al. [33] compare the results of three different hybrid MCDM methods in the context of the planning of sewage treatment plants in consideration of the sustainability aspects. It is found that the technology selection is method-dependent. Consequently, they recommend not basing a decision on a single MCDM method but comparing the results of different methods. In addition, the authors consider both hard, easily quantifiable, and soft, only qualitative, decision criteria.

Muqimuddin and Singgih [34], among others, deal with the risks resulting from disruptions in the production process. The authors carry out a failure mode and effects analysis (FMEA) with three different risk priority numbers, which are weighted using an AHP. Identified faults are then prioritized with the help of TOPSIS. By combining the methods, risks are prioritized depending on the subjective assessments of the decision-maker. Turskis et al. [35] also place their method for risk assessment in the context of disruptions that endanger the information technology (IT) security of critical infrastructure. This is a holistic risk management method, which includes both risk identification and risk assessment. Wang et al. [36] consider service risks in hospitals with the aim of increasing service quality. Since service quality is primarily measured using non-quantifiable parameters, the combination of FMEA and fuzzy MCDM developed by the authors is particularly suitable in this context.

2.3. Categorization of Measures

Risk treatment requires suitable measures to compensate for the effects of risks. Since production systems usually have very specific characteristics, various approaches to the generalization and categorization of measures can be found in the literature. Pielmeier [12] assigns measures that have a direct influence on production control to the levels of in-house production planning and control of the Aachen PPC model, which is described in Schuh and Stich [5]. The author emphasizes that event-specific control strategies can be selected through the classification. Furthermore, the classified measures are linked with target values using a cause–effect matrix. VDI 5207 Part 1 [37] assigns previously introduced energy flexibility measures to the level model of production. The energy flexibility measures cover a broad time horizon from a few seconds to several hours. The classification enables a targeted consideration of individual measures, depending on the current state of knowledge, and thus an efficient implementation of the measures.

Verhaelen et al. [38] consider reactive fault management in the context of global production processes. The methodology developed by the authors enables a flexible and quick response in the event of a malfunction. Faults are classified into a three-level categorization system based on their causes and linked to appropriate measures. Furthermore, a protocol for the description of the fault is developed that facilitates the prioritization of faults and the associated initiation of measures. Schwartz and Voß [7], on the other hand, clearly differentiate in their work between prevention strategies and reaction strategies for fault management in production. The use of the two strategies is tested using a simulation model, and the effects of the measures are assessed using efficiency and instability measures. The methodology is assigned to machine utilization planning. Hernández-Chover et al. [39] do not consider the planning of preventive and reactive measures directly, but they also weigh up between predictive maintenance and repairs that become necessary due to malfunctions. In doing so, they consider the critical infrastructure in an empirical case study and identify the proposed method as the optimal relationship between forward-looking investments and subsequent repairs.

2.4. Need for Research

The investigation of the relevant scientific subject areas led to the following research gaps, on the basis of which the need for action for the creation of the approach is derived:

1. Energy-oriented PPC mostly neglects operational fault and risk management or is based on complex algorithms with little practical suitability. In order to implement the selected measures, a sensible method of integrating measures is required.
2. A large number of methods exists for the individual weighting of alternatives, but there is no adaptation of MCDM to energy-oriented PPC. Furthermore, no approach offers the consideration of risk-specific criteria in energy-flexible production systems.
3. In the literature, there is either a subdivision into preventive and reactive measures or an assignment of measures to the Aachen PPC model. The process views of the Aachen PPC model are not divided into preventive and reactive sub-steps, which would enable both forward-looking direct adjustments to the plan and alternative strategies if faults occur.

In summary, an approach must be developed that selects relevant risks and measures to be considered in a structured manner. In addition, it should enable the subdivision into the preventive and reactive treatment of risks and finally contain the planning of the measures on the basis of the target values of the energy-oriented PPC.

3. Scientific Concept

The presented approach builds on the development and evaluation of risk treatment paths following Roth et al. [6]. It is based on the determination of interactions through interpretive structural modeling and the calculation of conditional probabilities using Bayesian networks. The result is a multitude of risk treatment paths that contain risks and measures that can occur in a production system under consideration.

The generation of the risk-treatment paths in Roth et al. [6] is subject to several assumptions, the most important of which are:

1. The regarded area of application is limited to the use of electrical energy in production systems,
2. Processes are formalized in discrete production timetables,
3. Experts are available to supply the necessary information to generate the Bayesian networks.

Any further limitations named in Roth et al. [6] equally apply to the approach in this paper. The following assumptions apply specifically to the approach presented in this paper:

1. A discretized, energy-optimal production schedule is available,
2. For the risk-treated energy-oriented PPC, the planned energy schedules including the maximum allowances for peak loads are available,
3. A detailed risk inventory including measures is available, wherein process-specific parameters in the dimensions of time and energy consumption are known for every risk and every measure.

Figure 2 depicts a graphical overview of the approach proposed in this paper. The boxed numbers in Figure 2 additionally show in which section one can find details on the respective process step. The main process steps of the approach are:

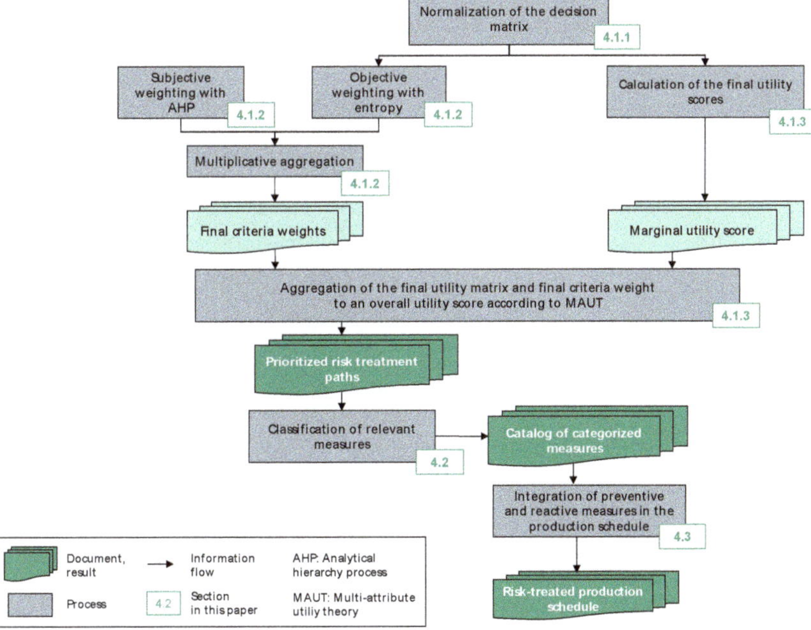

Figure 2. Overview of the approach.

1. Calculation of the final criteria weights by combining AHP and Shannon entropy,
2. Calculation of the marginal utility scores according to the problem-specific utility functions,
3. Calculation of the final utility score and subsequent prioritizing the risk-measure path profiles in descending order,

4. Classification of the relevant measures into preventive, reactive, and non-distinguishable in accordance with [37],
5. Integration of the measures into the process schedule depending on the prior classification, resulting in a risk-treated process schedule.

The application of the approach produces three overarching results: the prioritized risk treatment paths, a catalog of categorized measures, and as the final result, the risk-treated production schedule.

4. Description of the Approach

This section introduces the details of the developed approach. Section 4.1 summarizes the measure selection process, which is based on a hybrid MCDM approach with AHP and entropy for criteria weighting and MAUT for the prioritization of the paths. Then, Section 4.2 outlines the process for measure classification into preventive and reactive by utilizing the categorization of energy flexibility measures in the VDI 5207 standard [37]. Finally, Section 4.3 integrates the selected measures in the production schedule with the help of a mixed-integer linear problem (MILP) and a branch-and-bound optimization, minimizing the cost of additionally purchased energy. The integration of measures is understood as the creation of new, risk-treated scheduling that, through preventive measures, contains less risk potential. In addition, reactive measures are integrated so that backup schedules are created that are used when faults occur.

4.1. Prioritization of the Risk Treatment Paths with a Hybrid MCDM Approach

For the risk treatment, measures must be integrated into the existing production schedule. To identify which measures to integrate, the risk treatment paths that contain different measures are ranked, and, subsequently, measures are extracted from the chosen path. As the risk treatment paths are evaluated in multiple dimensions, the resulting decision problem is a multi-criteria decision-making problem and thus requires complex decision-making methods to identify the best alternative from all the available paths. This paper introduces a hybrid MCDM approach that weights the criteria combining AHP [23] and Shannon entropy [40] and subsequently ranks the alternatives according to the rules of MAUT [41].

4.1.1. Normalization of the Decision Matrix

The model developed by Roth et al. [6] evaluates each path according to the trilemma of cost, energy flexibility, and a risk priority number. In this research paper, the trilemma is modified to fulfill the requirements of criteria selection in MCDM problems, as stated by Wang et al. [42]. The authors define five principles to be obeyed when identifying suitable criteria: (1) the systemic principle, (2) the consistency principle, (3) the independency principle, (4) the measurability principle, and (5) the comparability principle. The systemic principle demands a holistic assessment of the regarded problem. In every PPC system, costs are one of the key decision factors [12] and therefore need to be considered in the decision problem, hereinafter referred to as cost. Reinhart and Schultz [43] propose energy flexibility as a key indicator in the energy-oriented PPC, as it incorporates several energy-related variables in one indicator, hereinafter referred to as energy flexibility. Furthermore, this research is positioned in risk and fault management, and thus a risk-related dimension needs to be added to the problem to describe it holistically. The risk-related indicator is hereinafter referred to as risk reduction. The expert weighting of the three criteria additionally ensures that the criteria of the decision problem are in line with the decision-maker's goals; thus, criteria (1) and (2) are considered fulfilled. To demonstrate that principles (3) to (5) are fulfilled, a closer look at how the criteria are calculated is necessary.

The costs are calculated according to the cost model introduced by Roth et al. [6], which distinguishes between, e.g., production costs, logistic costs, and delay costs, and it divides these in turn into respective sub-categories, such as material or personnel costs. Costs are defined as the change in costs in the modified production schedule in relation to the unmodified production schedule. The deviation is the sum of deviations in the above-mentioned cost categories. For every cost component, only measure-induced costs are regarded, resulting in the measure-induced change in costs $\Delta C_{tot}{}^M$, where $\Delta C_{tot}{}^M$ is negative for cost savings, positive for increased costs, and "0" for unchanged costs [6]. Thus, for $\Delta C_{tot}{}^M$, the lower the better. The cost thus indicates the costs incurred through the integration of the measures on the path. In practice, they therefore often correspond to the deviation from cost-optimal scheduling, which does not consider any risk effects.

The energy flexibility indicates the remaining flexibility available in the system to react to disruptions influencing the production system's energy consumption. The regarded dimensions of energy flexibility are the potential to change loads or so-called energy flexibility states EF_{states}, the potential to shift the time of consumption EF_t, and the potential of energy-flexibility technologies, e.g., the use of storage facilities and flexible on-site generation EF_{tech}. The overall energy flexibility EF is then calculated by multiplying the three dimensionless components:

$$EF = EF_t \Delta EF_{states} \Delta EF_{tech} \tag{1}$$

A value higher than one indicates the desired high energy flexibility, whereas a value lower than one indicates undesired low energy flexibility.

Risk reduction indicates how much the risk is reduced by the integration of the measures into the production schedule, such as a change in production sequence or a shift of production starts. Risk reduction can thus be understood as the added value of the risk treatment approach and is defined as the absolute difference between the risk potential of the original plan and the resulting risk in the risk-threatened plan:

$$Risk\ reduction = |risk\ potential - resulting\ risk| \tag{2}$$

The risk potential of a path, on the other hand, is defined as the product of the probability of occurrence of the risks P_{Risk} it contains and the costs caused by the risks $\Delta C_{tot}{}^R$, i.e., the damage [44–46]:

$$Risk\ potential = P_{Risk} \cdot \Delta C_{tot}{}^R \tag{3}$$

The resulting risk considers all path elements and thus consists of the product of the probability of occurrence of the risks and measures P and the total risk and measure costs of the path ΔC_{tot}:

$$Resulting\ risk = P \cdot \Delta C_{tot} \tag{4}$$

A high rating for the risk reduction is desired whereas the worst possible outcome is a risk reduction of zero, implying that no measures were integrated, and thus no potential risk was reduced.

Thus, all three criteria are in line with criteria (3), namely, independency. As for costs, a differentiation between measure and risk-induced costs is introduced, and the energy flexibility only considers cost-independent factors. Furthermore, all three criteria are quantified evaluations of the regarded system and thereby fulfill principle (4). To achieve comparability across the criteria that utilize different scales, normalization of the criteria ratings r_{ij} is necessary (criteria (5)). Depending on the direction of the criteria, Equation (5) or Equation (6) is applied [25], resulting in a normalized rating r_{ij}^*, with a value of one corresponding to the best possible alternative i and zero being the worst possible alternative i for the respective criterion j:

$$r_{ij}^* = \frac{r_{ij} - \min(r_{ij})}{\max(r_{ij}) - \min(r_{ij})}, for\ maximizing\ criteria \tag{5}$$

$$r_{ij}^* = 1 + \frac{\min(r_{ij}) - r_{ij}}{\max(r_{ij}) - \min(r_{ij})}, for\ minimizing\ criteria \tag{6}$$

The cost indicator $\Delta C_{tot}{}^M$ is to be minimized, and the energy flexibility indicator EF and the risk indicator risk reduction are to be maximized. The three criteria can now be graphically represented in a trilemma and are summarized in the decision matrix D with r_{ij}^* being the normalized rating for alternatives $i = 1 \ldots n$ with respect to the criterion $j = 1 \ldots m$:

$$D = \begin{bmatrix} r_{11}^* & \cdots & r_{1m}^* \\ \vdots & \ddots & \vdots \\ r_{n1}^* & \cdots & r_{nm}^* \end{bmatrix} \tag{7}$$

4.1.2. Subjective and Objective Criteria Weighting

For the subjective expert weighting, the three trilemma criteria AHP is applied. The AHP [23] is widely used in the literature and practice due to its straightforward application and high reliability, even in uncertain decision situations [47]. It is based on the principle of pairwise comparison, usually done by experts, and it serves as the subjective weighting method in the approach. The pairwise comparison results in an $n \times n$ decision matrix A with the elements a_{ij}, each of which indicates the relative weighting of two criteria, i and j. Values on the diagonal are equal to one. The remaining comparisons are filled with values from one to nine, and their inverse fractions for opposite dependencies [48]:

$$A = \begin{bmatrix} 1 & \cdots & a_{1n} \\ \vdots & \ddots & \vdots \\ \frac{1}{a_{1n}} & \cdots & 1 \end{bmatrix}, \; a_{ii} = 1, \; a_{ji} = \frac{1}{a_{ij}}, \; a_{ij} \neq 0 \tag{8}$$

The following applies for the pairwise comparison: the higher the chosen value for criterion i, the higher the preference of criterion i over criterion j. For the comparison, a scale from one to nine is introduced, as it is effective in expressing preferences with sufficient precision and does not overwhelm the human decision-maker [49]. A verbal explanation of the nine levels is given in Table 2.

Table 2. Descriptive explanation of the AHP pair-wise comparison scale [49].

Intensity of Importance on an Absolute Scale	Definition	Explanation
1	Equal importance	Two activities contribute equally to the objective
3	Moderate importance of one over another	Experience and judgment slightly favor one activity over another
5	Essential or strong importance	Experience and judgment strongly favor one activity over another
7	Very strong importance	An activity is strongly favored and its dominance demonstrated in practice
9	Extreme importance	The evidence favoring one activity over another is of the highest possible order of affirmation
2, 4, 6, 8	Intermediate values between the two adjacent judgments	When compromise is needed
Reciprocals	If activity i has one of the above numbers assigned to it when compared with activity j, then j has the reciprocal value when compared with i.	

The choice of the exact preference level within the range is the decision-maker's. As outlined in Section 4.1.1, the three relevant criteria to compare are cost, EF, and risk reduction. The regarded factors, which influence the pairwise comparison, are, e.g., a client's importance, sustainability goals, risk attitude, and time criticality. Some examples of decisions are listed below:

- Risk reduction is more important than cost: the orders are particularly time-critical and for particularly important customers. Additional costs are therefore accepted for the reduction of risks.

- Energy flexibility is more important than risk reduction: the information available in the planning period is vague because rush orders could arrive. High remaining energy flexibility is required in the system.
- Cost is more important than risk reduction: the forecasts for penalties for deviations are low, and load peaks are not expected in the period under consideration. There is no great need to incur costs for risk reduction measures.

These comparisons can serve as a guide for decision-makers. Individual preferences may exist in a specific production system, and thus individual decisions must take into account the circumstances and requirements of the production system in question.

The calculation of the criteria weights w_{AHP} for the decision matrix A is carried out by multiplying its largest eigenvalue λ_{max} by the respective eigenvector $\vec{v_i}$:

$$w_{AHP} = \lambda_{max}\, \vec{v_i},\ \ with\ \lambda_{max} = \max_i \lambda_i \tag{9}$$

Finally, the consistency of the expert judgments is checked with the consistency ratio (CR). The inputs are assumed to be consistent if CR $\leq$ 0.1 hold true; otherwise, the expert inputs must be revised. The CR is the fraction of the consistency index (CI) and random consistency index (RCI) as shown. The value of the RCI depends on the number of alternatives and was introduced by Dong [50]. Small inconsistencies are accepted, as human inputs are subject to error, especially with an increasing number of alternatives. Furthermore, if inconsistencies are small, they do not have a decisive influence on the result of the AHP.

In addition to the subjective weighting with AHP, an objective weighting takes place using the entropy (information theory) according to Wang et al. [42]. The aim of the entropy is to calculate a weighting that reflects the information and uncertainty contained in an individual criterion. The entropy e_j is calculated using Equation (10) and is based on the normalized decision matrix D (Equation (7)).

$$e_j = -k \sum_{i=1}^{n} r_{ij}^* ln\left(r_{ij}^*\right),\ j = 1 \ldots m \tag{10}$$

where $k = \frac{1}{\ln(n)}$ and n the number of alternatives per criterion. The higher the entropy, the higher the uncertainty of the criterion, and the lower the weighting w_j should be. This relationship is represented in the calculation of the objective weighting $w_{Entr,j}$ [32,40,42]:

$$w_{Entr,j} = \frac{f_j}{\sum_{j=1}^{m} f_j}\ with\ f_j = 1 - e_j \tag{11}$$

The final weighting w_j^{final} for a criterion is generated by multiplying the objective weighting $w_{Entr,j}$ and subjective weighting $w_{AHP,j}$ similar to [32,42]:

$$w_j^{final} = \frac{w_{Entr,j} \cdot w_{AHP,j}}{\sum_{j=1}^{n} w_{Entr,j} \cdot w_{AHP,j}} \tag{12}$$

The final weighting w_j^{final} combines the possibility of mapping individual preferences with the AHP and at the same time reduces the distortions by entropy due to the consideration of the contained information value of the different options. Since in this case there is sufficient data to calculate the entropy and an expert familiar with the use case is available, both weightings can easily be determined and combined.

4.1.3. Calculation of the Final Utility Scores and Aggregation into an Overall Utility Score

The MAUT assumes that every decision is based on maximizing one's own utility [24]. The normalized decision matrix D (Equation (7), Section 4.1.1) forms the basis for calculating the marginal utility score of alternative i with criteria j $u_{ij}\left(r_{ij}^*\right)$. A universal utility function does not exist, but rather the chosen functions highly depend on the decision-maker. In general, the distinction between linear and exponential utility functions is widespread [24,25,51]. Figure 3 shows a depiction of the different utility functions. If small changes in the criteria values in the lower third are rated as significant, a concave utility function should be selected. If small changes in the upper third of all values are rated as significant, a convex utility function should be selected.

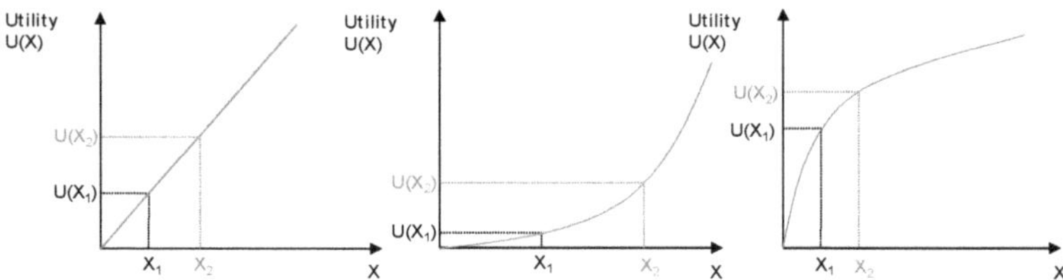

Figure 3. Utility functions (linear, convex, concave).

Equation (13) depicts the generalized form of the exponential utility function for the marginal utility score u_{ij}, where x depends on the decision-maker's utility with $x < 1$ for a concave function and $x > 1$ for a convex function [25].

$$u_{ij}\left(r_{ij}^*\right) = \frac{exp\left((r_{ij}^*)^x\right) - 1}{exp(1) - 1} \tag{13}$$

If the distribution is even, a linear utility function should be selected:

$$u_{ij}\left(r_{ij}^*\right) = r_{ij}^* \tag{14}$$

To identify a decision-maker's underlying utility function, direct methods or indirect methods can be applied [24]. With direct methods, the decision-maker answers direct questions about his or her preferences through ratings or preferences on lotteries, etc. Indirect methods can be versions of additive utility methods, which are based on linear programming [52]. If it is not possible to define the decision-maker's risk preference for a trilemma criterion, a neutral utility function should be chosen. As a result, a new decision matrix containing the marginal utility scores is created:

$$U_{ij}\left(r_{ij}^*\right) = \begin{pmatrix} U\left(r_{11}^*\right) & \cdots & U\left(r_{1m}^*\right) \\ \vdots & \ddots & \vdots \\ U\left(r_{n1}^*\right) & \cdots & U\left(r_{nm}^*\right) \end{pmatrix} \tag{15}$$

Eventually the final utility scores U_i are calculated for each risk treatment path, i.e., each alternative, with the marginal utility scores $U_{ij}\left(r_{ij}^*\right)$ and the final weighting w_j^{final}:

$$U_i = \sum_{j=1}^{m} U_{ij}\left(r_{ij}^*\right) \cdot w_j^{final}, \quad i = 1 \ldots n \tag{16}$$

Arranging the risk treatment paths descending from $U_{i,max}$ to $U_{i,min}$ results in a prioritized list of all paths and thus in the selection of measures that in the next steps are categorized and integrated into the production schedule.

4.2. Classification of Relevant Measures

To integrate the identified measures into the production schedule, it is first necessary to distinguish between preventive and reactive measures in order to create a catalog of categorized measures. The literature suggests that the main differentiator is the timing of the measure considering the occurring damage [7,25,29].

- Reactive measures are only used when damage has already occurred. The aim is to keep the resulting damage and the associated costs as low as possible. Due to the immediate implementation, short-term changes in the production schedule must be expected.
- Preventive measures are used to avoid potential damage and its financial impact as well as the reduction of the likelihood of occurrence. They are implemented at an early stage before the fault occurs. This excludes the possibility of changes to the production schedule with short notice.

In addition to differentiation according to the time of implementation of a measure, economic aspects must be considered. Preventive measures should generally be preferred in the case of high expected costs for reactive measures [53]. The advantage of taking both categories of measures into account in this approach is that the preventive measures reduce the impact and likelihood of potential faults. At the same time, the planning of reactive measures creates an information base for the reactions if faults still occur, so that a solution does not have to be sought under time pressure.

A generalized and thorough overview of relevant measures for energy-flexible production is given in VDI 5207 Part 1 [37]. The distribution of the measures within the three implementation levels of the energy-flexible factory serves as a reference point for selecting relevant measures for operational risk management (Figure 4).

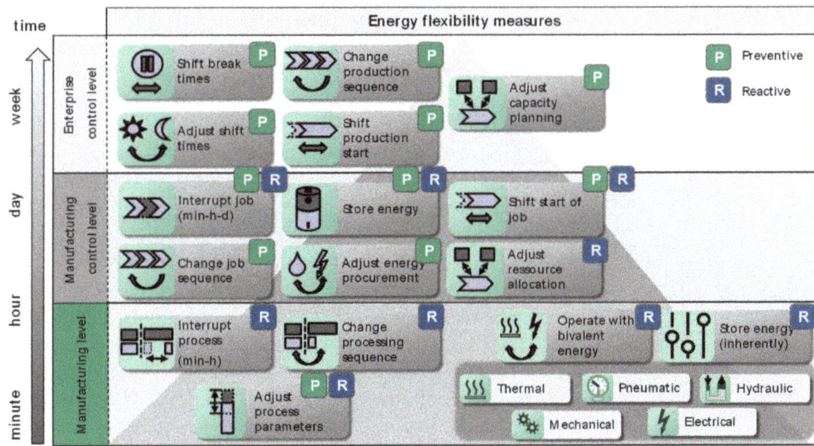

Figure 4. Energy-flexibility measures and their classification into preventive (P) and reactive (R) measures. Figure based on VDI 5207 Part 1 [37]. Reproduced with permission of the Verein Deutscher Ingenieure e. V.

The different implementation levels in Figure 4 imply different time horizons that serve as orientation for dividing the measures into preventive and reactive. Still, for some measures it is not possible to assign them to only one category without knowing the specific production context. Thus, these measures are marked as preventive and reactive. At the short-term manufacturing level, all measures are considered reactive except for the adjustment of process parameters, which can also be an activity planned in advance. At the manufacturing control level, most of the measures available are preventive, as the regarded time horizon ranges from hours to days and thus implies longer advance planning. The manufacturing control's measures further influence the measures at the lower level, and preventive planning of these should be done whenever possible. Nonetheless, in the event of severe disruption, a job may be interrupted reactively, and, if available, energy storage will be used. The enterprise control level only consists of preventive measures due to the necessary longer planning horizon of days to weeks. Finally, it should be noted that this division into preventive and reactive measures is a general orientation because in the wide variety of different industries with specific planning and production systems, the respective measures may differ.

4.3. Integration of Preventive and Reactive Measures in the Production Schedule

Once measures have been successfully divided into reactive and preventive, implementation of the measures into the production schedule needs to be planned. The rescheduling of reactive measures is considered to be segment-based rescheduling, similar to Toba [54]. This means that for reactive measures, only the segments after the potential risk occurrence are affected and rescheduled, whereas for preventive measures, the measure execution must be prior to the expected time of disruption to be effective. Therefore, the implementation of preventive measures may affect the entire production schedule.

The production schedule needs to be modified so that the previously selected measures are integrated as well as they can be, taking into account the logistical goals of the production system and the boundary conditions for the planned energy consumption. Due to the differentiation into preventive and reactive measures, new production schedules must be generated. These are the modified production schedule with all preventive measures, which replaces the original plan. Furthermore, a backup plan is drawn up for each reactive measure implemented, as the reactive measures only come into effect when a disruption occurs.

The integration of measures is formalized as a MILP, which can be solved through branch-bound-and-cut optimization, e.g., in MATLAB® [54,55]. For the present problem, it is important to aim for short computation times to ensure the applicability of the approach in operational practice. This can be achieved by reducing complexity wherever the problem setting allows it. The goal of this optimization is to integrate the measures energy optimally and thereby create a risk-treated, energy-optimal production schedule by the addition of preventive and reactive measures.

In the course of this section, the term *production* schedule refers to a plan that aggregates all relevant jobs on all workstations for the respective production period under consideration of resources and sequence restrictions, whereby one *job* contains a product's production steps, i.e., the necessary workstations including durations, sequence restrictions, and resource consumptions. The workstations come with capacity restrictions, and not every job is necessarily processed on every workstation. The initial production schedule prior to risk treatment is assumed to be available and energy-optimal, hereinafter referred to as energy-oriented PPC.

Measures are thus either treated as jobs and are fixed in their allocation to one workstation or modify the load and time dimensions of jobs scheduled by the energy-oriented PPC. When scheduling the measures, the risk-treated energy-oriented PPC must still comply with logistical and energy-consumption target values. Consequently, two options remain for the risk treatment:

1. An extension of the original energy-oriented PPC by risk-specific target values and constraints leading to a detailed and comprehensive optimization problem.
2. Setting the results of the initial energy-oriented PPC as an input variable for an optimization problem that is limited to the implementation of measures.

Option (1) results in correspondingly higher computing times due to increased complexity. This also leads to more difficulties in understanding the solution process and thus lowers the acceptance of the approach for the end users. Option (2), on the other hand, results in a non-optimal solution, but with expected significantly shorter calculation times, thus increasing flexibility in the application of the approach. It is also advantageous that the energy-oriented PPC is not redesigned but expanded. This increases understanding and acceptance if the generated solution fulfills the end user's standards. Due to the predominant advantages of option (2), this will be pursued further below.

Thus, the optimization problem consists of an energy-optimal production schedule as input that can be generated using different approaches, e.g., those described in Section 2.1. In addition, the measures of the chosen risk treatment path need to be scheduled to create the risk-treated production plans.

In order to meet the logistical goals and to avoid delays, the end times of the jobs scheduled in the energy-oriented PPC are fixed and block the workstations for the scheduling of measures in these time periods.

Usually, when creating an energy-oriented production plan, a cost-optimal result is sought after the variable price forecasts. The electricity demand planned and procured in this way should be consumed within tolerances in order to avoid high penalties. As part of the risk treatment, price fluctuations in the markets are no longer of central importance, as the plan is already generated, but the time-dependent penalties for deviations from the originally purchased electricity consumption have to be focused on now.

Thus, the objective function of the optimization problem minimizes the quantity deviation cost (QDC) QDC_t for each time unit t that arises from a deviation of the actual energy consumption from the forecasted load profile Δpow_t.

The QDC is substituted by the forecast for reBAP, which stands for "regelzonenübergreifender einheitlicher Bilanzausgleichsenergiepreis" and assigns a uniform price to the balancing energy that was necessary in the past. The reBAP is calculated in retrospect for every quarter-hour of a day. If no suitable forecasts are available for the reBAP, intraday market forecasts can also be used, as the amount gives an impression of the energy availability and the demand and thus the level of the penalty costs caused by deviations.

The objective of the optimization is to minimize the QDC, which arises due to the deviations from the planned load profile. This is shown in the target function with the deviation Δpow_t and the QDC_t at the respective point in time t.

$$Minimize \sum_{t \in T} (\Delta pow_t \cdot QDC_t) \tag{17}$$

To ensure logistic targets are met, a job must start early enough to not miss any due dates:

$$start_{ks} \leq end_{ks} - duration_{ks}, \ \forall i \in I, \ k \in K, \tag{18}$$

where $start_{ks}$ describes the start time of job k on working station s and end_{ks} the due date of the job k on station s. The duration of job k on station s is described by $duration_{ks}$.

To calculate the actual energy consumption $cons_t$, the binary x_{kit}, that is, one if job k is allocated to workstation s in time unit t and zero otherwise, is multiplied by the workstation and job-specific power consumption $Power_{ks}$:

$$cons_t = \sum_{i=1}^{m} \sum_{k=1}^{l} Power_{ks} \cdot x_{kst}, \ \forall t \in T \tag{19}$$

Additionally, it must be ensured that the measure integration does not lead to the peak loads being exceeded; thus, the total consumption $cons_t$ must be smaller than the maximum allowed peak load $Peak_{max}$ for every time unit

$$cons_t \leq Peak_{max}, \ \forall t \in T. \tag{20}$$

Finally, the deviation in energy consumption is calculated as the total consumption minus the planned consumption $Power_t^{planned}$.

$$\Delta pow_t = cons_t - Power_t^{planned}, \ \forall t \in T \tag{21}$$

The optimization is performed twice—once to generate the modified production schedule and once to generate the backup plan. Figure 5 depicts schematically how the optimization improves the handling of disruptions. In the above production schedule without prior risk treatment, a disruption leads to a spontaneous decision to post-process. This leads to the annual maximum load being exceeded, as shown in the adjacent diagram of the load profile. With the risk treatment shown in the lower area of Figure 5, the possible disruption is considered in advance with a backup plan, containing reactive measures for the case of the occurrence of the fault. The pause on station 4 enables post-processing to compensate for the disruption without exceeding the peak load. To create the risk-treated plan in the above-mentioned approach, only the affected jobs are rescheduled.

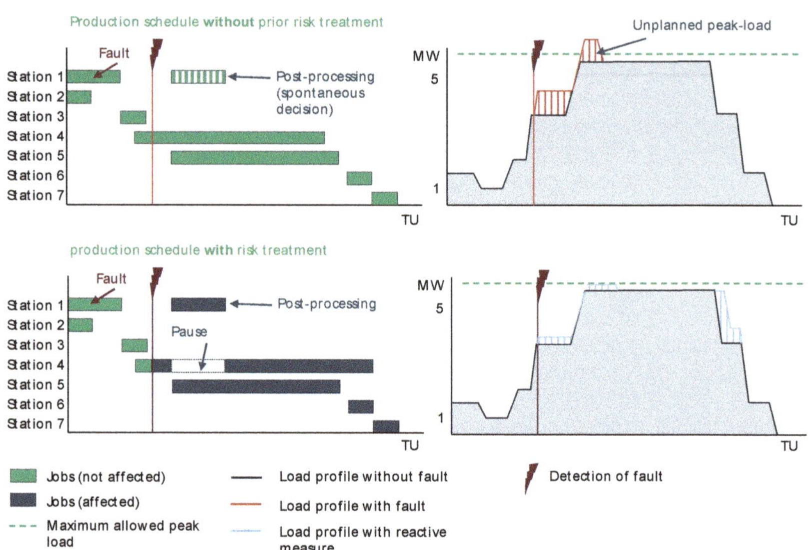

Figure 5. Schematic initial production schedule and the effects of disturbances on the load profile.

5. Application of the Approach

The use case for demonstrating the approach is located in the foundry industry. As the melting processes are considered especially energy-intensive, the foundry industry can significantly influence the power grid [56]. In the use case, four furnaces are used to melt raw iron, iron ore, and scrap iron. The ladles are transported to the casting fields via forklifts. Molds and cores are produced on site. The whole process of melting, molding, core preparation, casting, and post-processing is considered for a time period of 65 time units (TU), which is equivalent to 16.25 h. A detailed description of the use case can be found in Roth et al. [6].

The approach requires the risk treatment path profiles obtained in Roth et al. [6] as well as process-specific risk and measure data, which are gathered with the help of expert interviews and failure mode and effects analysis (FMEA). Six risks with their potential extents of damage and probability of occurrence were identified and are listed with the related nine different measures in Table 3. At this point, it is important to note that risks that are directly related to quality management processes were not considered, as they are not in the sphere of influence of the PPC.

Table 3. Risk inventory with the frequency of occurrence, extent of damage, and the related measures.

ID	Risk (Frequency)	Extent of Damage	ID	Measure
R1	Mold for a small part is faulty—exchange of mold pattern is possible (1 per week)	Time delay to casting and subsequent processes	M1	Switch to casting panel with identical material requirements
		Additional post-processing	M2	Casting with increased post-processing effort
R2	Mold for a small part is faulty—exchange of mold pattern is not possible (1 per week)	Additional order to be planned; changed load profile	M3	Preparation of an additional mold
		Additional post-processing	M2	Casting with increased post-processing effort
R3	Core for a small part is faulty—exchange of core box is possible (2 per week)	Time delay to casting and subsequent processes additional post-processing time	M1	Switch to casting panel with identical material requirements
			M2	Casting with increased post-processing effort
R4	Core for a small part is faulty—exchange of core box is not possible (2 per week)	Additional order to be planned; changed load profile	M4	Preparation of an additional core
		Additional post-processing	M2	Casting with increased post-processing effort
R5/R6/R7 *	Furnace failure (2 per year)	Delay melting	M5	Adjust furnace utilization if an unoccupied furnace is available
		Delay melting	M6	Adjust process start
		Delay melting; changed load profile	M7	Adjust parameter melting temperature and duration
			M8	Interrupt melting process (only possible for small TUs)
R6	Forklift failure (5 per week)	Time delay in follow-up processes	M9	Provision of a spare forklift
			M10	Switching to transport trolleys

* This risk applies to every furnace individually.

The measures are further subdivided into process-altering and supplementary process measures, e.g., M2: casting with increased post-processing effort alters the post-processing, and M3: preparation of an additional mold is a supplementary process to be planned in addition to the scheduled mold preparation processes. For every process-altering measure, the measure-induced deviation in duration and load profile, and for every supplementary process measure, the absolute duration and load profile is filed.

Additional input parameters for the scheduling are the planned load profile, the QDC, and the initial energy-optimal production schedule, including production-specific requirements.

In the following, the application of the approach is described. The section is structured based on the three main steps of the approach, with the calculation of the final criteria weights, the classification of the relevant measures, and, finally, the integration of the measures.

5.1. Prioritization of the Risk Treatment paths

For the measure selection, the AHP periodization matrix and utility functions for the three criteria of cost, risk reduction, and energy flexibility are needed. An interactive MATLAB® Live Script is created as the user interface. For optimization, MATLAB® offers an Optimization Toolbox™ that can solve MILPs efficiently. The approach is implemented in MATLAB® version 9.6.0.1472908 (R2019a). All input data are stored in Microsoft® Excel® and can therefore be easily modified.

Firstly, all ratings for the paths created in advance using the approach in Roth et al. [6] are normalized, resulting in trilemma criteria for every path. In Figure 6, the trilemma for three exemplary paths is depicted. To compare the three paths, their criteria are shown in the diagram below on the right.

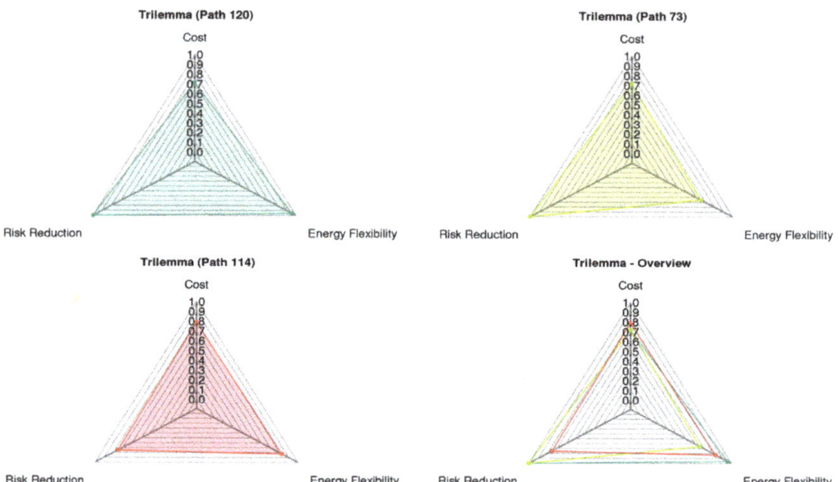

Figure 6. Trilemma of three risk treatment paths including an overview.

With the involvement of the experts from the production system, the approach to weighting the trilemma criteria was then carried out in order to be able to categorize the paths according to their suitability. For the decision matrix in Table 4, the principal eigenvector according to the AHP [23] determines the weighting of 0.0679 for cost, 0.7703 for energy flexibility, and 0.1618 for risk reduction, according to the assessment of the experts. This is due to a high need for the remaining energy flexibility in the system for reactive measures, since the orders are particularly tightly timed in the period under consideration, and failure to meet deadlines is associated with high penalties. Risk reduction is also given a higher weighting than cost; additional costs for measures are therefore accepted in favor of risk reduction.

Table 4. Decision matrix for the AHP.

	Cost	**Energy Flexibility**	**Risk Reduction**	**Final Weighting**
Cost	1	1/9	1/3	0.0679
Energy flexibility	9	1	6	0.7703
Risk reduction	3	1/6	1	0.1618

The objective weighting, applying Shannon entropy, results in a weighting of 0.0245 for cost, 0.0710 for energy flexibility, and 0.9044 for risk reduction, assuming all values are normalized according to Equation (5) or Equation (6). The proportionally high weighting of risk reduction in comparison to the two remaining criteria highly influences the final weighting obtained by multiplicative aggregation. The large deviation in values is not uncommon, as subjective and objective weighting are generated independently. Thus, the final weighting results in 0.0082 for cost, 0.2700 for energy flexibility, and 0.7218 for risk reduction.

For this use case, the utility functions based on the underlying data structure were selected, so that differences in ratings are amplified. This is achieved by choosing a convex utility function when most data points lie within the upper third of the scale. When most date points lie within the lower third of the scale, a concave utility function is applied. For the remaining third, a linear function was utilized. Thus, for cost and energy flexibility, a convex utility function was chosen and for risk reduction a linear utility function.

$$u_{ij}\left(r_{ij}^*\right) = \frac{exp\left(\left(r_{ij}^*\right)^2\right) - 1}{exp(1) - 1} \tag{22}$$

Applying the respective utility function to the decision matrix results in marginal utility scores for every alternative and every criterion. The final utility scores are calculated by weighted sums, and the paths, i.e., the alternatives, are ranked in descending order. Table 5 depicts the top three paths with their marginal and final utility scores under the assumption of the previously generated final weighting.

Table 5. Marginal and final utility scores of the three best performing paths.

Criteria	Marginal Utility Score			Final Utility Score
	Cost	Energy Flexibility	Risk Reduction	
Weight	0.0082	0.2700	0.7218	
No. 120	0.3930	0.8689	1.0000	0.9596
No. 73	0.3930	0.3107	1.0000	0.8089
No. 114	0.5028	0.5677	0.7507	0.6992

The top three paths contain the following measures: Path 120 contains R4 (core for a small part is faulty) and M4 (preparation of an additional core), path 73 contains R7 (delay melting) and M5–7 (adjust furnace utilization, adjust process start, adjust parameter melting temperature and duration), and path 114 contains only M7 (adjust parameter melting temperature and duration).

The selection of paths that contain little or no risk can be explained by the fact that the risk reduction criterion shows a strong accumulation in its assessments and is therefore generally avoided. The selection is therefore shifted to the criteria costs and EF, which are then rated higher. The distribution of the risk reduction depends crucially on the input variables of the extent of the damage and the probability of occurrence of the damage. During the application, the conscientious collection of this data is therefore of high relevance for the reliable selection of suitable paths. Furthermore, the measures contained in the prioritized paths must be compared and selected for integration into the production plan.

5.2. Classification of Relevant Measures

To showcase the potential of the risk reduction measures, the four measures (Table 6) from the prioritized paths were selected for the measures catalog to be integrated into the production schedule. The division of the measures into preventive and reactive is based on the allocation of the presented categorization of energy flexibility measures. For example, M5 can be assigned to the energy flexibility measure "adjust resource allocation" and is, therefore, a reactive measure.

Table 6. Catalog with four categorized measures.

ID	Description	Type	Effect on the Process
M4	Preparation of an additional core	Reactive	Altering
M5	Adjusting furnace utilization if an unoccupied furnace is available	Reactive	Supplementing
M6	Adjust process start furnace	Preventive	Supplementing
M7	Adjust parameter melting temperature and duration of furnace	Preventive	Supplementing

M7 is assigned to the energy flexibility measure "adjust process parameters" for which a decision must be made depending on the production situation. In the application, the process parameters should only be adjusted if a fault actually occurs and is therefore defined as a reactive measure.

5.3. Integration of Preventive and Reactive Measures in the Production Schedule

Figure 7 shows the result of the scheduling in the form of a risk-treated production plan for the allocation of orders on the workstations.

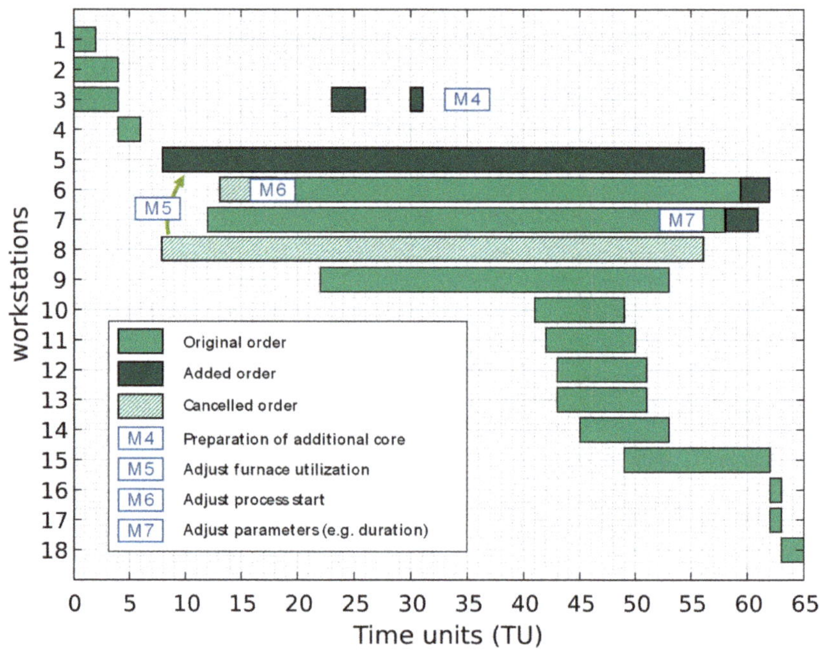

Figure 7. Risk-treated production schedule Gantt chart with marked reactive and preventive measures.

In order to counter R7, the failure of furnace 4, causing a delayed melting, the reactive measure M5 was implemented in the backup plan, which is only used when the risk arises. Measure M5, the utilization of another unoccupied furnace, can then be implemented by the switch from furnace 4 (workstation 8) to furnace 1 (workstation 5) if R7 occurs.

In addition, M6 and M7 were implemented in the plan as preventive measures, which means that they replace the original production plan.

1. M6 (adjustment of the process start) is implemented in the melting process on furnace 2 (workstation 6). This avoids possible warm-up times in furnace 2 due to a delayed casting time in the event of the occurrence of R7, as furnace 2 is used for the pre-melting for furnace 4.
2. M7 is implemented by the adjustment of temperature (decreased) and thus an extended duration of the melting process on furnace 3 (workstation 7). This is necessary in order to synchronize the termination of the melting processes on the furnaces for the casting.

As a reaction to a faulty core, which was identified as R4, the reactive measure M4, the preparation of an additional core, was scheduled in the backup plan to create a replacement core on workstation 3. Since the core creation can be carried out flexibly in a longer time interval, it was placed at times with lower reBAP prices.

Figure 8 shows the electrical load curve across the TU. The yellow dashed line shows the original load profile. The reBAP forecasts, which affect the timing of measures, are shown in red for orientation. The green load profile represents the risk-treated load. The short-term peaks for the creation of the additional core (M4), as well as the reduction of the load by adjusting parameters (M7), are the effects when the reactive measures are used. With the preventive measure M6, the load profile changes as an effect of the delayed start. Measure M5 with the changed utilization of the furnaces has no effect on the electrical load profile.

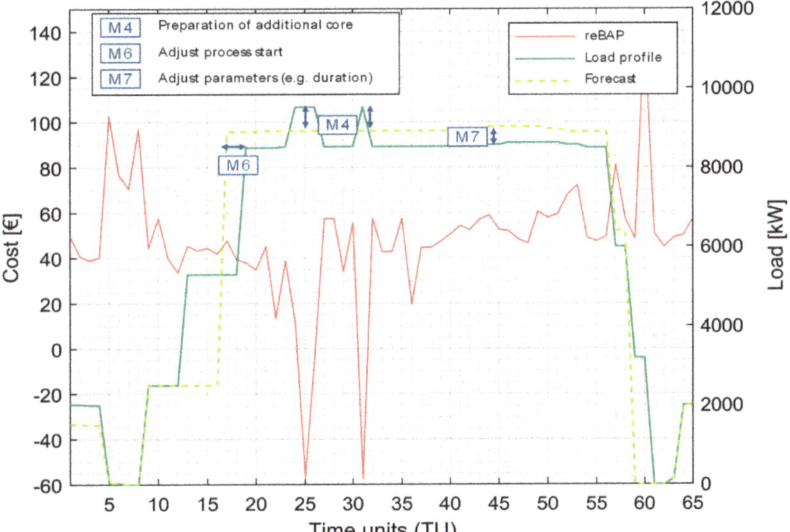

Figure 8. Load profile before (light green dashed line) and after the measure implementation (dark green line). For reference, the reBAP price is included.

5.4. Discussion

The application of the approach for the use case of a foundry resulted in a suitable risk treated energy-oriented production plan. The advantages over a situational decision in the event of faults as described in the state of the art lie in the consideration of the interactions of all known risks and measures in advance. In addition, preventive measures to reduce the probability of occurrence or the extent of damage are made possible.

The prerequisite for applicability is an acceptable level of effort in carrying out the risk treatment, especially since experts from different areas must be involved. By the introduction of this specific approach for the energy-oriented PPC, the effort required to select and adapt general approaches from the literature is reduced. Furthermore, the adaption to the energy-oriented PPC offers the possibility of automating the individual steps by the future development of software. The approach first requires a weighting of criteria for the selection of paths. Some manual steps are necessary here, which was supported by convenient graphical user interfaces for inputs. In addition, this weighting is only required occasionally, for example, if there have been changes in the production program or customer requirements. The classification of measures, which is also done manually, rarely needs to be adjusted after an initial assignment in most cases. The selection and integration of paths are then required more frequently, which is why these processes were more automated through the scheduling.

The approach was applied using the representative use case of a foundry. In further applications to use cases of other companies and sectors, the transferability of the approach should be tested and increased in order to take into account the inhomogeneity of the industrial sectors. In a long-term study, the effort of the application of the approach should be compared with the overall benefit through savings from avoided or reduced damage in the event of disruptions. This has remained open in the present use case due to a short period of application.

6. Conclusions and Outlook

As described in the introduction, industry is facing major challenges due to increasing energy prices. Energy flexibility measures can lead to cost reductions by adapting the load to the available energy. Faults in the often complex production system lead to high additional costs, so that risk treatment of production schedules is recommended.

This approach can offer significant added value with manageable effort through the structured and reliable consideration and treatment of risks. Applied in industrial practice, this can encourage the willingness of industry to be energy-flexible. Furthermore, risk treatment leads to the better handling of faults and lower subsequent costs for the company. Thus, planning security for the operation of the energy system with a high proportion of volatile feed-in increases.

To improve the approach in terms of ease of use and reliability, the following options have been identified:

- The approach assumes that the final selection of measures is monitored by a human decision-maker. It is also assumed that a manual definition of risk preferences and classification of measures is desired. As a result, the approach cannot be carried out fully automatically. This would require AHP, as well as the selection of the utility function to be replaced by data-based processes and machine learning.
- The optimization considered uses constraints to describe the restrictions of the production system. In the case of more complex production systems, it may be necessary to use meta-heuristics such as genetic algorithms in order to be able to map all boundary conditions and interactions.
- After the reactive measures have been carried out, the effects should be put in a feedback loop in order to take the findings into account when developing reactive measures in risk treatment. The approach should be expanded to include this functionality. The results of the feedback loop can, i.e., be used for the calculation of Bayesian networks described in Roth et al. [6].
- The input and output data of the approach can be adapted in such a way that interfaces to the common systems in industrial companies can be implemented more easily. For example, order data for scheduling can be transferred from enterprise resource planning systems and load profiles of workstations from the energy management system, and the risk-treated production plans can be visualized by the manufacturing execution system. Thereby, the effort required for the application can be further reduced.

Author Contributions: Conceptualization: S.R.; methodology: S.R. and M.H.; software: M.H.; validation: S.R. and M.H.; formal analysis: S.R. and M.H.; investigation: S.R. and M.H.; resources: S.R. and M.H.; data curation: S.R. and M.H.; writing—original draft preparation: S.R. and M.H.; writing—review and editing: J.S. and G.R.; visualization: S.R. and M.H.; supervision: J.S. and G.R.; project administration: S.R.; funding acquisition: G.R. All authors have read and agreed to the published version of the manuscript.

Funding: This research was funded by the German Federal Ministry of Education and Research (BMBF), grant number 03SFK3G1-2.

Acknowledgments: The authors gratefully acknowledge the financial support of the Kopernikus-Project "SynErgie" from the Federal Ministry of Education and Research (BMBF), and the supervision of the project by the Projektträger Jülich (PTJ) project management organization. Further thanks are due to Steffen Klan and Christoph Hartmann from Fraunhofer IGCV for their support with the use case of the foundry.

Conflicts of Interest: The authors declare no conflict of interest.

References

1. BDEW Bundesverband der Energie- und Wasserwirtschaft. BDEW-Strompreisanalyse Januar 2022—Haushalte und Industrie. 2022. Available online: https://www.bdew.de/media/documents/220124_BDEW-Strompreisanalyse_Januar_2022_24.01.2022_final.pdf (accessed on 20 April 2022).
2. Forschungsstelle für Energiewirtschaft. Deutsche Strompreise an der Börse EPEX Spot in 2020. Available online: https://www.ffe.de/veroeffentlichungen/deutsche-strompreise-an-der-boerse-epex-spot-in-2020/ (accessed on 3 January 2022).
3. EPEX SPOT. Market Data. Available online: https://www.epexspot.com/en/market-data (accessed on 3 January 2022).
4. Keller, F.; Schultz, C.; Braunreuther, S.; Reinhart, G. Enabling Energy-Flexibility of Manufacturing Systems through New Approaches within Production Planning and Control. *Procedia CIRP* **2016**, *57*, 752–757. [CrossRef]
5. Schuh, G.; Stich, V. *Produktionsplanung und—Steuerung 1: Grundlagen der PPS*, 4. Aufl.; Springer: Berlin, Germany, 2012; ISBN 978-3-642-25422-2.
6. Roth, S.; Kalchschmid, V.; Reinhart, G. Development and evaluation of risk treatment paths within energy-oriented production planning and control. *Prod. Eng. Res. Dev.* **2021**, *15*, 413–430. [CrossRef]
7. Schwartz, F.; Voß, S. Störungsmanagement in der Produktion—Simulationsstudien für ein hybrides Fließfertigungssystem. *Z. Plan. Unternehm.* **2004**, *15*, 427–447. [CrossRef]
8. Greve, J. Störungen im Industriebetrieb: Eine klassifizierende untersuchung der Störungen und Analyse des Störverhaltens betrieblicher Systeme unter Anwendung kybernetischer Betrachtungsweise. Ph.D. Thesis, Technische Hochschule, Darmstadt, Germany, 1970.
9. Schwartz, F. *Störungsmanagement in Produktionssystemen*; Shaker: Aachen, Germany, 2004; ISBN 3832230882.
10. Heil, M. *Entstörung Betrieblicher Abläufe*; Springer: Wiesbaden, Germany, 1995; ISBN 978-3-8244-6100-4.
11. Simon, D. *Fertigungsregelung durch Zielgrößenorientierte Planung und Logistisches Störungsmanagement*; Springer: Berlin, Germany, 1995; ISBN 978-3-540-58942-6.
12. Pielmeier, J. System zur Ereignisorientierten Produktionssteuerung. Ph.D. Thesis, Technical University of Munich, Munich, Germany, 2019.
13. Schultz, C. System zur Energieorientierten Produktionssteuerung in der Auftragsbezogenen Fertigung. Ph.D. Thesis, Technical University of Munich, Munich, Germany, 2018.
14. Rösch, M.; Lukas, M.; Schultz, C.; Braunreuther, S.; Reinhart, G. An approach towards a cost-based production control for energy flexibility. *Procedia CIRP* **2019**, *79*, 227–232. [CrossRef]
15. *DIN ISO 31000:2018*; Risikomanagement—Leitlinien. ISO: Geneva, Switzerland, 2018.
16. Oduoza, C.F. Framework for Sustainable Risk Management in the Manufacturing Sector. *Procedia Manuf.* **2020**, *51*, 1290–1297. [CrossRef]
17. Klöber-Koch, J.; Braunreuther, S.; Reinhart, G. Predictive Production Planning Considering the Operative Risk in a Manufacturing System. *Procedia CIRP* **2017**, *63*, 360–365. [CrossRef]
18. Abele, M.; Unterberger, E.; Friedl, T.; Carda, S.; Roth, S.; Hohmann, A.; Reinhart, G. Simulation-based evaluation of an energy oriented production planning system. *Procedia CIRP* **2020**, *88*, 246–251. [CrossRef]
19. Schultz, C.; Sellmaier, P.; Reinhart, G. An Approach for Energy-oriented Production Control Using Energy Flexibility. *Procedia CIRP* **2015**, *29*, 197–202. [CrossRef]
20. Golpîra, H. Smart Energy-Aware Manufacturing Plant Scheduling under Uncertainty: A Risk-Based Multi-Objective Robust Optimization Approach. *Energy* **2020**, *209*, 118385. [CrossRef]
21. Coca, G.; Castrillón, O.D.; Ruiz, S.; Mateo-Sanz, J.M.; Jiménez, L. Sustainable evaluation of environmental and occupational risks scheduling flexible job shop manufacturing systems. *J. Clean. Prod.* **2019**, *209*, 146–168. [CrossRef]
22. Simon, P.; Zeiträg, Y.; Glasschroeder, J.; Tomothy, G.; Reinhart, G. Approach for a Risk Analysis of Energy Flexible Production Systems. *Procedia CIRP* **2018**, *72*, 677–682. [CrossRef]
23. Saaty, T.L. The Analytic Hierarchy Process: Decision Making in Complex Environments. In *Quantitative Assessment in Arms Control*; Springer: Boston, MA, USA, 1984.
24. Ishizaka, A.; Nemery, P. *Multi-Criteria Decision Analysis: Methods and Software*; Wiley: Chichester, UK, 2013; ISBN 978-1-119-97407-9.
25. Alinezhad, A.; Khalili, J. *New Methods and Applications in Multiple Attribute Decision Making (MADM)*; Springer: Cham, Switzerland, 2019; ISBN 978-3-030-15009-9.
26. Vujicic, M.; Papic, M.; Blagojevic, M. Comparative analysis of objective techniques for criteria weighing in two MCDM methods on example of an air conditioner selection. *Tehnika* **2017**, *72*, 422–429. [CrossRef]
27. Wang, M.; Lin, S.-J.; Lo, Y.-C. The comparison between MAUT and PROMETHEE. In Proceedings of the 2010 IEEE International Conference on Industrial Engineering and Engineering Management, Macao, China, 7–10 December 2010; pp. 753–757. [CrossRef]
28. Zanakis, S.H.; Solomon, A.; Wishart, N.; Dublish, S. Multi-attribute decision making: A simulation comparison of select methods. *Eur. J. Oper. Res.* **1998**, *107*, 507–529. [CrossRef]
29. Zavadskas, E.K.; Turskis, Z.; Kildiene, S. State of art surveys of overviews on MCDM/MADM methods. *Technol. Econ. Dev. Econ.* **2014**, *20*, 165–179. [CrossRef]
30. Doczy, R.; AbdelRazig, Y. Green Buildings Case Study Analysis Using AHP and MAUT in Sustainability and Costs. *J. Archit. Eng.* **2017**, *23*, 5017002. [CrossRef]

31. Şahin, M. A comprehensive analysis of weighting and multicriteria methods in the context of sustainable energy. *Int. J. Environ. Sci. Technol.* **2020**, *8*, 591–1616. [CrossRef]
32. Feizi, F.; Karbalaei-Ramezanali, A.; Tusi, H. Mineral Potential Mapping Via TOPSIS with Hybrid AHP–Shannon Entropy Weighting of Evidence: A Case Study for Porphyry-Cu, Farmahin Area, Markazi Province, Iran. *Nat. Resour. Res.* **2017**, *26*, 553–570. [CrossRef]
33. Ren, J.; Liang, H.; Chan, F.T. Urban sewage sludge, sustainability, and transition for Eco-City: Multi-criteria sustainability assessment of technologies based on best-worst method. *Technol. Forecast. Soc. Chang.* **2017**, *116*, 29–39. [CrossRef]
34. Muqimuddin; Singgih, M.L. Integrated FMEA-MCDM For Prioritizing Operational Disruption in Production Process. *IOP Conf. Ser. Mater. Sci. Eng.* **2020**, *847*, 12028. [CrossRef]
35. Turskis, Z.; Goranin, N.; Nurusheva, A.; Boranbayev, S. Information Security Risk Assessment in Critical Infrastructure: A Hybrid MCDM Approach. *Informatica* **2019**, *30*, 187–211. [CrossRef]
36. Wang, L.-E.; Liu, H.-C.; Quan, M.-Y. Evaluating the risk of failure modes with a hybrid MCDM model under interval-valued intuitionistic fuzzy environments. *Comput. Ind. Eng.* **2016**, *102*, 175–185. [CrossRef]
37. VDI Department Factory Planning and Operation. *VDI Verein Deutscher Ingenieure e.V. VDI 5207 Energy-flexible factory, Part 1 Fundamentals*; VDI: Berlin, Germany, 2020.
38. Verhaelen, B.; Kehm, F.; Häfner, B.; Lanza, G. Reaktion auf Störungen globaler Produktionsanläufe. *ZWF* **2020**, *115*, 492–496. [CrossRef]
39. Hernández-Chover, V.; Castellet-Viciano, L.; Hernández-Sancho, F. Preventive maintenance versus cost of repairs in asset management: An efficiency analysis in wastewater treatment plants. *Process Saf. Environ. Prot.* **2020**, *141*, 215–221. [CrossRef]
40. Shannon, C.E.; Weaver, W. *The Mathematical Theory of Communication*; University of Illinois Press: Urbana, IL, USA, 1975; p. c1949. ISBN 0252725484.
41. Keeney, R.L.; Raiffa, H. *Decisions with Multiple Objectives*; Cambridge University Press: Cambridge, MA, USA, 2014; ISBN 9781139174084.
42. Wang, J.-J.; Jing, Y.-Y.; Zhang, C.-F.; Zhao, J.-H. Review on multi-criteria decision analysis aid in sustainable energy decision-making. *Renew. Sustain. Energy Rev.* **2009**, *13*, 2263–2278. [CrossRef]
43. Reinhart, G.; Schultz, C. Herausforderungen einer energieorientierten Produktionssteuerung. *Z. Wirtsch. Fabr.* **2014**, *109*, 29–33. [CrossRef]
44. Shakya, R.; Chauhan, P. Modelling of Risk Analysis in Production System. *IOP Conf. Ser. Mater. Sci. Eng.* **2019**, *691*, 12087. [CrossRef]
45. Romeike, F. *Risikomanagement*; Springer Gabler: Wiesbaden, Germany, 2018; ISBN 9783658139520.
46. Klöber-Koch, J.; Braunreuther, S.; Reinhart, G. Approach For Risk Identifikation And Assessment in A Manufacturing System. *Procedia CIRP* **2018**, *72*, 683–688. [CrossRef]
47. Velasquez, M.; Hester, P. An Analysis of Multi-Criteria Decision Making Methods. *Int. J. Oper. Res.* **2013**, *10*, 56–66.
48. Pereira, J.C.; Fragoso, M.D.; Todorov, M.G. Risk Assessment using Bayesian Belief Networks and Analytic Hierarchy Process applicable to Jet Engine High Pressure Turbine Assembly. *IFAC-PapersOnLine* **2016**, *49*, 133–138. [CrossRef]
49. Saaty, T.L. How to make a decision: The analytic hierarchy process. *Eur. J. Oper. Res.* **1990**, *48*, 9–26. [CrossRef]
50. Dong, Q.; Saaty, T.L. An analytic hierarchy process model of group consensus. *J. Syst. Sci. Syst. Eng.* **2014**, *23*, 362–374. [CrossRef]
51. Emovon, I.; Norman, R.; Murphy, A. Methodology of using an integrated averaging technique and MAUT method for failure mode and effects analysis. *J. Eng. Technol.* **2016**, *7*, 140–155.
52. Siskos, Y.; Grigoroudis, E.; Matsatsinis, N. Multiple criteria decision analysis: State of the art surveys. *Mult. Criteria Decis. Anal.* **2016**, *233*, 315–362. [CrossRef]
53. Foerster, F.; Nikelowski, L. Dynamic risk consideration of predicted maintenance needs regarding economic efficiency. *Procedia CIRP* **2020**, *93*, 915–920. [CrossRef]
54. Toba, H. Segment-based approach for real-time reactive rescheduling for automatic manufacturing control. In Proceedings of the 1999 IEEE International Symposium on Semiconductor Manufacturing Conference Proceedings (Cat No.99CH36314), Santa Clara, CA, USA, 11–13 October 1999; pp. 69–72. [CrossRef]
55. MathWorks. Mixed-Integer Linear Programming Algorithms. Available online: https://de.mathworks.com/help/optim/ug/mixed-integer-linear-programming-algorithms.html?searchHighlight=branch%20and%20bound&s_tid=srchtitle#btzwtmv (accessed on 6 April 2022).
56. Bosse, M.; Frost, E.; Hazrat, M.; Rhiemeier, J.-M.; Wolff, H. Ermittlung von branchenspezifischen Potentialen zum Einsatz von erneuerbaren Energien in besonders energieintensiven Industriesektoren am Beispiel der Gießerei-Industrie. Report number UBA-FB 00. 2013. Available online: https://www.bmuv.de/forschungsbericht/ermittlung-von-branchenspezifischen-potentialen-zum-einsatz-von-erneuerbaren-energien-in-besonders-energieintensiven-industriesektoren-am-beispiel-der-giesserei-industrie (accessed on 18 April 2022).

Article

Unfair and Risky? Profit Allocation in Closed-Loop Supply Chains by Cooperative Game Approaches

Ting Zeng [1] and Tianjian Yang [2,*]

[1] School of Economics and Management, Beijing University of Posts and Telecommunications, Beijing 100876, China; tingzeng@bupt.edu.cn
[2] School of Modern Post (School of Automation), Beijing University of Posts and Telecommunications, Beijing 100876, China
* Correspondence: frankytj@bupt.edu.cn

Abstract: Behavioral factors (i.e., risk aversion and fairness concern) are considered for profit allocation in a closed-loop supply chain. This paper studies a two-echelon closed-loop supply chain (CLSC) consisting of a risk-neutral manufacturer, a risk-averse fairness-neutral retailer, and a risk-neutral retailer having fairness concerns. Cooperative game analysis is used to characterize equilibriums under five scenarios: a centralized, a decentralized and three partially allied models. Analytical results confirm that even when factoring in retailers' risk aversion and fairness concern, the centralized model still outperforms decentralized. This paper makes a numerical study on the effects of risk aversion and fairness concern on profit distribution under these five models. It reveals that the impact of the risk aversion parameter and fairness concern parameter is dynamic, not always positive or negative. These research results provide helpful insights for CLSC managers to find out available choices and feasible ways to achieve fair profit allocations.

Keywords: risk aversion; fairness concern; cooperative game; closed-loop supply chain; utility maximization

Citation: Zeng, T.; Yang, T. Unfair and Risky? Profit Allocation in Closed-Loop Supply Chains by Cooperative Game Approaches. *Appl. Sci.* **2022**, *12*, 6245. https://doi.org/10.3390/app12126245

Academic Editor: Panagiotis Tsarouhas

Received: 9 May 2022
Accepted: 17 June 2022
Published: 19 June 2022

1. Introduction

Under the increasingly stressful environment, more and more enterprises remanufacture and recycle preowned goods, forming a closed-loop supply chain. Circular economy has been recognized as more efficient and competitive than linear economy [1]. Circular closed-loop supply chains (CLSCs) management provides a new perspective to the sustainability [2,3]. There remain challenges in managing circular supply chains effectively and various strategies are proposed by researchers [4–6].

One of the primary objectives of CLSCs is to maximize channel profits. Decision makers in CLSCs are usually not absolutely rational [7,8]. Behavioral factors such as risk aversion and fairness concern have impacts on operational decisions and profit distribution [9]. Under the uncertainties of market demand or recycling profit, CLSC decision makers tend to avoid potential risks. CLSC members are often concerned with the proportion of their profits in the total CLSC profit. Price strategy is a primary methodology to coordinate CLSC management.

This research addresses two themes: (1) derivation of equilibrated production quantities, prices and profits for a CLSC with both risk aversion and fairness concern under the five models; (2) evaluation of risk aversion and fairness concern's effect on profit allocations in cooperative game settings. Supply chains often have a single manufacturer and multiple retailers [10]. In this research, we incorporate retailers' risk aversion and fairness concern into a two-echelon CLSC consisting of a manufacturer (M), a risk-averse retailer (L) without fairness concern, and a risk-neutral retailer (F) having fairness concern. By cooperative game approaches, we examine five scenarios: (1) a centralized model where one central decision maker plans for all agents (CC); (2) a decentralized model where each agent makes decisions independently (DC); (3) L and F compose an alliance (LF); (4) M and L compose an alliance (ML) and (5) M and F compose an alliance (MF).

2. Literature Review

Considerable attention has been attracted to CLSCs in academia and practice. Remarkable studies have been conducted regarding two-echelon CLSCs [11–15] and three-echelon CLSCs [16–18]. Retailers of a CLSC frequently cooperate with manufacturers to collect preowned goods [19]. For instance, the Procter & Gamble company collects and resells used products through Loop, an online shopping platform which also sells new products. By forming strategic partnerships, enterprises in CLSCs can improve the profitability of individuals and distribution channels.

In past research on CLSCs, agents were assumed to be of complete rationality. In actual fact, decision making agents often show different social preferences such as risk aversion. Research on CLSC coordination with risk aversion has been carried out. Ke et al. [20] examined pricing and remanufacturing issues in a CLSC which is consisted of a dominant manufacturer and a risk-averse retailer. Zeballos et al. [21] constructed and analyzed a risk-averse multi-stage model of a CLSC which include several functional entities. Ma et al. [22] examined a bike-sharing operation network to study a multi-product, multi-agent, single-stage CLSC system under risk-averse criterion. Das et al. [23] studied a two-period risk-averse model of a CLSC for reusable packaging materials.

Fairness concerns also widely exist in CLSCs. Along this research line, Ma et al. [24] investigated reverse-channel CLSCs where retailers serve as collectors of preowned products while retailers have distributional fairness concern. Zheng et al. [25] incorporated fairness concern into a three-echelon CLSC composed of one manufacturer, one distributor and one retailer, and examined the impact of the retailer's fairness concern. Sarkar and Bhala [26] analyzed the coordination of a CLSC in the presence of fairness concerns with a constant wholesale price contract. Wang et al. [27] studied the impact of fairness concern in e-commerce CLSCs and showed that the fairness concern of the e-commerce platform reduces individual profit and systematic efficiency.

Limited research has been carried out on CLSCs regarding both risk aversion and fairness concern. He et al. [28] examined pricing strategies of a CLSC consisting of one manufacturer and one retailer considering the factors of risk aversion and fairness concern. Li et al. [29] investigated the price decisions in a CLSC with a risk-averse retailer with fairness concern and a risk-neutral manufacturer. Zhang and Zhang [30] studied a CLSC composed of two suppliers, a manufacturer, a risk-averse retailer and a fairness-concerned third-party under supply disruptions. These studies show that risk aversion and fairness concerns bring complexities to the coordination of a CLSC, and past research is limited to non-cooperative game setups. Differing from existing studies, this paper investigates the impact of risk aversion and fairness concern in a CLSC by cooperative game approaches. As summarized in Table 1, this paper differs from other existing CLSC literature in subject areas.

Table 1. An overview of characteristics of the literature and this paper.

References	Subject Areas			
	CLSCs	Risk Aversion	Fairness Concern	Cooperative Game Approach
Refs. [11–14,16–18]	Yes	No	No	No
Ref. [15]	Yes	No	No	Yes
Refs. [19–23]	Yes	Yes	No	No
Refs. [24,26,27]	Yes	No	Yes	No
Ref. [25]	Yes	No	Yes	Yes
Refs. [28–30]	Yes	Yes	Yes	No
This paper	Yes	Yes	Yes	Yes

3. Problem Description and Assumptions

This section describes a two-echelon supply chain model consisting of a risk-neutral manufacturer (M) and two competing retailers (L and F) who play a Stackelberg game. It is assumed that Retailer L has a higher market share than Retailer F, so that customers have higher willingness to purchase products from L than F. Given Retailer F's relatively weaker position and the wide recognition that disadvantaged agents often care about fairness issues [31,32], this research assumes that Retailer F has a fairness concern with the upper manufacturer. We assume Retailer L is risk-averse and attempts to avoid risks. The sales channels provided by the two retailers are available to all customers. Customers are risk neutral and make decisions to maximize their utility, and we consider one-period interactions among CLSC members.

To derive characteristic functions of the cooperative game, all potential alliances and the corresponding equilibriums are investigated. Five models are presented in Figure 1: the centralized model, CC, with a central decision maker, the decentralized model, DC, where agents make individual decisions independently, and three partial-cooperation models, LF, ML, and MF, where three partial alliances form and make centralized decisions within the respective coalition.

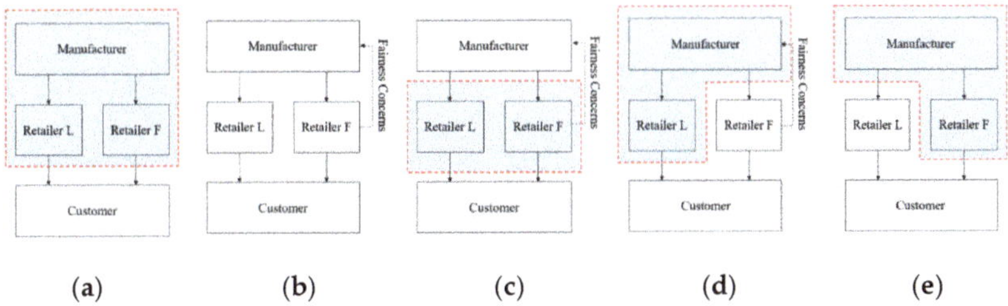

Figure 1. The cooperative, non-cooperative, and partial-cooperative models of the two-echelon CLSC consisting of one manufacturer and two retailers are: (**a**) Model CC; (**b**) Model DC; (**c**) Model LF; (**d**) Model ML; and (**e**) Model MF.

On the basis of the problem description, this paper employs the following notation throughout the research as shown in Table 2.

Table 2. Variables and parameters.

Notation	Definition
c_n, c_r	Unit production cost for a new or remanufactured product
w_n^L, w_r^L	Unit wholesale price for a new or remanufactured product offered by M to L
w_n^F, w_r^F	Unit wholesale price for a new or remanufactured product offered by M to F
p_n^L, p_r^L	Unit retail price for a new or remanufactured product offered by L
p_n^F, p_r^F	Unit retail price for a new or remanufactured product offered by F
q_n, q_r	Production quantity for new or remanufactured products
q_n^L, q_r^L	Quantity of new or remanufactured products transferred from M to L
q_n^F, q_r^F	Quantity of new or remanufactured products transferred from M to F
t_n^L, t_r^L	Utility of a consumer receiving from L for a new or remanufactured product
t_n^F, t_r^F	Utility of a consumer receiving from F for a new or remanufactured product
A	Exogenous unit cost for recycling a preowned product
λ	F's fairness concern parameter
η	L's risk aversion parameter, where $\eta \in (0,1)$
θ	Consumers' willingness-to-pay for a new product
δ	Consumers' value deduction for a remanufactured product, $\delta \in [0,1)$
ρ	The correlation coefficient between new products and remanufactured products
σ^2	The variance of demand uncertainty

Table 2. *Cont.*

Notation	Definition
$\frac{\alpha}{\beta}$	The maximum risk-aversion parameter to participate in the game
$U(L)$	The utility function of Retailer L
$\Pi(L)$	The profit function of Retailer L
$Var(L)$	The profit mean-variance function of Retailer L
π_j^i	Profit functions of alliance j of model i, $i \in$ {CC, DC, LF, ML, MF} and j = G (Model CC); M, L, F (Model DC); M, LF (Model LF); ML, F (Model ML); MF, L (Model MF), where G is the grand coalition.
u_y^x	Utility function of coalition y, $x \in$ {CC, DC, LF, ML, MF} and y = G (Model CC); M, L, F (Model DC); M, LF (Model LF); ML, F (Model ML); MF, L (Model MF), where G is the grand coalition.

In this CLSC, consumers' willingness to pay for a new product is assumed to be θ, which is a uniform distribution from 0 to 1. For a remanufactured product, consumers' willingness to pay is a portion δ of θ with $\delta \in [0, 1)$. The utility of a consumer receiving from Retailer L for a new product is $t_n^L(\theta) = \theta - p_n^L$, and the utility of a consumer receiving from Retailer L for a remanufactured product is $t_r^L(\theta) = \delta\theta - p_r^L$. Following the principle of utility maximization, if $t_n^L \geq \max\{t_r^L, 0\}$, consumers will purchase a new product, leading to a new product demand function $q_n^L(p_n^L, p_r^L) = 1 - \frac{p_n^L - p_r^L}{1-\delta}$. If $t_r^L \geq \max\{t_n^L, 0\}$, consumers will purchase a remanufactured product, resulting in a remanufactured product demand function $q_r^L(p_n^L, p_r^L) = \frac{p_n^L - p_r^L}{1-\delta} - \frac{p_r^L}{\delta}$ [25,33,34]. Likewise, the utility of a consumer receiving from Retailer F for a new product is $t_n^F(\theta) = \theta - p_n^F$, and for a remanufactured product is $t_r^F(\theta) = \delta\theta - p_r^F$. If $t_n^F \geq \max\{t_r^F, 0\}$, a new product demand function is $q_n^F(p_n^F, p_r^F) = 1 - \frac{p_n^F - p_r^F}{1-\delta}$. If $t_r^F \geq \max\{t_n^F, 0\}$, a remanufactured product demand function is $q_r^F(p_n^F, p_r^F) = \frac{p_n^F - p_r^F}{1-\delta} - \frac{p_r^F}{\delta}$.

Retailer F's fairness concern is a reaction towards adverse inequality related to the upstream agent [31]. Given this assumption, Retailer F's fairness concern becomes unrelated in the models of CC and MF due to the absence of financial transaction between M and F. However, in Models DC, LF, and ML, Retailer F shows fairness concerns with M.

With similar arguments in studies [35–40], it is assumed that a risk-averse retailer evaluates his or her profit on the basis of a mean–variance function. The utility function $U(L)$ of Retailer L considering its profit is presented as below:

$$\begin{aligned} U(L) = \Pi(L) - \;& \eta Var(L)/2 \\ = \;& \left(p_n^L - w_n^L\right) * q_n^L + \left(p_r^L - w_r^L\right) * q_r^L - \eta\sigma^2 \Big[\left(p_n^L - w_n^L\right)^2 \\ & + \left(p_r^L - w_r^L\right)^2 + 2\rho\left(p_n^L - w_n^L\right)\left(p_r^L - w_r^L\right)\Big]/2 \end{aligned} \tag{1}$$

where η (> 0) is the risk aversion parameter for Retailer L and σ^2 is the variance of demand uncertainty. A risk pooling effect exists [41] when the correlation coefficient between new products and remanufactured products is negative ($-1 < \rho < 0$). If the risk aversion parameter is too high, the risk-averse Retailer L does not participate in the market. We assume that $0 < \eta < \frac{\alpha}{\beta}$ to ensure L participates in the game.

4. The Equilibrium Analysis

The equilibrium analysis is carried out in the following and results are derived for the five base models, CC, DC, LF, ML, and MF.

4.1. The Centralized Case (Model CC)

In the centralized case, a central decision maker acts on behalf of all CLSC agents for maximization of the systematic profit as shown in Figure 1a. No financial transaction occurs between M and F, thus F's fairness concern does not exist. The centralized decision maker sells products to the end consumers directly. All CLSC members share risks, and L does not have risk aversion in this case. The profit function is formulated as:

$$\max_{p_n^L, p_r^L, p_n^F, p_r^F} \pi_G^{CC} = p_n^L q_n^L + p_n^F q_n^F - c_n\left(q_n^L + q_n^F\right) + p_r^L q_r^L + p_r^F q_r^F - c_r\left(q_r^L + q_r^F\right) - A\left(q_r^L + q_r^F\right) \tag{2}$$

Equation (2) presents the channel profit of this CLSC as two parts: the profit for new products and the profit for remanufactured products. The centralized decision maker determines the pricing for two types of products sold to two retailers, in order to achieve the highest channel profit.

Proposition 1. *After first order derivation, we get the optimal retail prices, the resulting sales quantities, and the optimal profit in the centralized model as following (Proof See Appendix A):*

$$p_n^{L*CC} = p_n^{F*CC} = \frac{1+c_n}{2}, \ p_r^{L*CC} = p_r^{F*CC} = \frac{c_r+A+\delta}{2}, \ q_n^{*CC} = q_n^{L*CC} + q_n^{F*CC}$$
$$= \frac{1-\delta-(c_n-c_r-A)}{1-\delta},$$
$$q_r^{*CC} = q_r^{L*CC} + q_r^{F*CC} = \frac{\delta c_n - c_r - A}{\delta(1-\delta)}, \ and \ \pi_G^{*CC} = \frac{(1-c_n)^2}{2} + \frac{(C_r+A-c_n\delta)^2}{2\delta(1-\delta)}$$

4.2. The Decentralized Case (Model DC)

In the decentralized case as shown in Figure 1b, a risk-neutral M and a risk-averse L are supposed to have no fairness concern, while F is fairness-caring with M. Besides its own profit, F also cares about its profit comparative to that of M. Similar to a large body of literature [42,43], the utility function of F is given as

$$u_F^{DC} = \pi_F^{DC} - \lambda\left(\pi_M^{DC} - \pi_F^{DC}\right) \tag{3}$$

where $\lambda \geq 0$ is F's fairness concern parameter.

In this model, M maximizes its profit while L and F pursue their utility maximization. M determines wholesale prices for L and F, then L and F decide their retail prices for products to end consumers. Therefore, the Stackelberg game composed of M, L, and F is expressed as

$$\max \pi_M^{DC} = \left(w_n^L - c_n\right)q_n^L + \left(w_n^F - c_n\right)q_n^F + \left(w_r^L - c_r - A\right)q_r^L + \left(w_r^F - c_r - A\right)q_r^F$$
$$s.t. \begin{cases} \max u_L^{DC} = \left(p_n^L - w_n^L\right) * q_n^L + \left(p_r^L - w_r^L\right) * q_r^L - \dfrac{\eta\sigma^2\left[\left(p_n^L-w_n^L\right)^2+\left(p_r^L-w_r^L\right)^2+2\rho\left(p_n^L-w_n^L\right)\left(p_r^L-w_r^L\right)\right]}{2} \\ \max u_F^{DC} = \pi_F^{DC} - \lambda\left(\pi_M^{DC} - \pi_F^{DC}\right) \end{cases} \tag{4}$$

where $\pi_F^{DC} = \left(p_n^F - w_n^F\right)q_n^F + \left(p_r^F - w_r^F\right)q_r^F$. The equilibrium results are obtained subsequently.

Proposition 2. *In the model DC, equilibrium wholesale prices are found as* $w_n^{*L} = \frac{c_n+1}{2}$, $w_r^{*L} = \frac{c_r+A+\delta}{2}$, $w_n^{*F} = \frac{\lambda+c_n+3\lambda c_n+1}{4\lambda+2}$, $w_r^{*F} = \frac{c_r+A+\delta+3\lambda c_r+3\lambda A+\lambda\delta}{4\lambda+2}$ *(Proof See Appendix B). The resulting equilibrium prices, sales quantities, and profits* p_n^{*L}, p_r^{*L}, p_n^{*F}, p_r^{*F}, q_n^{*L}, q_r^{*L}, q_n^{*F}, q_r^{*F}, π_M^{*DC}, u_L^{*DC}, u_F^{*DC} *can be obtained respectively. The corresponding results are not presented here because the expressions are too long.*

4.3. L and F form an Alliance (Model LF)

In the LF model as presented in Figure 1c, L and F form an alliance as one decision maker to decide p_n^L, p_r^L, p_n^F, and p_r^F. The coalition LF has fairness concerns with M, and it is also risk averse. Therefore, this LF model is formulated as

$$\max \pi_M^{LF} = \left(w_n^{LF} - c_n\right)q_n^{LF} + \left(w_r^{LF} - c_r - A\right)q_r^{LF}$$
$$s.t., \ \max u_{LF}^{LF} = \left(p_n^{LF} - w_n^{LF}\right) * q_n^{LF} + \left(p_r^{LF} - w_r^{LF}\right) * q_r^{LF}$$
$$- \eta\sigma^2\left[\left(p_n^{LF} - w_n^{LF}\right)^2 + \left(p_r^{LF} - w_r^{LF}\right)^2 + 2\rho\left(p_n^{LF} - w_n^{LF}\right)\left(p_r^{LF} - w_r^{LF}\right)\right]/2 - \lambda\left(\pi_M^{LF} - \pi_{LF}^{LF}\right)$$

where $\pi_{LF}^{LF} = \left(p_n^{LF} - w_n^{LF}\right) * q_n^{LF} + \left(p_r^{LF} - w_r^{LF}\right) * q_r^{LF}$. The equilibrium results are characterized in the following proposition.

Proposition 3. *In the model LF, equilibrium prices are found as*

$$p_n^{*LF}\left(w_n^{LF}, w_r^{LF}\right) = \frac{Kw_n^{LF} + \left(\eta\sigma^2\rho + \eta\sigma^2\delta\right)w_r^{LF} - \eta\sigma^2\lambda\delta^2 + \eta\sigma^2\rho\lambda c_r + \eta\sigma^2\rho\lambda A - \eta\sigma^2\lambda c_n\delta + \eta\sigma^2\lambda c_r\delta + \eta\sigma^2\lambda A\delta - \eta\sigma^2\rho\lambda c_n\delta + 4\lambda - 2\lambda c_n - 2\lambda^2 c_n + 2\lambda^2 - \eta\sigma^2\delta^2 + \eta\sigma^2\delta + \eta\sigma^2\lambda\delta + 2}{\eta^2\sigma^4\rho^2\delta^2 - \eta^2\sigma^4\rho^2\delta - \eta^2\sigma^4\delta^2 + \eta^2\sigma^4\delta + 4\eta\sigma^2\rho\lambda\delta + 4\eta\sigma^2\rho\delta + 2\eta\sigma^2\lambda\delta + 2\eta\sigma^2\lambda + 2\eta\sigma^2\delta + 2\eta\sigma^2 + 4\lambda^2 + 8\lambda + 4},$$

$$p_r^{*LF}\left(w_n^{LF}, w_r^{LF}\right) = \frac{\left(K + \eta\sigma^2\delta - \eta\sigma^2\right)w_r^{LF} + \left(\eta\sigma^2\delta + \eta\sigma^2\rho\delta\right)w_n^{LF} - \left(2\lambda + 2\lambda^2 + \eta\sigma^2\rho\lambda\delta + \eta\sigma^2\lambda\right)c_r + \left(\eta\sigma^2\lambda\delta + \eta\sigma^2\rho\lambda\delta\right)c_n + H}{\eta^2\sigma^4\rho^2\delta^2 - \eta^2\sigma^4\rho^2\delta - \eta^2\sigma^4\delta^2 + \eta^2\sigma^4\delta + 4\eta\sigma^2\rho\lambda\delta + 4\eta\sigma^2\rho\delta + 2\eta\sigma^2\lambda\delta + 2\eta\sigma^2\lambda + 2\eta\sigma^2\delta + 2\eta\sigma^2 + 4\lambda^2 + 8\lambda + 4}$$

where

$$K = 2 + 6\lambda + 4\lambda^2 + 2\eta\sigma^2 + \eta\sigma^2\delta + \eta^2\sigma^4\delta + 2\eta\sigma^2\lambda + 3\eta\sigma^2\rho\delta + 2\eta\sigma^2\lambda\delta + \eta^2\sigma^4\rho^2\delta^2 + 4\eta\sigma^2\rho\lambda\delta - \eta^2\sigma^4\delta^2 - \eta^2\sigma^4\rho^2\delta,$$

$$H = 2\delta + 4\lambda\delta + 2\lambda^2\delta - 2\lambda A - 2\lambda^2 A - \eta\sigma^2\lambda A - \eta\sigma^2\rho\delta + \eta\sigma^2\rho\delta^2 + \eta\sigma^2\rho\lambda\delta^2 - \eta\sigma^2\rho\lambda\delta - \eta\sigma^2\rho\lambda A\delta.$$

(Proof See Appendix C). The resulting equilibrium wholesale prices, sales quantities, and profits, $w_n^{*LF}, w_r^{*LF}, q_n^{*LF}, q_r^{*LF}, \pi_M^{*LF}, u_{LF}^{*LF}$, can be obtained, respectively. The specific results are not presented here because the expressions are too long.

4.4. M and L form an Alliance (Model ML)

In the ML model, as displayed in Figure 1d, M and L form an alliance and there is no financial transaction between M and L. Hence, M and L make decisions as one entity and share risks together, eliminating L's risk aversion in this model. The alliance of M and L, as the Stackelberg leader, interacts with F in a non-cooperative setup. Consequently, the ML model is expressed as

$$\max u_{ML}^{ML} = \left(p_n^{ML} - c_n\right)q_n^{ML} + \left(p_r^{ML} - c_r - A\right)q_r^{ML} + \left(w_n^F - c_n\right)q_n^F + \left(w_r^F - c_r - A\right)q_r^F$$
$$s.t.\ \max u_F^{ML} = \pi_F^{ML} - \lambda\left(u_{ML}^{ML} - \pi_F^{ML}\right)$$

where $\pi_F^{ML} = \left(p_n^F - w_n^F\right) * q_n^F + \left(p_r^F - w_r^F\right) * q_r^F$.

The alliance of M and L removes the competition between M and L as well as L's risk aversion. The ML coalition makes joint pricing decisions $\left(p_n^{ML}, p_r^{ML}\right)$ to the final customers. The equilibrium results are attained and described in Proposition 4.

Proposition 4. *The pricing equilibriums of Model ML are obtained as $p_n^{*ML} = \frac{c_n+1}{2}, p_r^{*ML} = \frac{c_r+A+\delta}{2}$, $w_n^{*F} = \frac{\lambda+c_n+3\lambda c_n+1}{4\lambda+2}$ and $w_r^{*F} = \frac{c_r+A+\delta+3\lambda c_r+3\lambda A+\lambda\delta}{4\lambda+2}$ (Proof See Appendix D). The resulting equilibrium prices, sales quantities, and profits, $p_n^{*F}, p_r^{*F}, q_n^{*F}, q_r^{*F}, q_n^{*ML}, q_r^{*ML}, u_F^{*ML}, u_{ML}^{*ML}$, can be obtained, respectively. The corresponding results are not presented here because the expressions are too long.*

4.5. M and F form an Alliance (Model MF)

In the MF model, M and F form an alliance as depicted in Figure 1e. There is no financial transaction as well as no profit transfer from M to F inside this alliance, making F's fairness concern ignorable in this model. L engages with the Stackelberg leader MF in a non-cooperative manner. As a result, Model MF is then formulated accordingly as

$$\max u_{MF}^{MF} = \left(w_n^L - c_n\right)q_n^L + \left(p_n^{MF} - c_n\right)q_n^{MF} + \left(w_r^L - c_r - A\right)q_r^L + \left(p_r^{MF} - c_r - A\right)q_r^{MF}$$
$$s.t.\ \max u_L^{MF} = \left(p_n^L - w_n^L\right) * q_n^L + \left(p_r^L - w_r^L\right) * q_r^L - \eta\sigma^2\left[\left(p_n^L - w_n^L\right)^2 + \left(p_r^L - w_r^L\right)^2 + 2\rho\left(p_n^L - w_n^L\right)\left(p_r^L - w_r^L\right)\right]/2$$

In this case, the coalition of M and F removes the competition between M and F so that F's fairness concern is irrelevant. A joint pricing decision $\left(p_n^{MF}, p_r^{MF}\right)$ is made by MF to the end consumers. The equilibrium results are represented in Proposition 5.

Proposition 5. *The pricing equilibriums of Model MF are obtained as $p_n^{*MF} = \frac{c_n+1}{2}, p_r^{*MF} = \frac{c_r+A+\delta}{2}$, $w_n^{*L} = \frac{c_n+1}{2}$ and $w_r^{*L} = \frac{c_r+A+\delta}{2}$ (Proof See Appendix E). The equilibrated prices, sales quantities,* *and utilities, p_n^{*L}, p_r^{*L}, q_n^{*L}, q_r^{*L}, q_n^{*MF}, q_r^{*MF}, u_L^{*MF}, u_{MF}^{*MF}, can be obtained, respectively. The corresponding results are not presented here because the expressions are too long.*

4.6. Analytical Comparison of Resulting Equilibriums

The resulting equilibriums in Propositions 1–5 are assessed comparatively and the conclusions are drawn as the following.

Proposition 6. *The wholesale pricing of the five models satisfy*

(1) $\left(w_n^{*L}\right)^{DC} = \left(w_n^{*L}\right)^{MF}$, $\left(w_r^{*L}\right)^{DC} = \left(w_r^{*L}\right)^{MF}$, $\left(w_n^{*F}\right)^{DC} = \left(w_n^{*F}\right)^{ML}$, $\left(w_r^{*F}\right)^{DC} = \left(w_r^{*F}\right)^{ML}$, *and* $\dfrac{\partial\left(w_n^{*L}\right)^i}{\partial\eta} = \dfrac{\partial\left(w_r^{*L}\right)^i}{\partial\eta} = 0$ *where* $i \in \{DC, MF\}$.

(2) *If* $A + \delta < 1$, $\left(w_n^{*L}\right)^i > \left(w_r^{*L}\right)^i$ *where* $i \in \{DC, MF\}$.

(3) *If* $(3\lambda + 1)(c_r + A - c_n) + \lambda(\delta - 1) + \delta < 1$, $\left(w_n^{*F}\right)^i > \left(w_r^{*F}\right)^i$ *where* $i \in \{DC, ML\}$.

Proposition 6(1) demonstrates that the wholesale price from M to L in Model DC equals that in Model MF. It also presents that the wholesale price from M to F in Model DC is equivalent to that in Model ML. This observation is explainable that M makes the same decision of wholesale price to the downstream single party (i.e., L or F). Thus we have $\left(w_n^{*L}\right)^{DC} = \left(w_n^{*L}\right)^{MF}$, $\left(w_r^{*L}\right)^{DC} = \left(w_r^{*L}\right)^{MF}$, $\left(w_n^{*F}\right)^{DC} = \left(w_n^{*F}\right)^{ML}$, and $\left(w_r^{*F}\right)^{DC} = \left(w_r^{*F}\right)^{ML}$. In addition, M considers L's risk aversion only when M and L form an alliance. Therefore, M makes decision of wholesale prices independently of L's risk aversion parameter η in the models of DC and MF. Therefore, we have $\dfrac{\partial\left(w_n^{*L}\right)^i}{\partial\eta} = \dfrac{\partial\left(w_r^{*L}\right)^i}{\partial\eta} = 0$, where $i \in \{DC, MF\}$.

Proposition 6(2) is a comparison of wholesale prices for new and remanufactured products in the DC and MF models. When $A + \delta < 1$, L has higher willingness to pay for new products than remanufactured products.

Proposition 6(3) compares the wholesale prices of new and remanufactured products of the DC and ML models. When $(3\lambda + 1)(c_r + A - c_n) + \lambda(\delta - 1) + \delta < 1$, F has lower willingness to pay for remanufactured products.

Proposition 7. *The retail pricing of new and remanufactured products in the five models satisfy*

(1) $p_n^{*ML} = p_n^{*MF} = p_n^{L*CC} = p_n^{F*CC}$, $p_r^{*ML} = p_r^{*MF} = p_r^{L*CC} = p_r^{F*CC}$, *and* $\dfrac{\partial p_n^{*i}}{\partial\eta} = \dfrac{\partial p_r^{*i}}{\partial\eta} = \dfrac{\partial p_n^{*i}}{\partial\lambda} = \dfrac{\partial p_r^{*i}}{\partial\lambda} = 0$ *where* $i \in \{ML, MF, CC\}$.

(2) $p_n^{L*CC} < p_n^{L*DC}$, $p_r^{L*CC} < p_r^{L*DC}$

(3) *If* $A + \delta < 1$, $p_n^{*i} > p_r^{*i}$ *where* $i \in \{ML, MF, CC\}$.

Proposition 7(1) shows that the retail prices of the ML coalition, MF coalition, and centralized model are the same. F's fairness concerns and L's risk aversion parameter do not affect the retail prices of the ML, MF, and MLF alliances. This is clear given that L or F partners with upstream agent M in the three models.

Proposition 7(2) concludes that the retail price of L in the CC model is always lower than that in the DC model. This is due to customers having higher willingness to purchase products from L than MLF.

Proposition 7(3) is a comparison of retail prices for new and remanufactured products in the three models. When $A + \delta < 1$, consumers have higher willingness to buy new products than remanufactured products.

Proposition 8. *The profits of the five models satisfy:*

$$\pi_G^{*CC} > \pi_M^{*DC} + u_L^{*DC} + u_F^{*DC}.$$

Proposition 8 compares the profits of the centralized case and decentralized case. With consideration of both fairness concern and risk aversion, the profit of the centralized model is higher than that of the decentralized model.

5. Numerical Experiment

In this section, we illustrate how L's risk aversion parameter η and F's fairness concern parameter λ affect profit distribution under the five models. To obtain the schemes of profit and utility in relation to parameters η and λ only, we set other variables and parameters at: the variance of demand uncertainty $\sigma^2 = 0.64$, the correlation coefficient between new products and remanufactured products $\rho = -0.5$, unit production cost for a new product $c_n = 1$, unit production cost for a remanufactured product $c_r = 0.5$, exogenous unit cost for recycling a preowned product $A = 0.1$, consumers' value deduction for a remanufactured product $\delta = 0.5$. By plugging these values into the profit and utility functions in Section 4, we obtain individual profit and utility functions with variables η and λ. The effect of L's risk aversion parameter η and F's fairness concern parameter λ on profit utility is graphically illustrated in Figures 2 and 3. In the DC model, the utility of F is negative, as shown in Figure 2c, due to $\frac{\pi_F^{DC}}{\pi_M^{DC}} < \frac{\lambda}{\lambda+1}$. Figure 2 demonstrates that in the DC model, L's risk aversion parameter η affects all parties in different ways. L's utility decreases linearly as η increases. M's profit and F's utility have non-linear relations with η and λ.

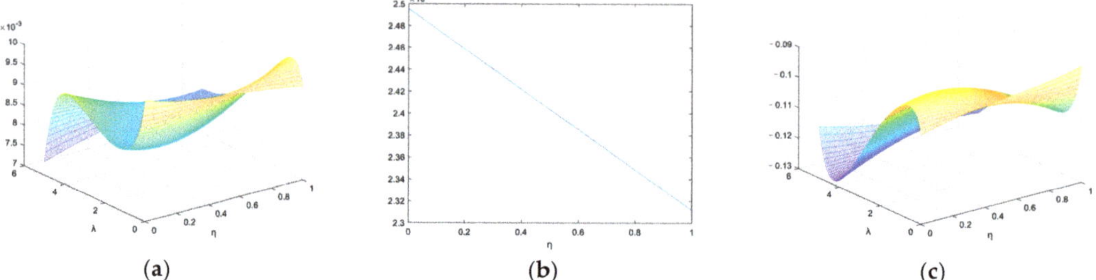

(a)   **(b)**   **(c)**

Figure 2. Individual utilities under the DC model are: (**a**) M's profit in the DC model; (**b**) L's utility in the DC model; (**c**) F's utility in the DC model.

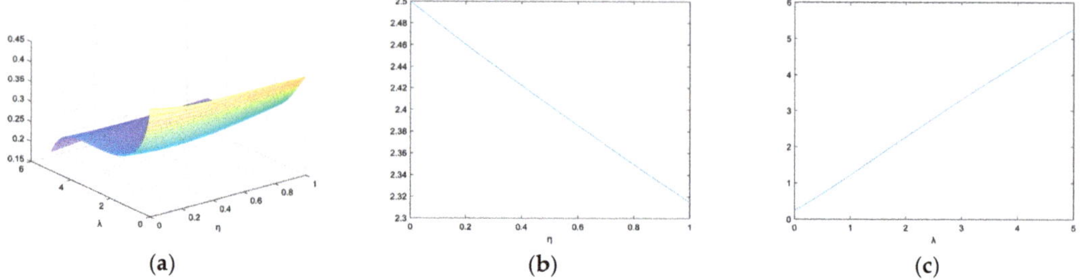

(a)   **(b)**   **(c)**

Figure 3. Individual utilities under the LF, MF, and ML models are: (**a**) M's profit in the LF model; (**b**) L's utility in the MF model; (**c**) F's utility in the ML model.

Figure 3 visually shows how η and λ affect individual profit and utility under different models. Figure 2a illustrates that M's profit in the LF model is affected by both η and λ. Figure 2b shows that L's profit utility in the MF model is in a negatively linear relation to its risk aversion parameter η. Figure 2c shows that F's profit utility in the ML model is in a positive linear relation to its fairness concern parameter λ. By comparing Figures 2a and 3a, it can be seen that M's profit is significantly higher in the LF model than in the DC model.

By comparing Figures 2b and 3b, it can be seen that L's utility is slightly higher in the MF model than in the DC model. By comparing Figures 2c and 3c, it can be seen that F's utility is higher in the ML model than in the DC model.

These numerical studies confirm that the decentralized case is the worst for profit optimization among all models. M, L, and F achieve higher optimal profit even when the other two parties form a coalition than when no coalition exists. In addition, it is further verified that the impact of risk aversion parameter η and fairness concern parameter λ is dynamic, not always positive or negative. The larger λ does not always lead to more profit being transferred to F. This phenomenon is different from what has been observed in supply chains [18] which consider only fairness concern but not risk aversion. The larger η, the less profit is transferred to L. It is understandable that lower risk endurance brings lower profit return.

6. Managerial Insights

The model analysis provides a framework for sustainable operational management of a CLSC. The operational sustainability of a CLSC relies on positive and consistent interactions among members. Through the incorporation of two irrational behavior factors and cooperative game approaches, CLSC participants can have a more comprehensive assessment towards pricing strategies. The cooperation or competition of members would determine the profit allocation and sustainability of a CLSC. The equilibrium analyses and numerical studies show that the profits of all members under the four cases (CC, LF, ML, MF) are higher than those of the decentralized case. This implies that members should make more efforts to maintain cooperation instead of competition. This research also demonstrates that the effects of irrational behavior factors vary due to whether they are related to wholesale prices or retail prices. The retail prices of alliances in the models of CC, ML, and MF are not affected by the two irrational behavior factors. According to past research, if only one irrational factor is considered, the higher the fairness concern level, the more profit is transferred from M to F. The higher the risk aversion parameter, the less profit is transferred from M to L. By incorporating two factors together, the effects of risk aversion and fairness concern on M's profit are complicated. As the degree of cooperation increases, the retail prices decrease. This indicates that the manufacturer should try to cooperate with retailers in order to maintain a stable market price without being affected by risk sensitivity and fairness concern level.

7. Conclusions and Future Research

On the basis of a two-echelon CLSC composed of M, L, and F, this research takes L's risk aversion parameter η and F's fairness concern parameter λ into account on five different occasions: a centralized (CC), a decentralized (DC), and three partially allied models (LF, ML, and MF). Analytical comparison of the resulting equilibriums reveals that the more decentralized the CLSC, the less profit it generates. In models LF and MF, L's risk aversion parameter η acts as a functional tool to reallocate profit between L and M. In models DC, LF, and ML where F's fairness concern is effective, the parameter λ plays a role in re-allocating the profits between F and M. Numerical studies are conducted to investigate how L's risk aversion parameter η and F's fairness concern parameter λ affect profit distribution under the five models. Numerical studies confirm that M, L, and F achieve higher optimal profit even when the other two parties form a coalition than when no coalition exists. The impact of risk aversion parameter η and fairness concern parameter λ is dynamic, not always positive or negative. The larger λ does not lead to more profit transferred to F. The larger η, the less profit is transferred to L. These research results provide helpful insights for CLSC managers to find out available choices and feasible ways to reach fair profit allocations. CLSC members should make more efforts to maintain cooperation instead of competition.

Great opportunities remain for future research. This research discusses a two-echelon CLSC with three agents. It would be meaningful to expand this study into a further complicated CLSC consisting of more agents and more echelons. In addition, this research assumes that one member has fairness concerns, and one other member has risk aversions. It would be very interesting to integrate more agents' risk aversions and fairness concerns into the CLSC and investigate their effects on profit allocations.

Author Contributions: Conceptualization and methodology, T.Z. and T.Y.; data curation, formal analysis, investigation, software, visualization, and writing—original draft, T.Z.; funding acquisition, project administration, resources, supervision, validation, and writing—review and editing, T.Y. All authors have read and agreed to the published version of the manuscript.

Funding: This research is funded by the National Natural Science Foundation of China, grant number 71001010.

Institutional Review Board Statement: Not applicable.

Informed Consent Statement: Not applicable.

Data Availability Statement: Not applicable.

Conflicts of Interest: The authors declare no conflict of interest.

Appendix A

Proof of Proposition 1. With a straight substitution of the demand functions $q_n^L(p_n^L, p_r^L)$, $q_r^L(p_n^L, p_r^L)$, $q_n^F(p_n^F, p_r^F)$ and $q_r^F(p_n^F, p_r^F)$ into the profit function π_G^{CC} and a partial derivation of the profit function concerning retail prices, we have $\frac{\partial^2 \pi_G^{CC}}{\partial p_n^{L2}} = -\frac{2}{1-\delta} < 0$, $\frac{\partial^2 \pi_G^{CC}}{\partial p_r^{L2}} = -\frac{2}{\delta(1-\delta)} < 0$, $\frac{\partial^2 \pi_G^{CC}}{\partial p_n^L \partial p_r^L} = \frac{2}{1-\delta}$, $\frac{\partial^2 \pi_G^{CC}}{\partial p_n^{F2}} = -\frac{2}{1-\delta} < 0$, $\frac{\partial^2 \pi_G^{CC}}{\partial p_r^{F2}} = -\frac{2}{\delta(1-\delta)} < 0$, and $\frac{\partial^2 \pi_G^{CC}}{\partial p_n^F \partial p_r^F} = \frac{2}{1-\delta}$. Therefore, $\frac{\partial^2 \pi_G^{CC}}{\partial p_n^{L2}} \frac{\partial^2 \pi_G^{CC}}{\partial p_r^{L2}} - \left(\frac{\partial^2 \pi_G^{CC}}{\partial p_n^L \partial p_r^L}\right)^2 = \frac{4}{\delta(1-\delta)} > 0$ and $\frac{\partial^2 \pi_G^{CC}}{\partial p_n^{F2}} \frac{\partial^2 \pi_G^{CC}}{\partial p_r^{F2}} - \left(\frac{\partial^2 \pi_G^{CC}}{\partial p_n^F \partial p_r^F}\right)^2 = \frac{4}{\delta(1-\delta)} > 0$, the profit function π_G^{CC} is strictly joint concave in p_n^L, p_r^L, p_n^F and p_r^F. This proves that there exists a unique optimal solution for profit maximization in Model CC.

By first-order settings $\frac{\partial \pi_G^{CC}}{\partial p_n^L} = 1 + \frac{c_n - c_r - 2(p_n^L - p_r^L)}{1-\delta} = 0$, $\frac{\partial \pi_G^{CC}}{\partial p_r^L} = \frac{c_r - 2p_r^L - \delta(c_n - 2p_n^L)}{\delta(1-\delta)} = 0$, $\frac{\partial \pi_G^{CC}}{\partial p_n^F} = 1 + \frac{c_n - c_r - 2(p_n^F - p_r^F)}{1-\delta} = 0$ and $\frac{\partial \pi_G^{CC}}{\partial p_r^F} = \frac{c_r - 2p_r^F - \delta(c_n - 2p_n^F)}{\delta(1-\delta)} = 0$, optimal pricing decisions are derived $p_n^{L*CC} = p_n^{F*CC} = \frac{1+c_n}{2}$ and $p_r^{L*CC} = p_r^{F*CC} = \frac{c_r + A + \delta}{2}$. The consequential optimal production decisions of new and remanufactured products are $q_n^{*CC} = q_n^{L*CC} + q_n^{F*CC} = \frac{1-\delta-(c_n-c_r-A)}{1-\delta}$ and $q_r^{*CC} = q_r^{L*CC} + q_r^{F*CC} = \frac{\delta c_n - c_r - A}{\delta(1-\delta)}$. Subsequently, the optimal total profit is determined as $\pi_G^{*CC} = \frac{(1-c_n)^2}{2} + \frac{(C_r + A - c_n \delta)^2}{2\delta(1-\delta)}$. □

Appendix B

Proof of Proposition 2. By substituting demand functions $q_n^L(p_n^L, p_r^L)$ and $q_r^L(p_n^L, p_r^L)$ into L's utility function u_L^{DC} and taking partial derivatives, we have $\frac{\partial^2 u_L^{DC}}{\partial p_n^{L2}} = -\frac{2}{1-\delta} - \eta\sigma^2 < 0$, $\frac{\partial^2 u_L^{DC}}{\partial p_r^{L2}} = -\frac{2}{1-\delta} - \frac{2}{\delta} - \eta\sigma^2 < 0$ and $\frac{\partial^2 u_L^{DC}}{\partial p_n^L \partial p_r^L} = \frac{2}{1-\delta} - \eta\sigma^2\rho > 0$. Then $\frac{\partial^2 u_L^{DC}}{\partial p_n^{L2}} \frac{\partial^2 u_L^{DC}}{\partial p_r^{L2}} - \left(\frac{\partial^2 u_L^{DC}}{\partial p_n^L \partial p_r^L}\right)^2 = \frac{4}{\delta(1-\delta)} + \frac{4\eta\sigma^2(1+\rho)}{1-\delta} + \frac{2\eta\sigma^2}{\delta} + \eta^2\sigma^4(1-\rho^2) > 0$ implies that L's utility function u_L^{DC} is strictly joint concave in p_n^L and p_r^L, and u_L^{DC} has a unique optimal solution. By first-order conditions $\frac{\partial u_L^{DC}}{\partial p_n^L} = 0$ and $\frac{\partial u_L^{DC}}{\partial p_r^L} = 0$, L's optimal response functions are obtained as $p_n^{L*}(w_n^L, w_r^L) = (2w_n^L - \eta\sigma^2\delta^2 + \eta\sigma^2\delta + 2\eta\sigma^2 w_n^L - \eta^2\sigma^4\delta^2 w_n^L + \eta\sigma^2\rho w_r^L + \eta\sigma^2\delta w_n^L + \eta\sigma^2\delta w_r^L + \eta^2\sigma^4\delta w_n^L - \eta^2\sigma^4\rho^2\delta w_n^L + 3\eta\sigma^2\rho\delta w_n^L + \eta^2\sigma^4\rho^2\delta^2 w_n^L + 2) / (\eta^2\sigma^4\rho^2\delta^2 - \eta^2\sigma^4\rho^2\delta - \eta^2\sigma^4\delta^2 + \eta^2\sigma^4\delta + 4\eta\sigma^2\rho\delta + 2\eta\sigma^2\delta + 2\eta\sigma^2 + 4)$ and $p_r^{L*}(w_n^L, w_r^L) = (2\delta + 2w_r^L + \eta\sigma^2 w_r^L - \eta^2\sigma^4\delta^2 w_r^L - \eta\sigma^2\rho\delta + \eta\sigma^2\delta w_n^L + 2\eta\sigma^2\delta w_r^L + \eta\sigma^2\rho\delta^2 + \eta^2\sigma^4\delta w_r^L - \eta^2\sigma^4\rho^2\delta w_r^L + \eta\sigma^2\rho\delta w_n^L + 3\eta\sigma^2\rho\delta w_r^L + \eta^2\sigma^4\rho^2\delta^2 w_r^L) /$

$(\eta^2\sigma^4\rho^2\delta^2 - \eta^2\sigma^4\rho^2\delta - \eta^2\sigma^4\delta^2 + \eta^2\sigma^4\delta + 4\eta\sigma^2\rho\delta + 2\eta\sigma^2\delta + 2\eta\sigma^2 + 4)$. By substituting $p_n^{L*}\left(w_n^L, w_r^L\right)$ and $p_r^{L*}\left(w_n^L, w_r^L\right)$ into demand functions $q_n^L\left(p_n^L, p_r^L\right)$ and $q_r^L\left(p_n^L, p_r^L\right)$, the optimal quantities $q_n^{L*}\left(w_n^L, w_r^L\right)$ and $q_r^{L*}\left(w_n^L, w_r^L\right)$ for L are obtained as functions composed of wholesale prices w_n^L and w_r^L.

By substituting demand functions $q_n^F\left(p_n^F, p_r^F\right)$ and $q_r^F\left(p_n^F, p_r^F\right)$ into F's utility function u_F^{DC} and taking partial derivatives, we have $\frac{\partial^2 u_F^{DC}}{\partial p_n^{F2}} = \frac{2(\lambda+1)}{\delta-1} < 0$, $\frac{\partial^2 u_F^{DC}}{\partial p_r^{F2}} = (\lambda+1)\left(\frac{2}{\delta-1} - \frac{2}{\delta}\right) < 0$, and $\frac{\partial^2 u_F^{DC}}{\partial p_n^F \partial p_r^F} = \frac{2(\lambda+1)}{1-\delta} > 0$. Then $\frac{\partial^2 u_F^{DC}}{\partial p_n^{F2}}\frac{\partial^2 u_F^{DC}}{\partial p_r^{F2}} - \left(\frac{\partial^2 u_F^{DC}}{\partial p_n^F \partial p_r^F}\right)^2 = \frac{4(\lambda+1)^2(1-\delta)}{\delta(\delta-1)^2} > 0$ implies that F's utility function u_F^{DC} is joint concave in p_n^F and p_r^F, thus u_F^{DC} has a unique optimal solution. With first-order conditions $\frac{\partial u_F^{DC}}{\partial p_n^F} = 0$ and $\frac{\partial u_F^{DC}}{\partial p_r^F} = 0$, F's optimal response functions are obtained as $p_n^{F*}\left(w_n^F, w_r^F\right) = \frac{\lambda + w_n^F - \lambda c_n + 2\lambda w_n^F + 1}{2(\lambda+1)}$, $p_r^{F*}\left(w_n^F, w_r^F\right) = \frac{\delta + w_r^F + \delta\lambda - \lambda c_r - \lambda A + 2\lambda w_r^F}{2(\lambda+1)}$. By substituting $p_n^{F*}\left(w_n^F, w_r^F\right)$ and $p_r^{F*}\left(w_n^F, w_r^F\right)$ into demand functions $q_n^F\left(p_n^F, p_r^F\right)$ and $q_r^F\left(p_n^F, p_r^F\right)$, the optimal quantities $q_n^{F*}\left(p_n^F, p_r^F\right)$ and $q_r^{F*}\left(p_n^F, p_r^F\right)$ for F are obtained as functions composed of wholesale prices w_n^F and w_r^F.

By plugging $q_n^{L*}\left(w_n^L, w_r^L\right)$, $q_r^{L*}\left(w_n^L, w_r^L\right)$, $q_n^{F*}\left(p_n^F, p_r^F\right)$ and $q_r^{F*}\left(p_n^F, p_r^F\right)$ into M's profit function π_M^{DC} with the first-order conditions $\frac{\partial \pi_M^{DC}}{\partial w_n^L} = 0$, $\frac{\partial \pi_M^{DC}}{\partial w_r^L} = 0$, $\frac{\partial \pi_M^{DC}}{\partial w_n^F} = 0$ and $\frac{\partial \pi_M^{DC}}{\partial w_r^F} = 0$, M's optimal decisions are obtained as $w_n^{L*} = \frac{c_n+1}{2}$, $w_r^{L*} = \frac{c_r+A+\delta}{2}$, $w_n^{F*} = \frac{\lambda + c_n + 3\lambda c_n + 1}{4\lambda+2}$ and $w_r^{F*} = \frac{c_r + A + \delta + 3\lambda c_r + 3\lambda A + \lambda\delta}{4\lambda+2}$. Subsequently, the channel and individual equilibrium profits are calculated, respectively, as π_M^{*DC}, π_L^{*DC}, π_F^{*DC}, and π_G^{*DC}.

Proposition 2 is therefore established. $\square$

Appendix C

Proof of Proposition 3. In the LF model, the alliance LF decides the sales prices to maximize the coalition utility. By substituting demand functions $q_n^{LF}\left(p_n^{LF}, p_r^{LF}\right)$ and $q_r^{LF}\left(p_n^{LF}, p_r^{LF}\right)$ into LF's utility function u_{LF}^{LF} and taking partial derivatives, we have $\frac{\partial^2 u_{LF}^{LF}}{\partial p_n^{LF2}} = -\eta\sigma^2 - \frac{2+2\lambda}{1-\delta} < 0$, $\frac{\partial^2 u_{LF}^{LF}}{\partial p_r^{LF2}} = -\eta\sigma^2 - \frac{2}{1-\delta} - \frac{2}{\delta} - \lambda\left(\frac{2}{1-\delta} + \frac{2}{\delta}\right) < 0$, and $\frac{\partial^2 u_{LF}^{LF}}{\partial p_n^{LF} \partial p_r^{LF}} = \frac{2+2\lambda}{1-\delta} - \eta\rho\sigma^2 > 0$. Then

$$\frac{\partial^2 u_{LF}^{LF}}{\partial p_n^{LF2}}\frac{\partial^2 u_{LF}^{LF}}{\partial p_r^{LF2}} - \left(\frac{\partial^2 u_{LF}^{LF}}{\partial p_n^{LF} \partial p_r^{LF}}\right)^2 = \frac{\left(-\eta\sigma^2(1-\delta)-2-2\lambda\right)\left(-\eta\sigma^2\delta(1-\delta)-2\delta-2(1-\delta)-2\lambda\right)-\delta\left(2+2\lambda-\eta\rho\sigma^2(1-\delta)\right)^2}{\delta(1-\delta)^2}$$

> 0 implies that LF's utility function u_{LF}^{LF} is joint concave in p_n^{LF} and p_r^{LF}, and u_{LF}^{LF} has a unique optimal solution. With first-order conditions $\frac{\partial u_{LF}^{LF}}{\partial p_n^{LF}} = 0$ and $\frac{\partial u_{LF}^{LF}}{\partial p_r^{LF}} = 0$, LF's optimal response functions are obtained as

$$p_n^{*LF}\left(w_n^{LF}, w_r^{LF}\right) = \frac{Kw_n^{LF} + \left(\eta\sigma^2\rho + \eta\sigma^2\delta\right)w_r^{LF} - \eta\sigma^2\lambda\delta^2 + \eta\sigma^2\rho\lambda c_r + \eta\sigma^2\rho\lambda A - \eta\sigma^2\lambda c_n\delta + \eta\sigma^2\lambda c_r\delta + \eta\sigma^2\lambda A\delta - \eta\sigma^2\rho\lambda c_n\delta + 4\lambda - 2\lambda c_n - 2\lambda^2 c_n + 2\lambda^2 - \eta\sigma^2\delta^2 + \eta\sigma^2\delta + \eta\sigma^2\lambda\delta + 2}{\eta^2\sigma^4\rho^2\delta^2 - \eta^2\sigma^4\rho^2\delta - \eta^2\sigma^4\delta^2 + \eta^2\sigma^4\delta + 4\eta\sigma^2\rho\lambda\delta + 4\eta\sigma^2\rho\delta + 2\eta\sigma^2\lambda\delta + 2\eta\sigma^2\lambda + 2\eta\sigma^2\delta + 2\eta\sigma^2 + 4\lambda^2 + 8\lambda + 4},$$

$$p_r^{*LF}\left(w_n^{LF}, w_r^{LF}\right) = \frac{\left(K + \eta\sigma^2\delta - \eta\sigma^2\right)w_r^{LF} + \left(\eta\sigma^2\delta + \eta\sigma^2\rho\delta\right)w_n^{LF} - \left(2\lambda + 2\lambda^2 + \eta\sigma^2\rho\lambda\delta + \eta\sigma^2\lambda\right)c_r + \left(\eta\sigma^2\lambda\delta + \eta\sigma^2\rho\lambda\delta\right)c_n + H}{\eta^2\sigma^4\rho^2\delta^2 - \eta^2\sigma^4\rho^2\delta - \eta^2\sigma^4\delta^2 + \eta^2\sigma^4\delta + 4\eta\sigma^2\rho\lambda\delta + 4\eta\sigma^2\rho\delta + 2\eta\sigma^2\lambda\delta + 2\eta\sigma^2\lambda + 2\eta\sigma^2\delta + 2\eta\sigma^2 + 4\lambda^2 + 8\lambda + 4}$$

where

$$K = 2 + 6\lambda + 4\lambda^2 + 2\eta\sigma^2 + \eta\sigma^2\delta + \eta^2\sigma^4\delta + 2\eta\sigma^2\lambda + 3\eta\sigma^2\rho\delta + 2\eta\sigma^2\lambda\delta + \eta^2\sigma^4\rho^2\delta^2 + 4\eta\sigma^2\rho\lambda\delta - \eta^2\sigma^4\delta^2 - \eta^2\sigma^4\rho^2\delta,$$
$$H = 2\delta + 4\lambda\delta + 2\lambda^2\delta - 2\lambda A - 2\lambda^2 A - \eta\sigma^2\lambda A - \eta\sigma^2\rho\delta + \eta\sigma^2\rho\delta^2 + \eta\sigma^2\rho\lambda\delta^2 - \eta\sigma^2\rho\lambda\delta - \eta\sigma^2\rho\lambda A\delta.$$

By substituting $p_n^{LF*}\left(w_n^{LF}, w_r^{LF}\right)$ and $p_r^{LF*}\left(w_n^{LF}, w_r^{LF}\right)$ into demand functions $q_n^{LF}\left(p_n^{LF}, p_r^{LF}\right)$ and $q_r^{LF}\left(p_n^{LF}, p_r^{LF}\right)$, the optimal quantities $q_n^{LF*}\left(p_n^{LF}, p_r^{LF}\right)$ and $q_r^{LF*}\left(p_n^{LF}, p_r^{LF}\right)$ for LF are obtained as functions composed of wholesale prices w_n^{LF} and w_r^{LF}. By plugging $q_n^{LF*}\left(p_n^{LF}, p_r^{LF}\right)$ and $q_r^{LF*}\left(p_n^{LF}, p_r^{LF}\right)$ into M's profit function π_M^{LF} with the first order conditions $\frac{\partial \pi_M^{LF}}{\partial w_n^{LF}} = 0$ and $\frac{\partial \pi_M^{LF}}{\partial w_r^{LF}} = 0$, M's optimal decisions w_n^{LF}, w_r^{LF} are obtained. $\square$

Appendix D

Proof of Proposition 4. In Model ML, coalition ML and F constitute a two-echelon Stackelberg game model with ML being the leader and F the follower. Plugging demand functions $q_n^F\left(p_n^F, p_r^F\right)$ and $q_r^F\left(p_n^F, p_r^F\right)$ into F's utility function and taking partial derivatives, we have $\frac{\partial^2 u_F^{ML}}{\partial p_n^{F2}} = \frac{2\lambda+2}{\delta-1} < 0$, $\frac{\partial^2 u_F^{ML}}{\partial p_r^{F2}} = \left(\frac{2}{\delta-1} - \frac{2}{\delta}\right)(\lambda+1) < 0$, and $\frac{\partial^2 u_F^{ML}}{\partial p_n^F \partial p_r^F} = \frac{2\lambda+2}{1-\delta} > 0$.

This implies that $\frac{\partial^2 u_F^{ML}}{\partial p_n^{F2}} \frac{\partial^2 u_F^{ML}}{\partial p_r^{F2}} - \left(\frac{\partial^2 u_F^{ML}}{\partial p_n^F \partial p_r^F}\right)^2 = \frac{4(\lambda+1)^2(1-\delta)}{\delta(1-\delta)^2} > 0$. Thus, F's utility function u_F^{ML} is strictly joint concave in p_n^F and p_r^F, and it has a unique optimal solution. With first-order conditions $\frac{\partial u_F^{ML}}{\partial p_n^F} = 0$ and $\frac{\partial u_F^{ML}}{\partial p_r^F} = 0$, F's optimal response functions are obtained

$$p_n^F\left(w_n^F, w_r^F\right) = \frac{\lambda+w_n^F-\lambda c_n+2\lambda w_n^F+1}{2\lambda+2}, \; p_r^F\left(w_n^F, w_r^F\right) = \frac{\delta+w_r^F-\lambda c_r-\lambda A+\lambda\delta+2\lambda w_r^F}{2\lambda+2}.$$

Plugging demand functions $q_n^{ML}\left(p_n^{ML}, p_r^{ML}\right)$ and $q_r^{ML}\left(p_n^{ML}, p_r^{ML}\right)$ into ML's utility function and taking partial derivatives, we have $\frac{\partial^2 u_{ML}^{ML}}{\partial p_n^{ML2}} = \frac{2}{\delta-1} < 0$, $\frac{\partial^2 u_{ML}^{ML}}{\partial p_r^{ML2}} = \frac{2}{\delta-1} - \frac{2}{\delta} < 0$, and $\frac{\partial^2 u_{ML}^{ML}}{\partial p_n^{ML}\partial p_r^{ML}} = \frac{2}{1-\delta} > 0$. So $\frac{\partial^2 u_{ML}^{ML}}{\partial p_n^{ML2}} \frac{\partial^2 u_{ML}^{ML}}{\partial p_r^{ML2}} - \left(\frac{\partial^2 u_{ML}^{ML}}{\partial p_n^{ML}\partial p_r^{ML}}\right)^2 = \frac{4(1-\delta)}{(1-\delta)^2\delta} > 0$ implies that ML's utility function is joint concave in p_n^{ML} and p_r^{ML}, and u_{ML}^{ML} has a unique optimal solution. With first-order conditions $\frac{\partial u_{ML}^{ML}}{\partial p_n^{ML}} = 0$ and $\frac{\partial u_{ML}^{ML}}{\partial p_r^{ML}} = 0$, ML's optimal responses are obtained as $p_n^{*ML} = \frac{c_n+1}{2}$, $p_r^{*ML} = \frac{c_r+A+\delta}{2}$.

By substituting $p_n^F\left(w_n^F, w_r^F\right)$, $p_r^F\left(w_n^F, w_r^F\right)$, p_n^{ML} and p_r^{ML} into demand functions $q_n^F\left(p_n^F, p_r^F\right)$, $q_r^F\left(p_n^F, p_r^F\right)$, $q_n^{ML}\left(p_n^{ML}, p_r^{ML}\right)$ and $q_r^{ML}\left(p_n^{ML}, p_r^{ML}\right)$, the optimal quantities are obtained as M's wholesale prices to F. By plugging the optimal quantities into ML's utility function with the first-order conditions $\frac{\partial u_{ML}^{ML}}{\partial w_n^F} = 0$ and $\frac{\partial u_{ML}^{ML}}{\partial w_r^F} = 0$, ML's optimal decisions $w_n^{*F} = \frac{\lambda+c_n+3\lambda c_n+1}{4\lambda+2}$ and $w_r^{*F} = \frac{c_r+A+\delta+3\lambda c_r+3\lambda A+\lambda\delta}{4\lambda+2}$ are obtained. $\square$

Appendix E

Proof of Proposition 5. In Model MF, alliance MF and L compose a two-echelon model with MF being the leader and L the follower in a Stackelberg game. By plugging demand functions $q_n^L\left(p_n^L, p_r^L\right)$ and $q_r^L\left(p_n^L, p_r^L\right)$ into L's utility function and calculating partial derivatives, we get $\frac{\partial^2 u_L^{MF}}{\partial p_n^{L2}} = -\eta\sigma^2 - \frac{2}{1-\delta} < 0$, $\frac{\partial^2 u_L^{MF}}{\partial p_r^{L2}} = -\eta\sigma^2 - \frac{2}{1-\delta} - \frac{2}{\delta} < 0$, and $\frac{\partial^2 u_L^{MF}}{\partial p_n^L \partial p_r^L} = \frac{2}{1-\delta} - \eta\sigma^2\rho > 0$.

Then $\frac{\partial^2 u_L^{MF}}{\partial p_n^{L2}} \frac{\partial^2 u_L^{MF}}{\partial p_r^{L2}} - \left(\frac{\partial^2 u_L^{MF}}{\partial p_n^L \partial p_r^L}\right)^2 = \left(\frac{2}{1-\delta} + \eta\sigma^2\right) * \left(\frac{2}{1-\delta} + \eta\sigma^2 + \frac{2}{\delta}\right) - \left(\frac{2}{1-\delta} - \eta\sigma^2\rho\right)^2 > 0$ implies that L's utility function u_L^{MF} is joint concave in p_n^L and p_r^L with a unique optimal solution. Through first-order conditions $\frac{\partial u_L^{MF}}{\partial p_n^L} = 0$ and $\frac{\partial u_L^{MF}}{\partial p_r^L} = 0$, we obtain L's optimal response functions

$$p_n^L\left(w_n^L, w_r^L\right) = \frac{2w_n^L-\eta\sigma^2\delta^2+\eta\sigma^2\delta+2\eta\sigma^2 w_n^L-\eta^2\sigma^4\delta^2 w_n^L+\eta\sigma^2\rho w_r^L+\eta\sigma^2\delta w_n^L+\eta\sigma^2\delta w_r^L+\eta^2\sigma^4\delta w_n^L-\eta^2\sigma^4\rho^2\delta w_n^L+3\eta\sigma^2\rho\delta w_n^L+\eta^2\sigma^4\rho^2\delta^2 w_n^L+2}{\eta^2\sigma^4\rho^2\delta^2-\eta^2\sigma^4\rho^2\delta-\eta^2\sigma^4\delta^2+\eta^2\sigma^4\delta+4\eta\sigma^2\rho\delta+2\eta\sigma^2\delta+2\eta\sigma^2+4},$$

$$p_r^L\left(w_n^L, w_r^L\right) = \frac{2\delta+2w_r^L+\eta\sigma^2 w_r^L-\eta^2\sigma^4\delta^2 w_r^L-\eta\sigma^2\rho\delta+\eta\sigma^2\delta w_n^L+2\eta\sigma^2\delta w_r^L+\eta\sigma^2\rho\delta^2+\eta^2\sigma^4\delta w_r^L-\eta^2\sigma^4\rho^2\delta w_r^L+\eta\sigma^2\rho\delta w_n^L+3\eta\sigma^2\rho\delta w_r^L+\eta^2\sigma^4\rho^2\delta^2 w_r^L}{\eta^2\sigma^4\rho^2\delta^2-\eta^2\sigma^4\rho^2\delta-\eta^2\sigma^4\delta^2+\eta^2\sigma^4\delta+4\eta\sigma^2\rho\delta+2\eta\sigma^2\delta+2\eta\sigma^2+4}$$

Plugging demand functions $q_n^{MF}\left(p_n^{MF}, p_r^{MF}\right)$ and $q_r^{MF}\left(p_n^{MF}, p_r^{MF}\right)$ into MF's utility function and taking partial derivatives, we have $\frac{\partial^2 u_{MF}^{MF}}{\partial p_n^{MF2}} = \frac{2}{\delta-1} < 0$, $\frac{\partial^2 u_{MF}^{MF}}{\partial p_r^{MF2}} = \frac{2}{\delta-1} - \frac{2}{\delta} < 0$, and $\frac{\partial^2 u_{MF}^{MF}}{\partial p_n^{MF}\partial p_r^{MF}} = \frac{2}{1-\delta} > 0$. Then $\frac{\partial^2 u_{MF}^{MF}}{\partial p_n^{MF2}} \frac{\partial^2 u_{MF}^{MF}}{\partial p_r^{MF2}} - \left(\frac{\partial^2 u_{MF}^{MF}}{\partial p_n^{MF}\partial p_r^{MF}}\right)^2 = \frac{4(1-\delta)}{(1-\delta)^2\delta} > 0$ implies that MF's utility function is joint concave in p_n^{MF} and p_r^{MF}, and u_{MF}^{MF} has a unique optimal solution. Through first-order conditions $\frac{\partial u_{MF}^{MF}}{\partial p_n^{MF}} = 0$ and $\frac{\partial u_{MF}^{MF}}{\partial p_r^{MF}} = 0$, MF's optimal responses are obtained as $p_n^{*MF} = \frac{c_n+1}{2}$, $p_r^{*MF} = \frac{c_r+A+\delta}{2}$.

By substituting $p_n^L\!\left(w_n^L, w_r^L\right)$, $p_r^L\!\left(w_n^L, w_r^L\right)$, p_n^{*MF} and p_r^{*MF} into demand functions $q_n^L\!\left(p_n^L, p_r^L\right)$, $q_r^L\!\left(p_n^L, p_r^L\right)$, $q_n^{MF}\!\left(p_n^{MF}, p_r^{MF}\right)$ and $q_r^{MF}\!\left(p_n^{MF}, p_r^{MF}\right)$, the optimal quantities are obtained as M's wholesale prices to L. By plugging the optimal quantities into MF's utility function with the first-order conditions $\frac{\partial u_{MF}^{MF}}{\partial w_n^L} = 0$ and $\frac{\partial u_{MF}^{MF}}{\partial w_r^L} = 0$, MF's optimal decisions $w_n^{*L} = \frac{c_n+1}{2}$ and $w_r^{*L} = \frac{c_r+A+\delta}{2}$ are obtained. $\square$

References

1. Farooque, M.; Zhang, A.; Thurer, M.; Qu, T.; Huisingh, D. Circular supply chain management: A definition and structured literature review. *J. Clean. Prod.* **2019**, *228*, 882–900. [CrossRef]
2. Lieder, M.; Asif, F.M.A.; Rashid, A.; Mihelič, A.; Kotnik, S. Towards circular economy implementation in manufacturing systems using a multi-method simulation approach to link design and business strategy. *Int. J. Adv. Manuf. Technol.* **2017**, *93*, 1953–1970. [CrossRef]
3. Acerbi, F.; Sassanelli, C.; Taisch, M. A conceptual data model promoting data-driven circular manufacturing. *Oper. Manag. Res.* **2022**, 1–20. [CrossRef]
4. Jabbour, C.J.C.; Jabbour, A.B.L.D.; Sarkis, J.; Filho, M.G. Unlocking the circular economy through new business models based on large-scale data: An integrative framework and research agenda. *Technol. Forecast. Soc. Chang.* **2019**, *144*, 546–552. [CrossRef]
5. Birkel, H.; Müller, J.M. Potentials of industry 4.0 for supply chain management within the triple bottom line of sustainability—A systematic literature review. *J. Clean. Prod.* **2021**, *289*, 125612. [CrossRef]
6. Taddei, E.; Sassanelli, C.; Rosa, P.; Terzi, S. Circular supply chains in the era of Industry 4.0: A systematic literature review. *Comput. Ind. Eng.* **2022**, 108268. [CrossRef]
7. Loch, C.H.; Wu, Y. Social Preferences and Supply Chain Performance: An Experimental Study. *Manag. Sci.* **2008**, *54*, 1835–1849. [CrossRef]
8. Davis, A.M.; Katok, E.; Santamaría, N. Push, Pull, or Both? A Behavioral Study of How the Allocation of Inventory Risk Affects Channel Efficiency. *Manag. Sci.* **2014**, *60*, 2666–2683. [CrossRef]
9. Tao, J.; Shao, L.; Guan, Z. Incorporating risk aversion and fairness considerations into procurement and distribution decisions in a supply chain. *Int. J. Prod. Res* **2020**, *58*, 1950–1967. [CrossRef]
10. Ghosh, P.K.; Manna, A.K.; Dey, J.K.; Kar, S. Supply chain coordination model for green product with different payment strategies: A game theoretic approach. *J. Clean. Prod.* **2021**, *290*, 125734. [CrossRef]
11. Mawandiya, B.K.; Jha, J.K.; Thakkar, J. Two-echelon closed-loop supply chain deterministic inventory models in a batch production environment. *Int. J. Sustain. Eng.* **2016**, *9*, 315–328. [CrossRef]
12. Mawandiya, B.K.; Jha, J.K.; Thakkar, J. Production-inventory model for two-echelon closed-loop supply chain with finite manufacturing and remanufacturing rates. *Int. J. Syst. Sci. Oper. Logist.* **2017**, *4*, 199–218. [CrossRef]
13. Modak, N.M.; Modak, N.; Panda, S.; Sana, S.S. Analyzing structure of two-echelon closed-loop supply chain for pricing, quality and recycling management. *J. Clean. Prod.* **2018**, *171*, 512–528. [CrossRef]
14. Modak, N.M.; Kazemi, N.; Cárdenas-Barrón, L.E. Investigating structure of a two-echelon closed-loop supply chain using social work donation as a Corporate Social Responsibility practice. *Int. J. Prod. Econ.* **2019**, *207*, 19–33. [CrossRef]
15. Sane Zerang, E.; Taleizadeh, A.A.; Razmi, J. Analytical comparisons in a three-echelon closed-loop supply chain with price and marketing effort-dependent demand: Game theory approaches. *Environ. Dev. Sustain.* **2018**, *20*, 451–478. [CrossRef]
16. Liu, Z.; Li, K.W.; Tang, J.; Gong, B.; Huang, J. Optimal operations of a closed-loop supply chain under a dual regulation. *Int. J. Prod. Econ.* **2021**, *233*, 107991. [CrossRef]
17. Ullah, M.; Asghar, I.; Zahid, M.; Omair, M.; AlArjani, A.; Sarkar, B. Ramification of remanufacturing in a sustainable three-echelon closed-loop supply chain management for returnable products. *J. Clean. Prod.* **2021**, *290*, 125609. [CrossRef]
18. Dabaghian, N.; Moghaddam, R.T.; Taleizadeh, A.A.; Moshtagh, M.S. Channel coordination and profit distribution in a three-echelon supply chain considering social responsibility and product returns. *Environ. Dev. Sustain.* **2022**, *24*, 3165–3197. [CrossRef]
19. Savaskan, R.C.; Van Wassenhove, L.N. Reverse Channel Design: The Case of Competing Retailers. *Manag. Sci.* **2006**, *52*, 1–14. [CrossRef]
20. Ke, H.; Li, Y.; Huang, H. Uncertain pricing decision problem in closed-loop supply chain with risk-averse retailer. *J. Uncertain. Anal. Appl.* **2015**, *3*, 14. [CrossRef]
21. Zeballos, L.J.; Méndez, C.A.; Barbosa-Povoa, A.P. Design and Planning of Closed-Loop Supply Chains: A Risk-Averse Multistage Stochastic Approach. *Ind. Eng. Chem. Res.* **2016**, *55*, 6236–6249. [CrossRef]
22. Ma, L.; Liu, Y.; Liu, Y. Distributionally robust design for bicycle-sharing closed-loop supply chain network under risk-averse criterion. *J. Clean. Prod.* **2020**, *246*, 118967. [CrossRef]
23. Das, D.; Verma, P.; Tanksale, A.N. Designing a closed-loop supply chain for reusable packaging materials: A risk-averse two-stage stochastic programming model using CVaR. *Comput. Ind. Eng.* **2022**, *167*, 108004. [CrossRef]
24. Ma, P.; Li, K.W.; Wang, Z. Pricing decisions in closed-loop supply chains with marketing effort and fairness concerns. *Int. J. Prod. Res.* **2017**, *55*, 6710–6731. [CrossRef]

25. Zheng, X.-X.; Liu, Z.; Li, K.W.; Huang, J.; Chen, J. Cooperative game approaches to coordinating a three-echelon closed-loop supply chain with fairness concerns. *Int. J. Prod. Econ.* **2019**, *212*, 92–110. [CrossRef]
26. Sarkar, S.; Bhala, S. Coordinating a closed loop supply chain with fairness concern by a constant wholesale price contract. *Eur. J. Oper. Res.* **2021**, *295*, 140–156. [CrossRef]
27. Wang, Y.; Yu, Z.; Guo, Y. Recycling decision, fairness concern and coordination mechanism in EC-CLSC. *J. Control Decis.* **2021**, *8*, 184–191. [CrossRef]
28. He, J.; Zhang, L.; Fu, X.; Tsai, F.-S. Fair but Risky? Recycle Pricing Strategies in Closed-Loop Supply Chains. *Int. J. Environ. Res. Public Health* **2018**, *15*, 2870. [CrossRef] [PubMed]
29. Li, C.; Guo, X.; Du, S. Pricing Decisions in Dual-Channel Closed-Loop Supply Chain Under Retailer's Risk Aversion and Fairness Concerns. *J. Oper. Res. Soc. China* **2021**, *9*, 641–657. [CrossRef]
30. Zhang, Y.; Zhang, T. Complex Dynamics in a Closed-Loop Supply Chain with Risk Aversion and Fairness Concerns under Supply Disruption. *Int. J. Bifurc. Chaos* **2021**, *31*, 2150132. [CrossRef]
31. Ho, T.; Su, X. Peer-Induced Fairness in Games. *Am. Econ. Rev.* **2009**, *99*, 2022–2049. [CrossRef]
32. Ho, T.; Su, X.; Wu, Y. Distributional and Peer-Induced Fairness in Supply Chain Contract Design. *Prod. Oper. Manag.* **2014**, *23*, 161–175. [CrossRef]
33. Örsdemir, A.; Kemahlıoğlu-Ziya, E.; Parlaktürk, A.K. Competitive Quality Choice and Remanufacturing. *Prod. Oper. Manag.* **2014**, *23*, 48–64. [CrossRef]
34. Yan, W.; Xiong, Y.; Xiong, Z.; Guo, N. Bricks vs. clicks: Which is better for marketing remanufactured products? *Eur. J. Oper. Res.* **2015**, *242*, 434–444. [CrossRef]
35. Robison, L.J.; Barry, P.J. *Competitive Firm's Response to Risk*; Macmillan: New York, NY, USA, 1987.
36. Gan, X.; Sethi, S.P.; Yan, H. Coordination of Supply Chains with Risk-Averse Agents. *Prod. Oper. Manag.* **2004**, *13*, 135–149. [CrossRef]
37. Xiao, T.; Choi, T. Purchasing choices and channel structure strategies for a two-echelon system with risk-averse players. *Int. J. Prod. Econ.* **2009**, *120*, 54–65. [CrossRef]
38. Choi, S.; Ruszczyński, A. A multi-product risk-averse newsvendor with exponential utility function. *Eur. J. Oper. Res.* **2011**, *214*, 78–84. [CrossRef]
39. Choi, T.; Cheng, A.; Bin, S.; Qi, S. Optimal pricing in mass customization supply chains with risk-averse agents and retail competition. *Omega* **2019**, *88*, 150–161. [CrossRef]
40. Zhang, J.; Sethi, S.P.; Choi, T.-M.; Cheng, T.C.E. Pareto Optimality and Contract Dependence in Supply Chain Coordination with Risk-Averse Agents. *Prod. Oper. Manag.* **2022**, 1–14. [CrossRef]
41. Xiao, T.; Xu, T. Pricing and product line strategy in a supply chain with risk-averse players. *Int. J. Prod. Econ.* **2014**, *156*, 305–315. [CrossRef]
42. Nie, T.; Du, S. Dual-fairness supply chain with quantity discount contracts. *Eur. J. Oper. Res.* **2017**, *258*, 491–500. [CrossRef]
43. Chen, J.; Zhou, Y.; Zhong, Y. A pricing/ordering model for a dyadic supply chain with buyback guarantee financing and fairness concerns. *Int. J. Prod. Res.* **2017**, *55*, 5287–5304. [CrossRef]

Article

Optimization Model for Selective Harvest Planning Performed by Humans and Robots

Ben Harel [1,*], Yael Edan [1] and Yael Perlman [2]

[1] Department of Industrial Engineering and Management, Ben-Gurion University of the Negev, Beer-Sheva 8410501, Israel; yael@bgu.ac.il
[2] Department of Management, Bar-Ilan University, Ramat Gan 5290002, Israel; yael.perlman@biu.ac.il
* Correspondence: benhare@post.bgu.ac.il

Abstract: This paper addresses the formulation of an individual fruit harvest decision as a nonlinear programming problem to maximize profit, while considering selective harvesting based on fruit maturity. A model for the operational level decision was developed and includes four features: time window constraints, resource limitations, yield perishability, and uncertainty. The model implementation was demonstrated through numerical studies that compared decisions for different types of worker and analyzed different robotic harvester capabilities for a case study of sweet pepper harvesting. The results show the influence of the maturity classification capabilities of the robot on its output, as well as the improvement in cycle times needed to reach the economic feasibility of a robotic harvester.

Keywords: nonlinear programming; agriculture; harvest planning; production

Citation: Harel, B.; Edan, Y.; Perlman, Y. Optimization Model for Selective Harvest Planning Performed by Humans and Robots. *Appl. Sci.* **2022**, *12*, 2507. https://doi.org/10.3390/app12052507

Academic Editor: Panagiotis Tsarouhas

Received: 31 January 2022
Accepted: 25 February 2022
Published: 28 February 2022

Publisher's Note: MDPI stays neutral with regard to jurisdictional claims in published maps and institutional affiliations.

1. Introduction

The supply chain for fresh fruit and vegetables involves a number of steps from crop selection to shipment to the customer, including harvesting, processing, packaging, and transporting the produce [1]. Operational research models have been developed for this supply chain at three decision levels: strategic, tactical, and operational [2,3]. The strategic level includes long-term horizon decisions, for example, deciding on farm and plot locations [4]. The tactical level involves medium-term decisions, such as crop selection, scheduling, and allocation [2,5]. The operational level involves short-term decisions, such as water allocation, land preparation, pricing decisions, and harvest scheduling [5–7].

Harvest planning includes several factors (e.g., scheduling and machinery capacity, allocation, and timing) that have been investigated at all three levels. At the strategic level, research has focused on topics such as the number of harvesting machines needed for filling a specific storage capacity [8] and the evaluation of different planting and harvesting alternatives using bio-economics characteristics [2]. At the tactical level, research has focused on allocating harvester machines between a number of fields to reduce costs [9] and on optimizing harvesting and loading sequences to minimize delays in moving products to storage [10]. At the operational level, research has been conducted on optimizing harvest data to maximize mill productivity, and on analyzing the labor effect on fruit quality and quantity at harvest [2,11,12]. However, there are no studies examining operational decisions related to the harvesting of individual fruit.

Selective harvesting of high-value crops, such as apples, tomatoes, and broccoli, is currently performed mainly by humans, rendering it one of the most labor-intensive and expensive agricultural tasks [13]. Any attempts to mechanize all or parts of the supply chain for these crops must take into account the decades-long research on the robotization of many different agricultural applications, such as transplanting, spraying, cultivating, trimming, and selective harvesting [14]. This research has indeed indicated the technical feasibility of using robotic harvesters in orchards (e.g., apples [15,16] and citrus [17,18]), greenhouses (e.g., tomatoes [19] and sweet peppers [20]) and open fields (e.g., melons [21] and asparagus [22]), but the robots developed to date still lack commercial applicability, since they have failed to reach the efficiency of human harvesters [13,23].

A robotic harvester must detect the fruit, reach the fruit, decide whether to harvest the fruit, and then grasp and disconnect the fruit from the branch. The harvest decision of a selective robotic harvester thus includes a number of aspects. First, methods for maturity classification must be developed [24,25]. Second, since many fruits are hidden or partially obscured, the best camera viewpoints for maturity classification must be selected [26]. Since additional viewpoints cost time, an intelligent decision as to the necessity for an additional viewpoint and its location is needed [27]. The third and last aspect investigates the operational considerations of maturity classification and harvesting, namely, deciding whether the fruit is ripe and whether to harvest it, based on a consideration of the above points. By addressing these aspects, the harvest decision of a selective robotic harvester can be improved.

To the best of our knowledge, no research has examined the decision making of a selective robotic harvester. This research lacuna may be attributed to the research focus on proving the technical feasibility of robotic harvesters [23], i.e., developing sensing and grasping capabilities. Once the technical feasibility of robotic harvesters has been proven, successful implementation and commercialization will depend on their integration into the production cycle [28], similarly to milking robots: after the technical feasibility of the milking robot had been successfully demonstrated, different operational research aspects were investigated at various levels, including optimal allocation in a robotic milking barn [29], the design and layout of an optimal barn [30], and prediction of the milking robot utilization [29].

One of the operational aspects of a harvesting robot is harvest planning, which includes scheduling, routing, and resource allocation to ensure high quality of the final products [1]. The agricultural environment is characterized by several variabilities and challenges, deriving from its unstructured and dynamic make-up and the biological nature of the product [27,31,32]. Fruit and vegetable crops vary widely in physiology and management practices, which may involve multiple harvests in a single season [33]. In particular, in high-value crops, harvesting is performed several times along a production season. These crops are generally non-staple crops, such as fruit, vegetables, ornamentals, and spices [34]. For such crops, the decision to harvest a specific fruit/vegetable depends on the fruit/vegetable maturity level, the harvesting capacity, and market demand.

In this work, we aimed to develop a model for selective harvesting based on the maturity of the fruit, where harvesting can be performed by human workers or robots, with each type of harvester having different abilities to identify maturity. The model formulation and analyses are demonstrated in numerical studies using sweet pepper harvesting as a case study.

We determined the optimal harvest production order to maximize farmers' profit. The production order defines how many peppers of each maturity level to harvest on each specific harvest day along the harvesting season. The production order defines the harvesting capacity needed, i.e., the number of workers/robots required to achieve this production. Usually, different types of workers are available for harvesting, and in the future, robotic harvesters might also be integrated into the harvesting process [14,23]. The factors that determine the type of worker include costs (salary), the number of peppers each worker can harvest each day, and the worker's ability to classify the peppers into different maturity classes. For example, an inexperienced worker is able to classify peppers into only two maturity classes—mature and immature. An experienced worker can classify the pepper into four maturity classes—immature, partly mature, mostly mature, and must be harvested [35]. Therefore, more experienced workers, who are better able to estimate the pepper maturity level, will harvest the fruit at the best fit time, resulting in an overall higher pepper weight (since the fruit will stay longer on the plant and continue to develop).

A robotic harvester will be able to non-destructively classify the exact maturity level [24,36–39] and will have a higher harvest capacity than a human worker, but will also have *higher costs* [23]. Only a few previous studies have investigated and analyzed the economic effect of robotic harvesting, focusing on the machine cost per area unit, compared to human workers [1,40]. This work extends the literature by *estimating the selective robotic harvester quality savings*, taking into consideration both the cost (comparing robot costs to manual costs) and the effect of maturity classification using a robotic harvester (the quality effect).

The remainder of this paper is organized as follows: Section 2 gives the relevant background and previous research. Section 3 describes in detail the basic formulation of the harvesting problem, and Sections 4 and 5, respectively, present expansions of the basic formulation and numerical studies. Finally, Section 6 concludes the paper.

2. Literature Review

Harvest yields tend to be uncertain regarding quantity, quality, and timing, due to dynamic and unpredictable weather, soil, and water conditions [1,41]. Moreover, the harvesting time of individual fruits or vegetables will influence the production yield and quality [11]. The harvest time is affected by many factors, such as cultivation practices, weather conditions, soil conditions, geographical location, rate of quality decay, storage capacity, the required processing needed, and transport considerations [1,11,41]. Given the yield and quality attributes and the transport and storage restrictions, mathematical optimization techniques, such as linear or integer programming, are often used to determine the best possible harvest time [8]. This type of programming *determines the best harvesting time* for different objective functions, such as maximum profit or minimum expenses [2,11]. To date, this *best harvesting time* has been considered *at the field level only for non-selective harvesting* (i.e., when to harvest a specific field with a single harvest).

Another decision that must be made before harvesting concerns the number of workers to recruit for the harvest season. Labor costs are a significant expense in agricultural production, specifically in the harvesting process [42]. Harvesting often involves migrant seasonal workers and requires recruitment processes that take time and require planning [43]. Furthermore, it is necessary to maintain a balance between ensuring a sufficient number of workers to harvest the ripe products (not missing harvestable produce due to a lack of workers) and ensuring that workers are fully utilized to reduce operational costs.

Usually, the harvest planning problem is modeled using a mathematical formulation and then solved by different operational research methods [2]. Integer linear programming, the most common method [2], has been used to find the optimal date for grape and apple harvesting, taking into consideration operational costs and fruit quality [44,45]. In these studies, the harvest date was set for non-selective harvesting, i.e., harvesting the whole field at once. A different planning model for wine grape harvesting involving uncertainty combined integer linear programming and robust stochastic optimization [26]. Other studies have used dynamic programming [10] and simulation models [46] to solve problems in harvest planning. However, all of these studies addressed *decisions related to the whole field* and *did not relate to individual fruits*.

Any study related to the modeling of harvest planning must take the following features into consideration, as reviewed in [1]:

- Time window constraints—The model should consider the optimal time for harvesting and the quality decay resulting from harvesting outside of that time window, as well as its effect on revenues.
- Resource limitations—These include capacity and productivity constraints, together with labor and machine availability.
- Yield perishability—The deterioration of fresh products during the post-harvest period must be taken into account. Yield perishability can be modeled in several ways, including continuous deterioration curves, a loss factor for each period after harvesting, and the effect of product deterioration on customer demand.

- Uncertainty—There is uncertainty in the harvest yield (quantity and quality), due to unknown weather conditions and the inherent variability of agricultural processes.
- Inventory control—The inventory should be considered in terms of holding costs, duration of keeping in the inventory constraints, or as a decision variable.

Integrating all of the above elements into a single model is complicated and may be impossible to solve in reasonable computational times. Therefore, heuristics algorithms are usually developed to derive an operational solution [47].

None of the studies covered in the review [1] included all the aforementioned features, and *only two included four out of the five features*. Maatman et al. created a multiperiod stochastic model to determine the cultivating strategy for crops under rainfall uncertainty [48]. This model did not include *time window constraints*, i.e., the optimal time for harvesting was not considered. Annetts and Audsley considered four out of the five features described above. Their study dealt with cultivating strategy and machinery selection [49], aiming to maximize profits while minimizing environmental impact. The developed model did not include the *uncertainty* feature; the researchers used historical weather data to determine the available working hours of the machines but did not consider the weather data effect on crop maturity, yield, and costs [49]. In addition, *none of the above models considered selective harvesting* (focusing on decisions related to individual fruit). *The current paper addresses this knowledge gap by focusing on decisions related to individual fruit and taking into consideration four features for selective harvesting.* The inventory control feature was excluded, because we assume that fresh fruit are supplied on the same day of harvest.

The literature review revealed three studies that solved a problem with similarities to *planning the harvest of a selective crop*. Arnaout and Maatouk developed a model for scheduling the harvest operation of grapes to maximize harvest quality with minimum cost. Their model used a quality decay function to measure the quality loss of the grapes and its effect on the profits corresponding to the harvest date [47]. However, in this case, the grape crop was *harvested once in a season* (not selectively harvested). In contrast, crops such as sweet peppers require *selective harvesting, once every few days. Each fruit could be harvested at different maturity levels over an extended period.* Therefore, when planning to address the harvest of a selective crop problem, the inventory level and product quality can be controlled via the right decision related to the harvested product's maturity level.

Golenko-Ginzburg et al. developed a multilevel decision-making system for human and machine cotton harvesters. They created a hierarchical system of three levels; the first level determines the harvesting speed, the second level re-allocates the resources to create balanced teams, and the third level re-allocates the teams among farms, aiming to maximize the probability of completing cotton harvesting by a due date [50]. The principle of integrating two types of resources (man–machine) is similar to the current work formulation. However, cotton, similar to grapes, is a non-selective crop and is *harvested all at once*. Therefore, *adjustment for a selective crop*, such as sweet peppers, was needed and was the aim of this paper.

Albornoz et al., suggested an approach for grape-selective harvest scheduling and planning via management zone delineation [51]. Their work minimized the total costs of harvest operations and established planning and scheduling for selective harvest of each selected management zone. Although dealing with selective harvesting, their article *did not consider the fruit growth function* or the *maturity status of individual fruit*.

In this paper, we present *a model for deciding on the selective harvest of an individual fruit*. Four features are considered in the model, which is based on an economic analysis that takes into account robot/worker capabilities and the growth model of the fruit. The model is demonstrated for sweet pepper harvesting, but it is general and can be applied to other crops requiring selective harvesting.

3. Problem Description and Formulation

3.1. Growth Function

Sweet pepper fresh weight, similar to other fruits, is determined in terms of days after anthesis (DAA). Anthesis is taken as time zero, after which the pepper continues to grow, and after d_m days from anthesis, if not harvested, the pepper is considered rotten and cannot be supplied to the customer. Let $W(d)$ denote the weight of a pepper d days after anthesis. $W(d)$ is determined via the pepper growth function. Several growth functions for sweet peppers can be found in the literature [52], and the following common growth function was used [52]:

$$W(d) = \frac{W_{max}}{\left(1 + e^{-a(d-d_m)}\right)} \tag{1}$$

where W_{max} is the maximum weight of the pepper, d_m is the maximal value of d, and a is a constant determining the curvature of the growth pattern [33].

The pepper maturity class is determined by d. In the future, a robotic harvester might be able to identify the exact d for the pepper. However, workers can distinguish only between pepper maturity classes (they cannot identify the exact d of a pepper), since it would require tagging each flower, which is an expensive and complicated task. Therefore, the maturity classes are modeled as follows; the lowest maturity class is defined as pepper DAA between $[d_{1start}, \ldots , d_{1end}]$, the second-lowest maturity level is defined as pepper DAA between $[d_{2start}, \ldots , d_{2end}]$, etc.

3.2. Model Formulation

Inputs

n—number of periods
m—number of maturity levels
d—number of days after anthesis
d_{jstart}—min days after anthesis of maturity level j
d_{jend}—max days after anthesis of maturity level j
$y_{d,1}$—number of peppers d DAA available to harvest in the first period
I_0—number of peppers at anthesis each day
C_h—harvester capablities in one period
p—price per kilogram of pepper fresh weight
$W(d)$—fresh weight (kilogram) of pepper d days after anthesis
S—harvester's salary/rental cost through all planning horizons
F—total fixed expenses through all of planning horizons, including utilities, land, water, fertilization, planting, and taxes

Variables

N—number of harvesters (hired workers or equivalent numbers of robots see Section 5.2.2) (*decision variable*)
$H_{j,t}$—peppers harvested from maturity class j in period t (*decision variable*)
THW_t—total harvest weight of the peppers in period t (fresh weight in kilograms)
$A_{j,t}$—peppers from class j available to harvest at the beginning of period t
$P_{d,t}$—peppers d days after anthesis that are harvested in period t
$Y_{d,t}$—peppers available to harvest d days after anthesis at the beginning of period t

The farmer has two decisions to make: *the number of harvesters to use* for the harvest season (N), and *the number of peppers to harvest from each maturity class at each period* ($H_{j,t}$). Since we are dealing with *selective harvesting*, the number of harvesters is calculated as the number of workers or the equivalent-performing robotic harvesters (see Section 5.2). The harvest season is n periods long, and the harvesters can distinguish between peppers in m maturity classes.

The objective function represents the profit to the farmer and includes the following components: income from the harvested peppers (calculated as the fresh weight of the harvested peppers (THW_t) multiplied by the price per kilogram (p)); variable expenses (calculated as the number of harvesters (N) multiplied by the workers' salaries or the robots' rental cost (S) through all planning horizons); the total fixed expenses (F) through all planning horizons (calculated by summing expenses due to utilities, land, water etc.):

$$Max\ Z = p \times \sum_{t=1}^{n} THW_t - N \times S - F \tag{2}$$

subject to: (the constraints are explained one by one below):

$$Y_{1,t} = I_0 \qquad \forall t = 2 \ldots n \tag{3}$$

$$Y_{d,t} = Y_{d-1,t-1} - P_{d-1,t-1} \qquad \forall t = 2 \ldots n,\ d = 2 \ldots d_m \tag{4}$$

$$Y_{d,t} \geq P_{d,t} \qquad \forall t = 1 \ldots n,\ d = 1 \ldots d_m \tag{5}$$

$$H_{j,t} = \sum_{d=djstart}^{djend} P_{d,t} \forall t = 1 \ldots n,\ j = 1 \ldots m \tag{6}$$

$$A_{j,t} = \sum_{d=djstart}^{djend} Y_{d,t} \forall t = 1 \ldots n,\ j = 1 \ldots m \tag{7}$$

$$A_{j,t} \geq H_{j,t} \forall t = 1 \ldots n,\ j = 1 \ldots m \tag{8}$$

$$\sum_{j=1}^{m} H_{j,t} \leq N \cdot C_h \forall t = 1 \ldots n \tag{9}$$

$$THW_t = \sum_{d=1}^{dm} (W(d) \times P_{d,t}) \forall t = 1 \ldots n \tag{10}$$

Equations (3)–(5) define the mechanism of pepper growth in the greenhouse using the variables $Y_{d,t}$ and $P_{d,t}$. The number of available peppers d DAA at the beginning of the planning horizon ($Y_{d,1}$) is given. Since the actual d of a pepper is unknown to the workers, the decision is made based on the maturity classes ($A_{j,t}$ and $H_{j,t}$), as described above. The composition of each maturity class is obtained from Equations (6) and (7). In addition, the number of peppers harvested from each maturity class is limited by the peppers available, seen in Equation (8), and the maximal number of peppers harvested is determined by the number of harvesters in Equation (9). The actual harvest weight (THW) is determined via the harvested pepper's d, as in Equation (10).

If we take into consideration the potential of a robotic harvester, the above set of constraints is somewhat different. If we assume that the robots can identify the exact DAA d, then $H_{j,t} = P_{d,t}$ and thus constraints 6–8 can be removed and $P_{d,t}$ is replaced by $H_{j,t}$ in the remaining set of constraints.

This model formulation contains $[(n \times m) + 1]$ variables and $[n \times (m + 1) + 2]$ constraints. We note that different methods for solving nonlinear programming differ in the computational effort expressed by the number of variables and constraints.

3.3. Dealing with Uncertainty

The complex aspect of this formulation is to update the value of $P_{d,t}$, which determines the harvested yield weight and the relation between $H_{j,t}$ and $P_{d,t}$. To update the values of $Y_{d,t}$ between periods, the value of $P_{d,t}$ must be known, as is shown in Equation (4).

To determine the exact value of $P_{d,t}$, the distribution of the harvested peppers ($H_{j,t}$) within the different DAAs must be known. Since it is impossible to know the actual DAA of a harvested pepper, a stochastic process must be considered. We considered four ways to deal with this problem:

- Distribute the harvested peppers uniformly among the different DAAs.
- Distribute the harvested peppers in proportion to the available peppers of each DAA.
- Create a worst-case scenario in which the peppers with the lowest weight (lower DAA) available in a maturity class will be harvested.
- Create a best-case scenario in which the peppers with the highest weight (highest DAA) available in a maturity class will be harvested.

For example, in a problem in which the second maturity level is defined as peppers of 5 to 7 DAAs, let us assume that in period 1, the peppers available are: $Y_{5,1} = 4$, $Y_{6,1} = 6$, $Y_{7,1} = 2$. This implies that $A_{2,1} = \sum_{k=5}^{7} Y_{k,t} = 4 + 6 + 2 = 12$. Now, let us assume the algorithm decided to harvest six peppers from level 2 in period 1: $H_{2,1} = 6$.

- The first method, distributing uniformly, will set $P_{5,1} = 2, YP_{6,1} = 2, P_{7,1} = 2$.
- The second method, distributing in proportion, will set $P_{5,1} = \frac{4}{12} \times 6 = 2$, $P_{6,1} = \frac{6}{12} \times 6 = 3$, $P_{7,1} = \frac{2}{12} \times 6 = 1$.
- The third method, the worst-case scenario, will set $P_{5,1} = 4, YP_{6,1} = 2, P_{7,1} = 0$.
- The fourth method, the best-case scenario, will set $P_{5,1} = 0, YP_{6,1} = 4, P_{7,1} = 2$.

All four ways suggested for dealing with uncertainty are a function of the decision variables and, therefore, create a nonlinear formulation.

4. Model Extensions

The suggested formulation (Section 3) is based on certain assumptions that enable a solution to be reached in reasonable computational time. This section discusses and offers modifications of the formulation to accommodate different aspects of harvest planning.

4.1. Limiting the Harvested Rows and Deciding on the Rows to Harvest

In the current formulation, all fruits in the greenhouse are available for harvesting in each period. The greenhouse is separated into rows. Usually, the harvesters harvest only a subset of the total rows in the greenhouse (block) each working day, spacing the time between one harvest cycle and the next. Such spacing enables fruits to continue to grow (increase weight) and creates a higher supply of grown peppers to harvest in the next harvest cycle of the block (the subset). Moreover, harvesting only a subset of rows saves transition times while advancing along the rows, creating a more productive environment.

An additional index, the block index, should be added to the formulation to apply this approach. However, such a change will increase the number of variables and constraints in the current formulation, making it difficult to solve in a reasonable computational time. Therefore, we applied an additional solution for this approach, which is easier to solve; we ran the formulation for each block of rows separately. By doing this, the definition of time between periods could be changed. Instead of representing one day, the time between periods could be the time between two harvest cycles of the same block. This change in the definition of time between periods requires scaling the growth function to the appropriate time, rather than days. For example, if each block is harvested every ten days, the growth function should be scaled from the change in fruit weight d DAA to change in fruit weight $10 \times d$ DAA.

4.2. Modeling the Change in Pepper Price

The price of a kilogram of fresh peppers varies between different periods along the harvesting season. While this price variation affects the decision of how many peppers are to be harvested in each period, the price for each period cannot be known in advance, since it relies on market supply and demand. Therefore, the price can only be estimated using post data. Once the prices are estimated, they can be added to the formulation as input.

Then, instead of multiplying the total harvest weight (THW) of all planning horizons for the same fixed price per kilogram (p), the THW of each period can be multiplied by the specific period.

5. Numerical Studies—Harvesters with Different Capabilities to Classify Pepper Maturity Levels

As described in the Introduction, workers may have different capabilities to classify peppers into different maturity classes. A robotic harvester may be able to identify the exact maturity level of a pepper, but renting a robot incurs a certain cost. Based on the different harvesting capabilities, farmers need to select harvesters. Three numerical studies were conducted representing problems faced by a typical farmer. The first (Section 5.1) explores different types of worker. The second (Section 5.2) analyzes the potential capabilities of a robotic harvester in comparison to human harvesters. Finally, an economic return-on-investment analysis for the robotic harvester was conducted (Section 5.3) based on the results of the second example. The model parameters were chosen based on [20,53,54] and data obtained from an interview with a sweet pepper farmer in Israel for a sweet pepper greenhouse of a size of 1.2 hectares (Table 1). Since the aim of the numerical studies was to demonstrate the practical implementation of the model, small/medium-size problems were analyzed, where the number of periods n was set to 20. The problems were solved via a Microsoft Excel solver add-in using the generalized reduced gradient nonlinear method, which derives a local optimum solution (solver's maximal size is 200 decision variables and 100 constraints).

Table 1. Model parameters for the numerical example analysis.

General		Workers		Yield		Growth Function	
n	20	S	3530	$y_{d,1}$	7000 $\forall d$	d_m	30
p	2	C_h	6750	I_{n0}	7000	W_{max}	250
						a	0.3

5.1. Analysis of the Type of Worker

This numerical example aims to analyze the effect on the objective function of the workers' ability to classify peppers into multiple classes. The workers were classified into three types based on their experience, where the assumption was that more experienced workers can classify fruit into a higher number of maturity levels. The data regarding the classifying ability of each worker is given in Table 2. Two versions of the model formulation were run for each worker type. The first version ran the exact formulation described in Section 3.2 to determine the optimal number of workers and how many peppers to harvest from each maturity class (Section 5.1.1). In the second version, the number of workers was fixed, and the only decision variable was the number of peppers harvested from each maturity class for a given number of workers (Section 5.1.2).

Table 2. Data for the ability of three types of worker to classify peppers into maturity levels (m).

Type A: $m = 2$			Type B: $m = 3$			Type C: $m = 4$		
Class	d Start	d End	Class	d Start	d End	Class	d Start	d End
1	31	45	1	31	40	1	31	40
2	46	60	2	41	50	2	41	45
			3	51	60	3	46	50
						4	51	60

5.1.1. Number of Workers Is a Decision Variable

The results in Table 3 show that the best performance for a variable number of workers was derived for the experienced workers (Type C). However, the mid-experienced workers (Type B) obtained a higher total harvest weight with only a slight difference (0.4% less) in the profit. It seems from the results that for each type of worker a solution with a different number of workers was derived. A possible reason for this result is that the generalized reduced gradient method produces a local optimum solution. The difference between the profit values signifies the maximal difference in workers' salaries that the farmer will agree to pay for the change in the type of worker. The difference resulted in 14.2%, 0.4%, and 13.5% improvements in profit when changing the workers from Type C to A, Type C to B, and Type B to A, respectively. The formulation results lead to two strategies for the harvest plan (Figure 1) based on the inputs in Table 1.

Table 3. Results of the numerical example for the analysis of worker performance.

	Variable Number of Workers			Fixed Number of Workers		
	N	**Profit**	**THW**	**N**	**Profit**	**THW**
Type A: $m = 2$	2	98,066.7	52,553.37	2	98,066.7	52,553.37
Type B: $m = 3$	4	111,593	62,836.68	2	107,918	57,478.77
Type C: $m = 4$	3	112,049	61,304.32	2	108,022	57,531.19

m = # maturity classes.

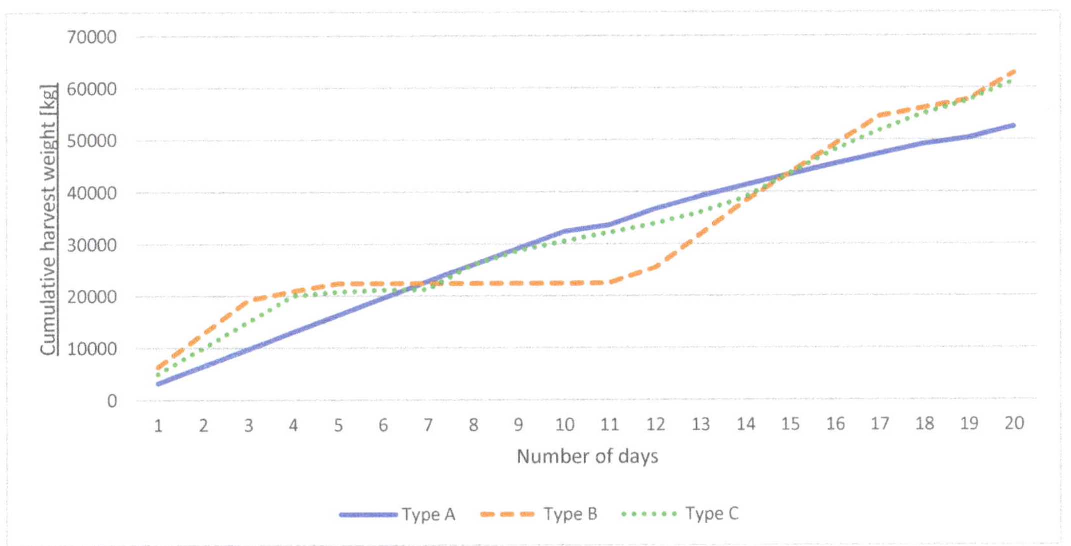

Figure 1. The cumulative total harvest weight (kilogram) for each worker type.

In the first strategy, workers of Types A and C harvest peppers at the same rate throughout the whole month. In the second strategy, workers of Type B harvest only at the beginning and at the end of the month. This strategy allows the fruit to keep gaining weight until the harvest at the end of the month.

The first strategy is simpler to implement, since the solution fully utilizes the workers/robots. The second strategy has the potential for better utilization, but requires tight planning. Since there are periods without harvesting, in those periods, the workers/robots can work in a different greenhouse (or different areas in a larger greenhouse) by synchronizing the harvest periods.

5.1.2. Fixed Number of Workers

This example demonstrates how workers' experience affects the total harvest weight. The formulation was executed for the three sub-cases with a fixed number of workers as input (the number of workers, N, was set to two instead of being a decision variable). The results in Table 3 show that there is a slight difference between the performances of Type B and Type C workers. However, there are more marked differences in the total harvest weight between Type A and Types B and C, with 9.3% and 9.5% improvements, respectively.

5.2. Analysis of the Robotic Harvester Capability

The assumption is that a robotic harvester can potentially classify fruits into multiple maturity classes, even to their exact DAA, by using AI methods [52]. Peppers are therefore harvested at the optimal time giving higher weights. In this numerical example, we compared different types of human harvesters (Table 1) with a robotic harvester.

5.2.1. Difference in the Total Harvest Weight between Robotic and Human Harvesters with the Same Capabilities

This analysis investigated the difference in the total harvest weight (THW) between a single robot and a number of workers of different types, where the robot and the total number workers had the same harvest capabilities C_h. This type of analysis can be applied to estimate the cost efficiency of a robotic harvester and the maximal price a farmer will agree to pay for such a robot for one month of operation. To evaluate how the robot's ability to classify peppers according to their DAA affects the harvest weight (THW) in one month, the THW of a robot was compared to the THW of a number of workers, where the number of workers was set to a number giving the same C_h as one robot (Table 1). The robot's THW was then compared to the THW of the workers for each worker type. Six workers ($N = 6$) were selected as the basis for comparison, with a total capacity of $C_h = 40{,}500$ fruits per working day being equivalent to the ability of one robot. Any number lower than this (produced by six workers) creates an imbalanced use case with a high supply of peppers in the greenhouse, resulting in the harvesting of peppers solely from the highest maturity class. Similarly, any number higher than $C_h = 40{,}500$ creates a use case in which all the peppers available in the greenhouse are harvested.

The results in Table 4 show the robotic harvester's high potential, with a 12.28, 7.33, and 4.9% increase in the THW when compared to workers of Types A, B, and C, respectively. The results show the importance of the ability to classify maturity for evaluating the feasibility of using a robotic harvester. Since the results are presented in terms of harvest weight, the higher the pepper price, the greater the advantage of the robot when compared to the workers. From Table 4, the difference in THW between the robot and the workers can be calculated as quality savings [55] (pp. 683–687) measured for the planning horizon period. For example, the quality savings between the robot and Type B workers is 2716 kg/month, resulting in overall savings of 23,765 kg/year. The quality savings can then be used to calculate return-on-investment measures. This analysis also reveals the importance of decreasing the robot's cost to increase its economic feasibility, as expected.

Table 4. The total harvest weight (kilogram) for a harvester robot and for different types of workers with the same C_h for one month and for an estimated growing period of 35 weeks.

Harvester Type	Robot	6 Type A Workers	6 Type B Workers	6 Type C Workers
THW (kg) in one month	37,069.03	32,885.64	34,352.45	35,253.99
THW (kg) in one year (35 harvest weeks)	324,354	287,749.35	300,583.93	308,472.41

5.2.2. Required Cycle Time for the Harvester Robot

This analysis aimed to assess the required cycle time for the harvester robot to reach the same THW of the human harvesters based on the results of the analysis in 5.2.1 (Table 4). The required robot harvest capabilities (C_h) to achieve the same THW were searched manually using the "goal seek" method. The results show that robot harvest capabilities (C_h) of 34,000, 36,070, and 37,370 kg in one working day are needed to reach the same THW as Type A, B, and C workers, respectively. However, it should be remembered that the robot's capabilities were compared to the capabilities of six workers. Therefore, assuming that the robot can work 20 h a day [56], Table 5 shows the equivalent number of workers for different harvester robot cycle times. It has been reported that harvester robots achieved cycle times between 5.5 and 24 s to harvest a single fruit [20], which according to this analysis, is equal to 0.5 to 2.3 workers (Table 5). The results show the importance of the robot harvest capabilities in one period and the improvement in cycle times needed for a robotic harvester to reach economic feasibility.

Table 5. The equivalent number of workers for different harvester robot cycle times.

Robot Cycle Time	Type A Workers	Type B Workers	Type C Workers
0.5	25.4	24.0	23.1
1	12.7	12.0	11.6
2.5	5.1	4.8	4.6
5.5	2.3	2.2	2.1
7.5	1.7	1.6	1.5
10	1.3	1.2	1.2
12.5	1.0	1.0	0.9
15	0.8	0.8	0.8
20	0.6	0.6	0.6
25	0.5	0.5	0.5

5.3. Payback Period and Rate-of-Return Analysis

Two return-on-investment measures for the harvester robot were calculated, similar to the analysis suggested by [55] (pp. 683–687): the payback period and the internal rate of return (IRR) in five years. The payback period is the number of years required for incoming cash flows to balance cash outflows. The IRR is the annual growth rate that an investment is expected to generate [55]. The analysis required cost estimation of the investment in the robot and the savings via its use. It includes the following data based on SWEEPER project estimations [56]:

- The cost of the robot is set at 100,000, 130,000, or 160,000 €
- The harvest season lasts 35 weeks per year
- Manual harvesting requires 3 s/pepper
- Robot harvesting requires 10 s/pepper
- Manual harvesters work five months, five days/week, 8 h/day
- A robotic harvester works 20 h/day, six days/week
- Manual harvest hourly rates of 16.5 (corresponding to rates in the Netherlands) and 9.97 € (corresponding to rates in Israel) result in 23,100 and 14,000 €/year, respectively

In addition, the analysis estimates robot operation costs of 20,000 € for the first two years and 10,000 € for the years thereafter. The use tax rate and the depreciation were based on the United States Internal Revenue Service's Modified Accelerated Cost Recovery System (MACRS), as in ref. [55] (p. 684).

The analysis included the following data, based on the corresponding use case (Section 5.2.1); the use case considers four harvest weeks of planning, multiplied by 8.75

to represent the number of plannings per year. The formulation result was used to estimate the robot quality savings. Type B workers were chosen for the analysis, resulting in 23,769.4 kg/year of quality savings.

The results in Table 6 show that, considering the highest robot cost (160,000 €), for pepper prices lower than 1.25 and 1.75 €/kg, it will take more than five years to recover the robot investment in both the Netherlands and Israel. The p of sweet peppers in the Netherlands varies between 1.2–1.3 €/kg, resulting in 4.72 payback periods and 1.9% IRR. The p of sweet peppers in Israel varies between 1.7–2.2 €/kg, resulting in 3.98 payback periods and 7.8% IRR.

Table 6. Payback periods (years) and IRR in five years in the Netherlands and Israel for different pepper p and robot costs.

Pepper p (€)/ Robot Cost (€)	The Netherlands						Israel					
	Payback Periods			IRR in Five Years (%)			Payback Periods			IRR in Five Years (%)		
	100K	130K	160K	100K	130K	160K	100K	130K	160K	100K	130K	160K
1	3.72	4.58	5.50	10.5	2.9	0.0	4.89	>6	>6	0.0	0.0	0.0
1.25	3.21	3.99	4.72	16.2	7.8	1.9	4.05	5.00	>6	7.3	0.0	0.0
1.5	2.89	3.52	4.19	21.1	12.4	5.9	3.46	4.29	5.08	13.3	5.2	0.0
1.75	2.63	3.17	3.76	25.7	16.6	9.8	3.05	3.76	4.46	18.5	9.9	3.8
2	2.42	2.91	3.40	30.3	20.3	13.6	2.76	3.33	3.98	23.3	14.5	7.8
2.25	2.24	2.70	3.14	34.7	24.0	16.8	2.53	3.04	3.58	27.9	18.3	11.6
2.5	2.08	2.52	2.93	39.0	27.6	19.8	2.33	2.81	3.27	32.4	22.1	15.1
2.75	1.95	2.36	2.75	43.2	31.0	22.8	2.16	2.61	3.03	36.7	25.7	18.2
3	1.82	2.23	2.59	47.4	34.5	25.8	2.02	2.45	2.84	41.0	29.2	21.3

For a lower robot cost, a 100,000 € return on investment will reduce to a payback of 3.21 and 2.71 years for the Netherlands and Israel, respectively. As expected, the return-on-investment measures are sensitive to both the pepper price and the robot cost.

6. Conclusions

This paper presents a formulation for decisions related to the *selective harvesting of individual fruit*. The formulation serves as the basis for *the operational decision of the production order* and can assist farmers of selective crops in deciding *how many fruit to harvest from each maturity class* and *the required number of harvesters (workers to hire for the harvesting season or equivalent performing robot)*. It takes into consideration four features, namely, time window constraints, resource limitations, yield perishability, and uncertainty. We suggest ways to deal with *agricultural product uncertainty* and adopted the formulation to different variations of the problem. The formulation was demonstrated here for a case study of sweet pepper harvesting. The case study presents the required cycle times for a robotic harvester and return-on-investment measures for different pepper and robot costs. The research formulation also *enables the estimation of robot quality savings,* a critical factor in a return on investment analysis and provides an economic analysis for different robot capabilities. The numerical examples analyzed the difference between types of worker and showed the potential added value of a robotic harvester, which can improve maturity classification and harvesting capabilities. This work focused on demonstrating use cases for farmers based on solutions obtained with personal computer-based systems. The aim was to show how our model can assist the individual farmer to decide on the production order. Specifically, we showed how this analysis can be used to *decide on the number of workers* and *type of workers*. Furthermore, the formulation results have proven the *quality savings of a robotic harvester* and reveal the importance of reducing its cycle times. A practical tool for the farmer can be developed based on this formulation and could be part of an IT package provided when renting a robotic system and/or part of a production management software.

The current *operational model for selective harvesting* can be developed further. The current formulation does not enable a combination of different types of workers/robots due to formulation complexity. This problem can be solved by developing a simulation model to simulate different combinations of workers and robots to determine the best combination. Despite this limitation, running the formulation separately for each available type of worker can contribute to the decision as to the best single type of worker to hire.

The current formulation does not take into consideration maturity classification mistakes of the robot and manual harvesters; thus, future research should consider adding such an analysis along with addressing inventory control aspects. Additional aspects of the robotic harvester that influence performance should also be considered, such as detection and grasping rates. These can influence the individual operational decision for a single fruit (whether to harvest or not) and not the overall production order (the number of workers to hire and peppers to harvest from each maturity level), which was the main goal of this paper. These future developments will be important once robotic harvesters will have become available and if human–robot collaborative harvesting is considered. The overall economic justification for robotic harvesting presented here indicates the need for further development of such harvesters. It demonstrates the importance of ensuring the high-quality maturity classification of such a robot and the reducing of its cycle time. The formulation of harvesting decisions related to the *individual fruit* is an important contribution to the field, providing a basic model in production and operations management of selective harvesting.

Author Contributions: Conceptualization: B.H., Y.E., Y.P.; Methodology: B.H., Y.E., Y.P.; Software: B.H.; Data analysis: B.H.; Interpretation of the data: B.H., Y.E., Y.P.; Writing—original draft preparation: B.H.; Writing—critical review and editing: B.H., Y.E., Y.P.; supervision: Y.E., Y.P.; project administration: Y.E.; funding acquisition: Y.E. All authors have read and agreed to the published version of the manuscript.

Funding: Partially funded by the European Commission (SWEEPER GA No. 66313) and by Ben-Gurion University of the Negev through the Helmsley Charitable Trust, the Agricultural, Biological, and Cognitive Robotics Initiative, the Marcus Endowment Fund, and the Rabbi W. Gunther Plaut in Manufacturing Engineering.

Institutional Review Board Statement: Not applicable.

Informed Consent Statement: Informed consent was obtained from all subjects involved in the study.

Data Availability Statement: Not applicable.

Conflicts of Interest: The authors declare no conflict of interest.

References

1. Kusumastuti, R.D.; Van Donk, D.P.; Teunter, R. Crop-related harvesting and processing planning: A review. *Int. J. Prod. Econ.* **2016**, *174*, 76–92. [CrossRef]
2. Soto-Silva, W.E.; Nadal-Roig, E.; González-Araya, M.C.; Pla-Aragones, L.M. Operational research models applied to the fresh fruit supply chain. *Eur. J. Oper. Res.* **2016**, *251*, 345–355. [CrossRef]
3. Perlman, Y. Establishing a dual food supply chain for organic products in the presence of showrooming—A game theoretic analysis. *J. Clean. Prod.* **2021**, *321*, 128816. [CrossRef]
4. Ahumada, O.; Villalobos, J.R. Operational model for planning the harvest and distribution of perishable agricultural products. *Int. J. Prod. Econ.* **2011**, *133*, 677–687. [CrossRef]
5. Ahumada, O.; Villalobos, J.R. Application of planning models in the agri-food supply chain: A review. *Eur. J. Oper. Res.* **2009**, *196*, 1–20. [CrossRef]
6. Perlman, Y.; Ozinci, Y.; Westrich, S. Pricing decisions in a dual supply chain of organic and conventional agricultural products. *Ann. Oper. Res.* **2019**, 1–16. [CrossRef]
7. Ozinci, Y.; Perlman, Y.; Westrich, S. Competition between organic and conventional products with different utilities and shelf lives. *Int. J. Prod. Econ.* **2017**, *191*, 74–84. [CrossRef]

8. Grisso, R.D.; McCullough, D.; Cundiff, J.S.; Judd, J.D. Harvest schedule to fill storage for year-round delivery of grasses to biorefinery. *Biomass Bioenergy* **2013**, *55*, 331–338. [CrossRef]
9. Carpente, L.; Casas-Méndez, B.; Jácome, C.; Puerto, J. A model and two heuristic approaches for a forage harvester planning problem: A case study. *TOP* **2010**, *18*, 122–139. [CrossRef]
10. Starbird, S.A. Optimal loading sequences for fresh-apple storage facilities. *J. Oper. Res. Soc.* **1988**, *39*, 911–917. [CrossRef]
11. Higgins, A.J.; Muchow, R.C.; Rudd, A.V.; Ford, A.W. Optimising harvest date in sugar production: A case study for the Mossman mill region in Australia I. Development of operations research model and solution. *Field Crops Res.* **1998**, *57*, 153–162. [CrossRef]
12. Higgins, A.J.; Neville, D.W. Australian sugar mills optimize harvester rosters to improve production. *Interfaces* **2002**, *32*, 15–25. [CrossRef]
13. Kootstra, G.; Wang, X.; Blok, P.M.; Hemming, J.; van Henten, E. Selective Harvesting Robotics: Current Research, Trends, and Future Directions. *Curr. Robot. Rep.* **2021**, *2*, 95–104. [CrossRef]
14. Edan, Y.; Adamides, G.; Oberti, R. Agriculture automation. In *Handbook of Automation*; Springer: Cham, Switzerland, 2022.
15. De-An, Z.; Jidong, L.; Wei, J.; Ying, Z.; Yu, C. Design and control of an apple harvesting robot. *Biosyst. Eng.* **2011**, *110*, 112–122. [CrossRef]
16. Silwal, A.; Davidson, J.R.; Karkee, M.; Mo, C.; Zhang, Q.; Lewis, K. Design, integration, and field evaluation of a robotic apple harvester. *J. Field Robot.* **2017**, *34*, 1140–1159. [CrossRef]
17. Mehta, S.S.; Burks, T.F. Vision-based control of robotic manipulator for citrus harvesting. *Comput. Electron. Agric.* **2014**, *102*, 146–158. [CrossRef]
18. Hu, X.; Yu, H.; Lv, S.; Wu, J. Design and experiment of a new citrus harvesting robot. In Proceedings of the International Conference on Control Science and Electric Power Systems (CSEPS), Shangai, China, 28–30 May 2021; pp. 179–183. [CrossRef]
19. Feng, Q.; Zou, W.; Fan, P.; Zhang, C.; Wang, X. Design and test of robotic harvesting system for cherry tomato. *Int. J. Agric. Biol. Eng.* **2018**, *11*, 96–100. [CrossRef]
20. Arad, B.; Balendonck, J.; Barth, R.; Ben-Shahar, O.; Edan, Y.; Hellström, T.; Hemming, J.; Kurtser, P.; Ringdahl, O.; Tielen, T.; et al. Development of a sweet pepper harvesting robot. *J. Field Robot.* **2020**, *37*, 1027–1039. [CrossRef]
21. Edan, Y.; Rogozin, D.; Flash, T.; Miles, G.E. Robotic melon harvesting. *IEEE Trans. Robot. Autom.* **2000**, *16*, 831–835. [CrossRef]
22. Leu, A.; Razavi, M.; Langstadtler, L.; Ristic-Durrant, D.; Raffel, H.; Schenck, C.; Graser, A.; Kuhfuss, B. Robotic green asparagus selective harvesting. *IEEE/ASME Trans. Mechatron.* **2017**, *22*, 2401–2410. [CrossRef]
23. Bac, C.W.; van Henten, E.J.; Hemming, J.; Edan, Y. Harvesting robots for high-value crops: State-of-the-art review and challenges ahead. *J. Field Robot.* **2014**, *31*, 888–911. [CrossRef]
24. Harel, B.; Parmet, Y.; Edan, Y. Maturity classification of sweet peppers using image datasets acquired in different times. *Comput. Ind.* **2020**, *121*, 103274. [CrossRef]
25. Li, B.; Lecourt, J.; Bishop, G. Advances in non-destructive early assessment of fruit ripeness towards defining optimal time of harvest and yield prediction—A review. *Plants* **2018**, *7*, 3. [CrossRef]
26. Harel, B.; van Essen, R.; Parmet, Y.; Edan, Y.; van Essen, R.; Parmet, Y.; Edan, Y. Viewpoint Analysis for Maturity Classification of Sweet Peppers. *Sensors* **2020**, *20*, 3783. [CrossRef]
27. Kurtser, P.; Edan, Y. Statistical models for fruit detectability: Spatial and temporal analyses of sweet peppers. *Biosyst. Eng.* **2018**, *171*, 272–289. [CrossRef]
28. Džidiæ, A.; Halachmi, I.; Havranek, J.L. Prediction of Milking Robot Utilization Predvidanje iskoristenja robota za strojnu muznju. *Agric. Conspec. Sci.* **2001**, *66*, 137–143.
29. Halachmi, I.; Metz, J.H.M.; van't Land, A.; Halachmi, S.; Kleijnen, J.P.C. Case Study: Optimal facility allocation in a robotic milking barn. *Trans. ASAE* **2002**, *45*, 1539–1546. [CrossRef]
30. Halachmi, I.; Metz, J.H.M.; Maltz, E.; Dijkhuizen, A.A.; Speelman, L. Designing the optimal robotic milking barn, Part 1: Quantifying facility usage. *J. Agric. Eng. Res.* **2000**, *76*, 37–49. [CrossRef]
31. van Herck, L.; Kurtser, P.; Wittemans, L.; Edan, Y. Crop design for improved robotic harvesting: A case study of sweet pepper harvesting. *Biosyst. Eng.* **2020**, *192*, 294–308. [CrossRef]
32. Edan, Y.; Engel, B.A.; Miles, G.E. Intelligent control system simulation of an agricultural robot. *J. Intell. Robot. Syst.* **1993**, *8*, 267–284. [CrossRef]
33. Johnson, L.K.; Bloom, J.D.; Dunning, R.D.; Gunter, C.C.; Boyette, M.D.; Creamer, N.G. Farmer harvest decisions and vegetable loss in primary production. *Agric. Syst.* **2019**, *176*, 102672. [CrossRef]
34. Temu, A.E.; Temu, A.A. High value agricultural products for smallholder markets in sub-saharan Africa: Trends, opportunities and research priorities. In Proceedings of the High Value Agricultural Products Workshop, Cali, Columbia, 3–5 October 2005; pp. 1–37.
35. Kapach, K.; Barnea, E.; Mairon, R.; Edan, Y.; Ben-Shahar, O. Computer vision for fruit harvesting robots-state of the art and challenges ahead. *Int. J. Comput. Vis. Robot.* **2012**, *3*, 4–34. [CrossRef]
36. Landahl, S.; Terry, L.A. Non-destructive discrimination of avocado fruit ripeness using laser Doppler vibrometry. *Biosyst. Eng.* **2020**, *194*, 251–260. [CrossRef]
37. Saranwong, S.; Sornsrivichai, J.; Kawano, S. Prediction of ripe-stage eating quality of mango fruit from its harvest quality measured nondestructively by near infrared spectroscopy. *Postharvest Biol. Technol.* **2004**, *31*, 137–145. [CrossRef]
38. Azarmdel, H.; Jahanbakhshi, A.; Mohtasebi, S.S.; Muñoz, A.R. Evaluation of image processing technique as an expert system in mulberry fruit grading based on ripeness level using artificial neural networks (ANNs) and support vector machine (SVM). *Postharvest Biol. Technol.* **2020**, *166*, 111201. [CrossRef]

39. Ratprakhon, K.; Neubauer, W.; Riehn, K.; Fritsche, J.; Rohn, S. Developing an Automatic Color Determination Procedure for the Quality Assessment of Mangos (Mangifera indica) Using a CCD Camera and Color Standards. *Foods* **2020**, *9*, 1709. [CrossRef] [PubMed]
40. Lowenberg-DeBoer, J.; Huang, I.Y.; Grigoriadis, V.; Blackmore, S. Economics of robots and automation in field crop production. *Precis. Agric.* **2020**, *21*, 278–299. [CrossRef]
41. Allen, S.J.; Schuster, E.W. Controlling the risk for an agricultural harvest. *Manuf. Serv. Oper. Manag.* **2004**, *6*, 225–236. [CrossRef]
42. Zion, B.; Mann, M.; Levin, D.; Shilo, A.; Rubinstein, D.; Shmulevich, I. Harvest-order planning for a multiarm robotic harvester. *Comput. Electron. Agric.* **2014**, *103*, 75–81. [CrossRef]
43. Amaruchkul, K. Planning migrant labor for green sugarcane harvest: A stochastic logistics model with dynamic yield prediction. *Comput. Ind. Eng.* **2021**, *154*, 107016. [CrossRef]
44. Ferrer, J.C.; Mac Cawley, A.; Maturana, S.; Toloza, S.; Vera, J. An optimization approach for scheduling wine grape harvest operations. *Int. J. Prod. Econ.* **2008**, *112*, 985–999. [CrossRef]
45. Plà-Aragonés, L.M. *Handbook of Operations Research in Agriculture and the Agri-Food Industry*; Springer: Cham, Switzerland, 2015; ISBN 9781493924820.
46. Ampatzidis, Y.G.; Vougioukas, S.G.; Whiting, M.D.; Zhang, Q. Applying the machine repair model to improve efficiency of harvesting fruit. *Biosyst. Eng.* **2014**, *120*, 25–33. [CrossRef]
47. Arnaout, J.P.M.; Maatouk, M. Optimization of quality and operational costs through improved scheduling of harvest operations. *Int. Trans. Oper. Res.* **2010**, *17*, 595–605. [CrossRef]
48. Maatman, A.; Schweigman, C.; Ruijs, A.; Van Der Vlerk, M.H.; Maatman, A.; Schweigman, C.; Ruijs, A.; Van Der Vlerk, M.H. Modeling farmers' response to uncertain rainfall in Burkina Faso: A stochastic programming approach. *Oper. Res.* **2002**, *50*, 399–414. [CrossRef]
49. Annetts, J.E.; Audsley, E. Multiple objective linear programming for environmental farm planning. *J. Oper. Res. Soc.* **2002**, *53*, 933–943. [CrossRef]
50. Golenko-Ginzburg, D.; Sinuany-Stern, Z.; Kats, V. A multilevel decision-making system with multiple resources for controlling cotton harvesting. *Int. J. Prod. Econ.* **1996**, *46–47*, 55–63. [CrossRef]
51. Albornoz, V.M.; Araneda, L.C.; Ortega, R. Planning and scheduling of selective harvest with management zones delineation. *Ann. Oper. Res.* **2021**. [CrossRef]
52. Maaike Wubs, A.; Ma, Y.T.; Heuvelink, E.; Hemerik, L.; Marcelis, L.F.M. Model selection for nondestructive quantification of fruit growth in pepper. *J. Am. Soc. Hortic. Sci.* **2012**, *137*, 71–79. [CrossRef]
53. Elkoby, Z.; Van Ooster, B.; Edan, Y. Simulation analysis of sweet pepper harvesting. In Proceedings of the IFIP International Conference on Advances in Production Management Systems (APMS), Ajaccio, France, 20–24 September 2014; pp. 441–448.
54. Melamed, Z. Analysis of Human-Robot Harvesting Operations in Sweet Pepper Greenhouses. Master's Thesis, Ben Gurion University of the Negev, Beersheba, Israel, 2016.
55. Nof, S.Y. *Handbook of Industrial Robotics*; John Wiley & Sons, Ltd.: Hoboken, NJ, USA, 1999.
56. Sweeper Project Workpackages Overview. Available online: http://www.sweeper-robot.eu/workpackages (accessed on 1 December 2021).

Article

Human–Machine Systems Reliability: A Series–Parallel Approach for Evaluation and Improvement in the Field of Machine Tools

Rosa Ma Amaya-Toral [1], Manuel R. Piña-Monarrez [2], Rosa María Reyes-Martínez [1], Jorge de la Riva-Rodríguez [1], Eduardo Rafael Poblano-Ojinaga [1], Jaime Sánchez-Leal [3] and Karina Cecilia Arredondo-Soto [4,5,*]

[1] Division of Graduate Studies and Research Tecnologico Nacional de Mexico/Instituto Tecnologico de Ciudad Juarez, Ave. Tecnologico #1340, El Crucero, Ciudad Juarez 32500, Chihuahua, Mexico; rosa.at@chihuahua2.tecnm.mx (R.M.A.-T.); rosa.rm@cdjuarez.tecnm.mx (R.M.R.-M.); jriva@itcj.edu.mx (J.d.l.R.-R.); jefatura_depi@itcj.edu.mx (E.R.P.-O.)

[2] Institute of Engineering and Technology, Autonomous University of Ciudad Juarez; Ciudad Juarez 32310, Chihuahua, Mexico; manuel.pina@uacj.mx

[3] Industrial, Manufacturing and System Engineering Department, College of Engineering, University of Texas at El Paso (UTEP), El Paso, TX 79968, USA; jsanchez21@utep.edu

[4] Chemical Sciences and Engineering Faculty, Universidad Autonoma de Baja California, Tijuana 22390, Baja California, Mexico

[5] Department of Industrial Engineering, Tecnologico Nacional de Mexico/Instituto Tecnologico de Tijuana, Tijuana 22414, Baja California, Mexico

* Correspondence: karina.arredondo@uabc.edu.mx or karina.arredondo@tectijuana.edu.mx

Citation: Amaya-Toral, R.M.; Piña-Monarrez, M.R.; Reyes-Martínez, R.M.; de la Riva-Rodríguez, J.; Poblano-Ojinaga, E.R.; Sánchez-Leal, J.; Arredondo-Soto, K.C. Human–Machine Systems Reliability: A Series–Parallel Approach for Evaluation and Improvement in the Field of Machine Tools. *Appl. Sci.* **2022**, *12*, 1681. https://doi.org/10.3390/app12031681

Academic Editor: Panagiotis Tsarouhas

Received: 31 December 2021
Accepted: 29 January 2022
Published: 6 February 2022

Publisher's Note: MDPI stays neutral with regard to jurisdictional claims in published maps and institutional affiliations.

Abstract: Machine workshops generate high scrap rates, causing non-compliance with timely delivery and high production costs. Due to their natural characteristics of a low volume, high-mix production batches, and serial and parallel configurations, generally the causes of their failure are not well documented. Thus, to reduce the scrap rate, and evaluate and improve their reliability, their system characteristics must be considered. Based on them, our proposed methodology allows us to evaluate the system, subsystem, and component–subsystem relationship by using either the Weibull and/or the exponential distribution. The strategy to improve the system performance includes reliability tools, expert interviews, cluster analysis, and root-cause analysis. In the application case, the failure sources were found to be mechanical and human errors. The component maintenance/setup, institutional conditions/attitude, and subsystem process/operation were the machine factors that presented the lowest reliability indices. The improved activities were monitored based on the Weibull β and η parameters that affect the system reliability. Finally, by using a life–effort analysis, and the method of comparative analysis of two sequential periods, we identified the causes that generated a change in the Weibull parameters. The contribution of this methodology lies in the grouping of the tools in the proposed application context.

Keywords: reliability; FMEA; life–strength model; statistical evaluation model; Weibull and exponential distribution; machine tools; root-cause analysis; cluster analysis; performance evaluation

1. Introduction

Reliability theory allows us to analyze failures that occur in the components of a system over time; therefore, its application crosses diverse areas of knowledge, including health sciences [1–3], social sciences [4,5], engineering and technology [6–8], and industry in general [9]. In the industrial sector, the concept of reliability is used in maintenance, where it is necessary to keep machines in good working condition, under statistical control, and even avoiding variation. However, complex tools are needed to achieve these goals due to random variables.

Moreover, in manufacturing where computer numerical control (CNC) equipment plays an essential role in production, high reliability is desirable. Unfortunately, determining the system's reliability is complex. Existing research studies mainly focus on univariate degradation data or multiple correlated data, which might not be accurate in equipment reliability assessments [10]. Li et al. [11] investigated the early failure of the Main Drive Systems of Computerized Numerical Control machine tools and developed a Bayesian network model to conduct a comprehensive reliability analysis. Li et al. [12] adopted a Bayesian network predictive analysis to model and analyze the reliability characteristics, such as failure probability, failure rate, and mean time to failure, of a floating offshore wind turbine. Bobbio et al. [13] proposed improving the analysis of dependable systems by mapping fault trees into Bayesian networks. Guo et al. [14] introduced a degradation-analysis-based reliability assessment method for CNC machine tools under performance testing by considering unit non-homogeneity. Langeth and Portinale [15] discussed the properties of the modeling framework that makes BNs particularly well suited for reliability applications and pointed to ongoing research that is relevant to practitioners in reliability. Mi et al. [16] focused on the reliability analysis of a complex multi-state system (MSS) based on the Bayesian network (BN) method using the Dempster–Shafer (DS) evidence theory to express the epistemic uncertainty in the system, considering a complex redundant system and introducing a modified β parametric factor for the reliability analysis.

Huai-Wei et al. [17] proposed a novel Failure Mode and Effects Analysis (FMEA) model based on multi-criteria group decision-making that integrates a rough best–worst method for machine tool risk analysis. Adamyan and He [18] presented a methodology that can be used to identify the failure sequences and assess the probability of their occurrence in a manufacturing system. Bai et al. [19] proposed a modified fatigue damage accumulation model to precisely predict lifetimes of aero-engine materials according to an interaction factor that is able to quantitatively map damage interactions introduced by high cycle fatigue and low cycle fatigue loads.

Dundulis et al. [20] present an overall framework for the estimation of the failure probability of pipelines based on the results of a deterministic–probabilistic structural integrity analysis, the corrosion rate, the number of defects, and failure data. A method for estimating the reliability of a wind power system based on modeling and analysis was developed by Erylmaz and Kan [21]. In Fan et al. [22], the evaluation of a small sample was applied in a CNC grinding machine evaluation based on Bayes theory. In terms of the production of CNC machine tools, due to the use of big data technology, CNC machine tools have been rapidly produced on a large scale, which has played an important role in the upgrading of CNC machine tools and the improvement of their quality and efficiency [23].

In Li et al. [24], the fuzzy theory was employed to represent uncertainties involved in prediction. Reliability prediction under fuzzy stress with and without fuzzy strength is conducted by using a dynamic stress–strength interference model that takes types of cycles of aero-engines into consideration. Taking into account the fuzziness associated with the failure probability of hydraulic systems, a fuzzy fault tree was proposed by integrating fuzzy set theory into the conventional fault tree analysis method.

Mi et al. and Vineyard et al. [25,26] describe failures and repair rate characteristics as well as distribution data for a typical flexible manufacturing system (FMS) that is in use in a manufacturing plant in the United States. Data included mechanical, hydraulic, electrical, electronic, software, and human failures as well as repairs. The data were also fit with appropriate theoretical distributions.

For this case, we present a methodology to evaluate and improve the reliability of the man–machine system in a machine tool workshop by considering the system characteristics, particularly those presented by the machine shops of Chihuahua city. Workshops of this metal-mechanic sector work with a low-volume production system and high-mix batches. These, by using the same machine, can manufacture parts of different materials for different customers. Therefore, a great variety of cutting tools and the use of different operations are needed. Consequently, failures that occur in the machine tool can be generated by several factors, which in our proposed methodology are conveniently addressed as a component, a component–system relationship, or a subsystem, to reduce the scrap rate and to improve the system reliability.

Although the man–machine system's evaluation is complex due to the interaction of the system's elements, its reliability evaluation consists of determining the reliability of each component-subsystem relationship by applying the probability rules that apply according to the series–parallel configuration, and then by computing the whole system's reliability. The application of the methodology to a regional man–machine system enables us to use the Weibull and exponential distributions to determine the reliability of the addressed component–subsystem combinations and the subsystems themselves. Finally, by implementing the proposed methodology the causes of critical failures were addressed and improved. This was done by considering that failures of mechanical and human origin are the major contributors to the system's failure (by a case study), and that the component maintenance/setup and institutional conditions/attitude are subcauses with a greater impact on the presence of failures in the machine tool. These components interacting with the subsystem process/operation and the machine were the factors that presented the lowest reliability indices. Once they were improved, the improvement actions were monitoring using their related Weibull β and η parameters.

2. Materials and Methods

Machining shops in the metal-mechanical sector operate as batch production systems with a high mix and a low volume. Conventional and CNC machine tools are included in the system, which presents as a common problem the generation of high scrap rates, which affect the ability to comply with the delivery times committed to customers and produce an increase in production costs. Another characteristic of this machine shop is that there is no culture of recording incidents related to the defects or failures that occur; therefore, there is no systematic analysis and follow-up of them. The interaction of the elements of the system makes it complex and affects its reliability.

The evaluation of the reliability of the man–machine system consists of determining the appropriate reliability by modeling each component of the system, applying the rules of probability according to its configuration, and computing the system's reliability. The components within a system can be related in the following two primary ways: serial configuration and parallel configuration. In serial reliability, all components must function for the system to operate correctly. In the case of a parallel (redundant) configuration, at least one component must be functioning for the system to operate correctly [27].

For this case, the methodological design consists of four phases: an operational context, a human–machine system design, the reliability evaluation model, and some improvement methods. They are presented in Figure 1. The four phases for the evaluation and improvement of the reliability of the human–machine system from Figure 1 range from the system approach and production system definition to the human–machine system design, reliability assessment, and improvement methods. These phases were developed using a system approach, reliability engineering concepts, root-cause analysis techniques, and some probability and statistical concepts. Some other methodologies, such as Bayesian networks, fuzzy theory, rates of failures and repairs, and data distributions, have also been found to be useful to evaluate the reliability of different kinds of systems. However, they have not focused on evaluating the reliability of a complex and flexible system with a series–parallel configuration using some statistical models with the same configuration, such as the one presented in this publication. A description of these four phases will be presented in the following subsections.

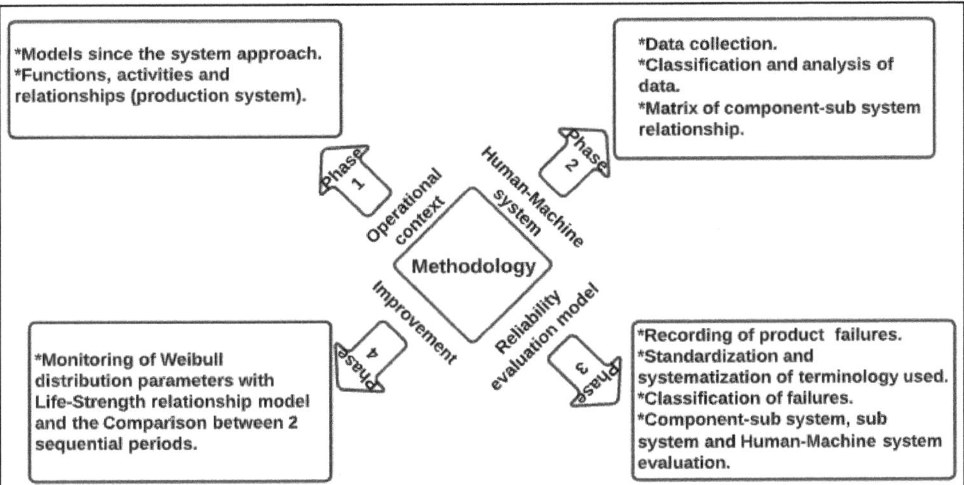

Figure 1. The four phases for the evaluation and improvement of the reliability of the human–machine system.

3. Case Study. The Applied Evaluation Processes

3.1. Phase 1: Operational Context

Preliminary research was conducted to identify the different kinds of Conventional and Numerical Control machines used in machine shops and the required knowledge and responsibilities. Table 1 shows some of the machines, knowledge, and responsibilities considered in this research. In addition to identifying the machinery, knowledge, and responsibilities, it was necessary to investigate the functions and activities developed to manufacture parts. Some of them are shown in Table 2. Then, based on functions and relationships, a Production System of the machine tools model was developed.

Table 1. Some of the machines, knowledge, and responsibilities considered.

Machine	Knowledge	Responsibility
Milling	Measurement	Take care of yourself
Griding	Geometric tolerances	Machine care
Machining center	Machine and tools	Order
Turnstile	Materials	Precision

Table 2. Some functions and activities considered for the production system.

Function	Activity
Production control	Plan the quantity of parts to be manufactured.
Shipment	Packing and shipping of parts
Quality control	Inspection of material and parts manufactured
Quotes	Quote orders

Functions, activities, and relationships (production systems). Using interviews with experts from the machine shop, the Production System of the machine tools was developed based on the workshop functions, activities, and relationships. This model is presented in Section 3. Some other models were developed using a system approach.

Models for the system approach. A system approach was used to analyze the production system as an open system. The inputs of the system and the outputs were identified in order to understand how it operates. Figure 2 shows the production system with an open system approach. Figure 2 shows how the customer requirements or inputs activate functions and activities and relationships between them. The influence of internal and external factors affects the functioning of the system, as well as the entropy or tendency to disorder. The outputs of the system are the manufactured pieces. Some quality and reliability indices were required to generate the guideline to improve the system.

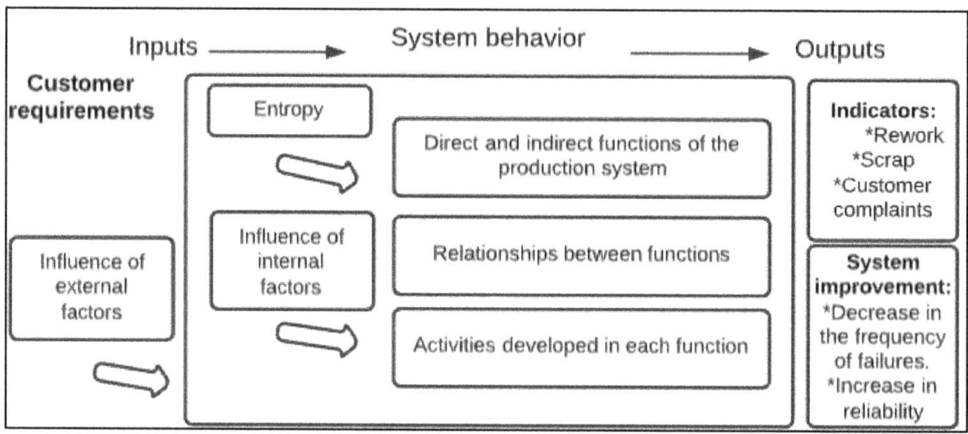

Figure 2. The production system with an open system approach.

Activities and functions and the relationships between them were helpful in establishing the Production System of the human–machine system presented in Figure 3.

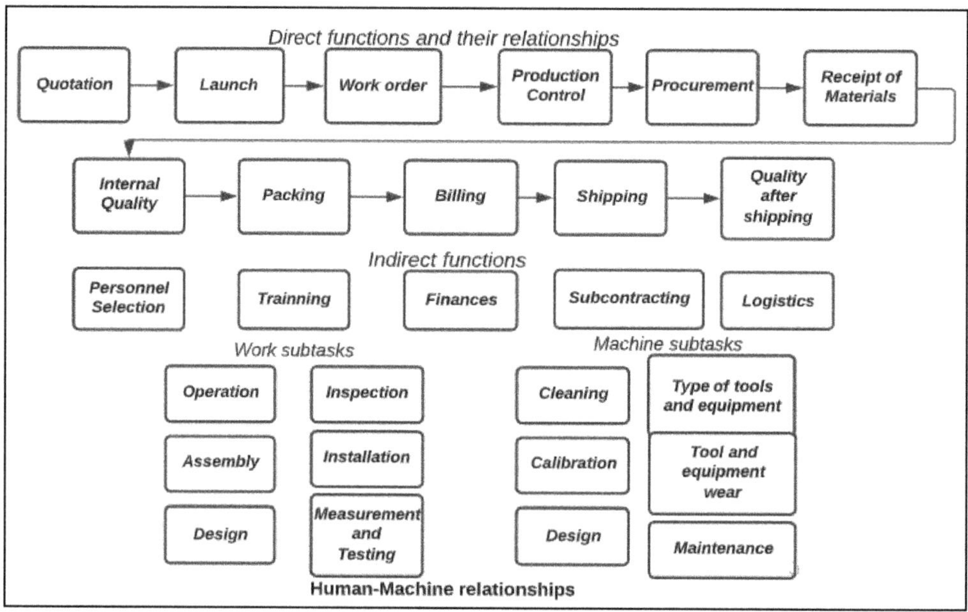

Figure 3. The production system of the human–machine system.

3.2. Phase 2: Human–Machine System

A human–machine system for the machine tool was designed in this phase. The procedure was as follows:

Data collection. To obtain uniform and consistent information, first, a standardization of the terminology used to identify the failures that occurred and the assignment of their causes and sub-causes were performed using the following materials and methodology:

a. An awareness interview was conducted with operators, technicians, designers, programmers, and managers to establish a diagnosis that facilitated the survey's design.
b. The survey was designed and applied to identify the terminology used in the presence of failures, when they occurred in the area, the causes and sub-causes that originated them, and the places where they occurred.
c. Based on the standardized terminology and the survey's results, a relationship matrix was created. Based on the name of the failure and the assigned causes and sub-causes in the different areas, the used terminology was standardized and used to create a catalog of visual aids.
d. A database was designed in excel that records the failures, causes, and sub-causes. They were recorded according to their (standardized) name, production date, part number, and standardized type of failure.

Classification and analysis of data. Once the information was collected in the database and it was standardized, for the analysis of the failures, a classification was performed to identify the causes, classify them, determine the place where they were detected, and determine the sub-causes of the failure. Each failure's classification was carried out with the information from the three participating machining workshops. To carry out this activity, the participation of production, quality, and engineering personnel from the three machine shops considered was required. The procedure followed is described below.

Detection of failures that occurred in the machine tool area. Detection of the failures in the machine tool area in the three machine shops was conducted in order to use only standardized failures. Some examples of these failures are great height, pores, large groove dimension, groove out of tolerance, bad grips, and chopped materials.

Identification of causes and sub-causes of failures and the places where the failures occurred. The failures that occurred in the machining area were analyzed based on their causes, sub-causes, and where they occurred using the root-cause analysis (RCA) [28–31]. The sub-causes of failure were listed and classified by groups or clusters and the place where the failure occurred. As a result, the six sub-causes used in this research are called components and the six places where the failures occurred are called subsystems. With them, a proper and adequate classification was obtained to develop a relationship matrix.

Matrix of component–subsystem relationships. Using the classification of faults through the cluster technique and root-cause analysis (RCA), a relationship matrix between components and subsystems was constructed, identifying the component and subsystem relationship. This matrix is presented in Table 3, showing that Admission Profile and Institutional Conditions/Attitude are present in every one of the six subsystems. Maintenance/Setup, Supplier, Calibration, and Information/Communication only influenced some blocks.

Table 3. Matrix of the component–subsystem relationship.

Subsystems	Sub-Causes/Components					
	Maintenance/Setup	Admission Profile	Institutional Conditions/ Attitude	Information/ Communication	Calibration	Supplier
Machine	X	X	X	X		X
Tool	X	X	X	X		
Process/Operation	X	X	X	X		
Measurement/ Instrumentation	X	X	X	X	X	
Programming/Design		X	X	X		
Material		X	X			X

This matrix resulting from the component–subsystem relationship is the basis of the model of the human–machine system for the machine tool given in Figure 4. Considering Figure 4, a serial–parallel system approach as an adequate configuration to evaluate the system's reliability was developed. It is important to mention that, for this research, the term "reliability" refers to the probability of finding a product without failures for which it must be scrapped.

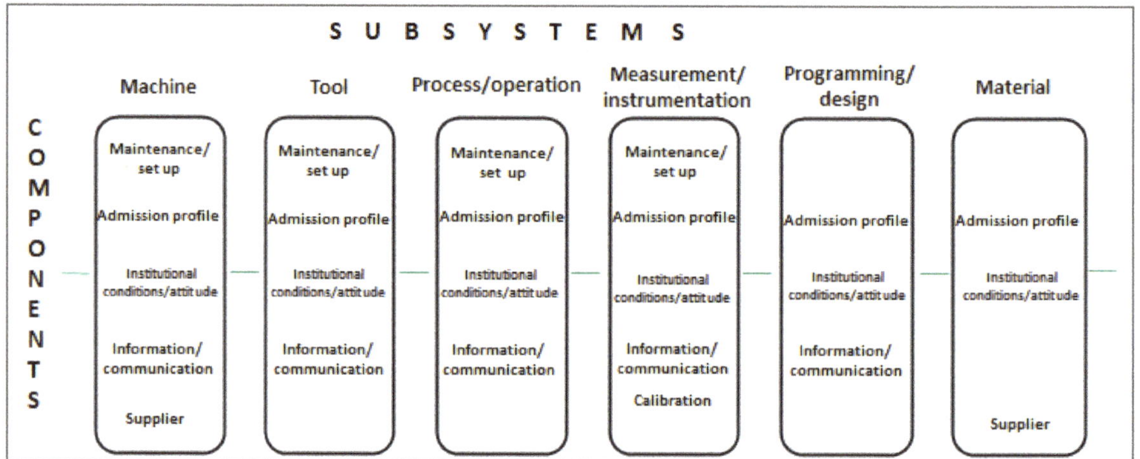

Figure 4. Human–machine system of the machine tool.

3.3. Phase 3: Reliability Evaluation Models

To evaluate the reliability of the human–machine system, some statistical models were formulated using a serial–parallel system approach, some concepts of reliability engineering, and statistics. According to these statistical models, it was necessary to have a database that recorded failures that occurred over time.

Recording of product failures. When a failure is recorded, the sub-cause of failure and the place where the failure occurred are also analyzed. A list of the six components (sub-causes) and subsystems (where the failure occurred) is conveniently included to facilitate this analysis. Table 4 shows the database used to collect the product failure information and its respective analysis. The database from Table 4 was obtained from the scrap report by a machine, and it was helpful in failure control and the identification of component–subsystem relationships.

Standardization and systematization of terminology. The process of standardization of the terminology used to feed the database consisted of the following:

a. An awareness interview was conducted with operators, designers, and managers to establish a diagnosis to help design the survey.

b. The survey was designed and applied in each one of the three machining workshops to identify the terminology used by each person involved in the different functions, the failures found in the products in each area, the causes and sub-causes of the failures, and the places where failures occurred.

c. A matrix of relationships was constructed between the areas involved using the terminology identified in the application of the surveys in the different areas of the machine shops, reducing terms to obtain a standardization of the terminology used.

It is important to ensure (through the commitment of the main actors in each machine shop) that the information-gathering process is carried out periodically and without interruption, and that the analysis of the collected data is constantly monitored (as required by the designed database). It is the responsibility of each machine shop that participates in this activity.

Table 4. Database used to collect and analyze failures that occurred in machine tools.

Date	Machine	Part Number	Failure #	QTY. Failures/Day	Total Failures/Day	Time to Failure	Cause	Subsystem	Component
jan/06	MT	138	37	5			Operation	3	1
jan/06	MT	138	19	2	7	0.73684	Adjustment	3	1
jan/07	MT	138	33	1			Adjustment	3	1
jan/07	RS3	591	16	1			Program setting	3	1
jan/07	TMO	18	33	1			Adjustment	3	1
jan/07	RAB	584	24	1	4	0.42105	Adjustment	3	1
jan/08	RS3	591	16	2	2	0.21053	Program setting	3	1
jan/09	LOP	82	6	1	1	0.10526	Tool selection	3	1
jan/12	MT	954	19	1			Adjustment	3	1
jan/12	RS3	563	19	1	2	0.21053	Adjustment	3	1
jan/13	LM	131	4	1	1	0.10526	Tool selection	2	1
jan/14	R2	121	8	1			Adjustment	3	1
jan/14	R2	246	8	1			Adjustment	3	1
jan/14	TUI	563	21	1	3	0.31579	Mat handling	3	3

Determination of the periodicity of the analysis of the information. We previously selected the periodicity with which the analysis of the information collected would be carried out (monthly, weekly, daily, etc.), depending on the response time we wanted to obtain and the amount of data available for the analysis and evaluation of reliability.

Evaluation of the reliability using a statistical model. For this research, reliability is understood as the probability that the man–machine system works appropriately or does not produce defective parts after some time. The basis of the reliability evaluation is the defective parts that have been manufactured in the machine tool area. The subsystems considered are machines, tools, processes, measurement/instrumentation, programming/design, and materials, and the components considered are maintenance, admission profile, institutional conditions/attitude, information/communication, calibration, and supplier.

Evaluation of the reliability of a component–subsystem combination. The first step in evaluating the reliability of a component–subsystem combination is to perform a goodness-of-fit test to determine the theoretical distribution that best describes the failure's behavior and its parameters. Here, the parameter estimation was performed by using the Weibull ++ Software. For data that follow a Weibull distribution, the reliability of each component–subsystem combination is evaluated by using Equation (1a). For data that follow an exponential distribution, their reliability is evaluated by using Equation (2).

$$R(t) = e^{-\left\{\frac{t}{\eta}\right\}^{\beta}} \tag{1a}$$

In the Weibull reliability function given in Equation (1a), $R(t)$ is the reliability index, β is the shape parameter, and η is the scale parameter. In Equation (1b) is given the instantaneous Weibull time-dependent risk function.

$$h(t) = \frac{\beta}{\eta}\left(\frac{t}{\eta}\right)^{\beta-1} \tag{1b}$$

From Equation (1b), notice that because, for the Weibull distribution, the instantaneous risk is dependent on time, then any failure mode whose probability depends on time should be modeled by using the Weibull distribution. In particular, observe that the time-dependent behavior implies that the event is generated by a non-homogeneous Poisson process (see [Rine] for details), and, consequently, the exponential distribution should not be used to represent time-dependent events. The exponential distribution does not depend on time, and it is generated by a homogeneous Poisson process. In Equation (2), the exponential reliability is given, where λ is the non-time-dependent risk function.

$$R(t) = e^{-\lambda t} \tag{2}$$

Evaluation of the reliability of a subsystem. The evaluation of the reliability of a subsystem affected by one or more components that participated in the failure of the product assumes that components have a series configuration within the subsystem, as they are independent and not mutually exclusive. The statistical model of Equation (3) is used to evaluate the reliability of a subsystem with a series configuration. In Equation (3), Rs is the reliability of the subsystem, and R_i is the reliability of the component–subsystem relationship.

$$R_s = \prod_{i=1}^{n} R_i \tag{3}$$

Evaluation of the reliability of the human–machine system. To evaluate the reliability of the human–machine system, the assumption is made that the subsystems have a parallel configuration since they are independent and mutually exclusive. Using the statistical model of Equation (4), the reliability index of the human–machine system is evaluated. It is worth mentioning that the subsystems that do not have any component affecting them will have a reliability of 1 and will not affect the calculation of the system's reliability, given by

$$R_S = 1 - \prod_{i=1}^{n}(1 - R_i) \tag{4}$$

In Equation (4), R_S is the reliability of the man–machine system, and R_i is the reliability of the subsystem. The equation for its calculation assumes a parallel system. The reliability index obtained by the statistical models represented by Equations (1)–(4) represents the probability of not finding defective parts to be scrapped after a period t. Depending on the reliability obtained, it may be necessary to identify the lowest reliability presented in the system and perform improvement actions, such as those presented in the next section.

3.4. Phase 4: Improvement

In this phase, two methods that can help to improve the reliability of the human–machine system in the machine tool area are applied. These methods depend on the type of origin of the failures that occurred and their effect on the system's reliability. Failures were analyzed to identify their origin. A case study was used for this purpose, so the data that appear were obtained from a machine shop.

Origin and contribution of failures. The analysis consists of a classification of kinds of failures by origin using a cluster analysis to detect the kinds of origins of failures in the machine tool area according to similar characteristics of each identified failure that occurred in the machine tool.

Table 5 shows this classification. Table 5 shows that the failure origins identified were mechanical, human error, heat treatment, material, and unexpected event and that failures of a mechanical origin were most common in the machine tools.

Table 5. Classification of kinds of failure by origin.

Mechanical	Mechanical	Mechanical	Human Error	Heat Treatment	Material	Unexpected Event
3	15	31	1	43	26	42
4	18	35	2	49	30	45
5	19	37	10	50	33	
6	20	38	16	51	34	
7	21	40	17	52	58	
8	22	41	27	53		
9	23	46	32	55		
11	24	47	36	56		
12	25	48	39	57		
13	28		44			
14	29		54			

The frequency of each kind of failure by origin is presented in Table 6. Through the Pareto Diagram shown in Figure 5, we identified the sources of failure that most affected their appearance and made the greatest contribution to or had the highest impact on the reliability of the man–machine system. Table 6 shows that the origin with the highest number of failures is the mechanical type, with 31 out of 58 (53.54%). Using data from Table 6, a Pareto Diagram of the origins of failures and their contribution from Figure 5 was drawn. In Figure 5, failures of mechanical and human-error origin are vital causes, with 72.4% of the contribution. In contrast, failures of heat treatment, material, and other origin were identified as trivial causes, with a 27.6% contribution to the failures in the products of the machine tool area. After identifying that the failures of both mechanical and human-error origin contribute most to the appearance of failures in the products, two improvement methods, namely monitoring the parameters β and η of the Weibull distribution for time-dependent failures, such as those of a mechanical origin, and monitoring the parameter λ of the exponential distribution for non-time-dependent failures, were developed.

Table 6. Frequency of kinds of failures by origin.

Origin	Qty. of Failures	Relative	Cumulative
Mechanical	31	53.45%	53.45%
Human Error	11	18.97%	72.42%
Heat treatment	9	15.51%	87.93%
Material	5	8.62%	96.55%
Unexpected event	2	3.45%	100.00%
Total	58	100.00%	

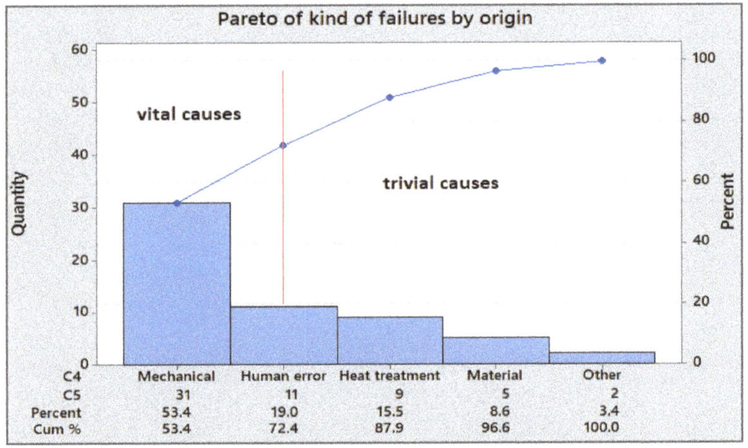

Figure 5. Pareto diagram of the origins of failures and their contribution.

Monitoring of Parameters between Two Sequential Periods

The objective of monitoring the parameters is to identify the lowest reliabilities in the human–machine system during two sequential periods. A case study from a machine shop was carried out to develop this method. The steps were as follows:

Component–subsystem relationships present in failures. This refers to the time to failure generated by a specific component–subsystem combination from a data distribution, either Weibull or exponential, with their corresponding parameters. Tables 7 and 8 show the parameters for each component–subsystem relationship presented in the failure analysis for the period from 25 January to 30 January 2021 and for the period from 17 May to 28 May 2021, respectively. These parameters were helpful in evaluating the reliability of each component–subsystem relationship and they were obtained using the software Weibull [++].

Table 7. Component–subsystem relationships presented in the failure analysis and their parameters from 25 to 30 January 2021.

Combination	Weibull		Exponential
	β	η	Λ
1, 1	1	3.1667	
1, 6	1	9.5	
2, 1	1	9.5	
3, 1	1.0688	3.1268	

Table 8. Component–subsystem relationships presented in the failure analysis and their parameters from 17 May to 28 May 2021.

Combination	Weibull		Exponential
	β	η	Λ
1, 1	3.4615	7.9737	
1, 6	2.18399	7.19726	
2, 1	1.52633	5.32846	
3, 1	1.90918	2.34728	
3, 3			1/9.5

Evaluation of the reliability of component–subsystem relationships. To evaluate the reliability from the period of 25 January to 30 January 2021, the statistical model represented by Equation (1) (Weibull distribution) was used because only failures of mechanical origin were present. To evaluate the reliability from the period of 17 May to 28 May 2021, both the Weibull distribution (Equation (1a)) and the exponential distribution (Equation (2)) were used because there were failures of mechanical and human-error origin. Table 9 shows the reliability indices for the component–subsystem relationships present in the machine tool during the period of 25 January to 30 January 2021 for five time periods considered within the 9.5-h shift.

Table 9. Reliability for the component–subsystem relationships during five time periods from 25 to 30 January 2021.

Periods	$R_{1,1}$	$R_{1,6}$	$R_{2,1}$	$R_{3,1}$	$R_{3,3}$
1 h	0.7292	0.9	0.9	0.645	0.7292
3 h	0.3877	0.7292	0.7292	0.2683	0.3877
5 h	0.2061	0.5907	0.5907	0.1111	0.2061
7 h	0.1096	0.4786	0.4786	0.0464	0.1096
9 h	0.0583	0.3877	0.3877	0.0193	0.0583

Table 10 shows the reliability indices for the component–subsystem relationships present in the machine tool during the period of 17 to 28 May 2021 for five time periods considered within the 9.5-h shift.

Table 10. Reliability for the component–subsystem relationships during five time periods from 17 to 28 May 2021.

Periods	$R_{1,1}$	$R_{1,6}$	$R_{2,1}$	$R_{3,1}$	$R_{3,3}$
1 h	0.6478	0.7382	0.7509	0.4437	0.9000
3 h	0.2718	0.4023	0.4234	0.0871	0.7292
5 h	0.1141	0.2193	0.2387	0.0171	0.5907
7 h	0.0478	0.1195	0.1346	0.0033	0.4786
9 h	0.0200	0.0651	0.0759	0.0006	0.38776

Evaluation of the reliability of the subsystems and the man–machine system. Using Equation (3), which considers a series configuration, the reliabilities of the subsystems that contributed to the failures were evaluated. Considering Equation (4), which considers a parallel configuration, the reliability of the man–machine system was evaluated. Table 11 shows the reliability indices of the subsystems that contributed to the failures and the reliability indices of the man–machine system for five time periods considered within the 9.5-h shift during the period from 25 to 30 January 2021.

Table 11. Reliability of the subsystems and the man–machine system during five time periods for the period from 25 to 30 January 2021.

Periods	R_1	R_2	R_3	R_S
1 h	0.6563	0.9	0.645	0.9878
3 h	0.2827	0.7292	0.2683	0.8579
5 h	0.1218	0.5907	0.1116	0.6807
7 h	0.0524	0.4786	0.0464	0.5289
9 h	0.0226	0.3877	0.0193	0.4131

Table 11 shows that, in the 7-h period after the start of the shift, in subsystem 1 (machine), the probability that there were no failures for scrap was 5.24%, while in subsystem 2 (tool) it was 47.86% and in subsystem 3 (process/operation) it was 4.64%. The lowest reliability was obtained for subsystem 3. Regarding the reliability of the man–machine system, the analysis showed a probability of 52.89% that no scrap failures would occur in the 7-h period after the shift started.

Table 12 shows the reliability indices of the subsystems present in the failure analysis and the reliability indices of the man–machine system for five time periods considered within the 9.5-h shift during the period from 17 May to 28 May 2021. Table 12 shows that in the 7-h period after the start of the shift, in subsystem 1 (machine), the probability that there were no failures for scrap was 0.57%, while in subsystem 2 (tool) it was 13.46% and in subsystem 3 (process/operation) it was 0.16%.

Table 12. Reliability of subsystems and the man–machine system during five time periods for the period from 17 to 28 May 2021.

Periods	R_1	R_2	R_3	R_S
1 h	0.4782	0.7509	0.3990	0.9219
3 h	0.1094	0.4234	0.0635	0.5191
5 h	0.025	0.2387	0.0101	0.2653
7 h	0.0057	0.1346	0.0016	0.1409
9 h	0.0013	0.0759	0.0002	0.0773

The lowest reliability was obtained for subsystem 3. Regarding the reliability of the man–machine system, the analysis showed a probability of 14.09% that no scrap failures would occur in the 7-h period after the shift started. Once the lowest reliabilities of the subsystems (process/operation) and the components (maintenance/setup) were identified, to determine which component or components had an influence on the low reliability, the causes that generated it, the origin of these causes, and the actions to be taken, we performed a FMEA analysis. The recommended actions were documented as a work instruction. This action is described below.

Failure Mode and Effect Analysis (FMEA). The analysis developed in the previous section for the case study yielded the causes that generated the low reliability of the subsystem (process/operation) and the components (maintenance/setup and institutional conditions/attitude). With this information as an input, we generated a FMEA analysis with a work team of two operators, one maintenance technician, and the production supervisor. Table 13 shows the FMEA analysis used to control and monitor the activities that affect the reliability of the man–machine system.

Table 13. FMEA analysis used to control and monitor activities.

Activity	Potential Failure Mode	Potential Effect of Failure	Sev	Potential Causes/Failure Mechanism	Oc	Current Process (Detection)	Det	RPN
Check the condition of the tool.	Tool in bad condition/damaged	Excess of burrs	3	Inability to calculate tool life expectancy.	2	None	2	12
			3	The machinery for sharpening is not available	2	None	5	30
Perform set-up of tools.	Wrong tool	Scrap	7	Operator's error	2	None	3	42
Load the material into the machine.	Incorrect material	Scrap	7	Warehouseman's error	1	None	7	49
	Incorrect location	Scrap	6	Operator's error	2	None	2	24
Assembly, welding, and polishing.	Pores	Noncompliance with specifications (error/scrap)	7	Material contamination	5	None	2	70
			9	Tungsten contamination	5	None	2	90
			7	Incorrect gas flow	3	None	2	42
			7	Fixture dirt	5	INS204 base assembly and fabrication	2	70

Table 13 presents some activities identified as being of a mechanical and human-error origin. The corresponding activity for assembling, welding, and polishing, whose defect was the presentation of porous pieces due to contaminated tungsten, was the one that presented the highest NPR value. This particular action was monitored and controlled, as a work instruction, to prevent recurrence as shown in Figure 6. The work instruction presented in Figure 6 was used to reduce or eliminate failures due to this failure mode and evaluate the system's improvement.

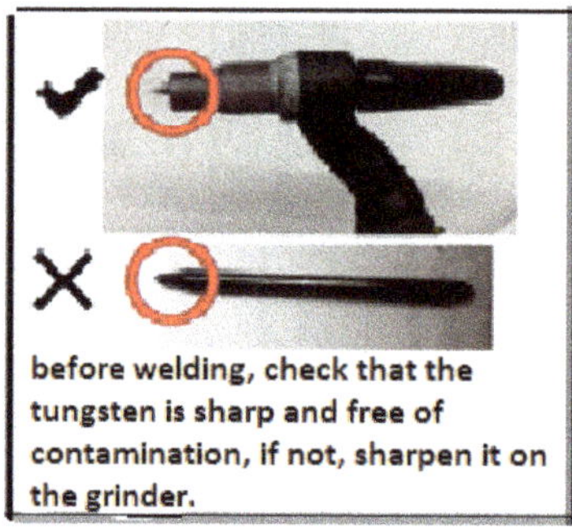

Figure 6. Work instruction to avoid the recurrence of the failure. Source: SOMSAC.

Monitoring of the parameters β and η and the parameter λ for activities with critical failures. The relationship between the reliability index and the Weibull β and η parameters shown in Equation (1a) is unique; for fixed t and $R(t)$ values, unique β and η values exist. For activities that depend on time, addressed here as failures generated by a mechanical source, we propose to monitor the system's reliability by monitoring the corresponding β and η parameters. Therefore, since the parameter β represents the dispersion of the logarithm of the failure times, then when its value is lower compared with the previous period, the log-dispersion among the failure times will increase, implying that no improvement was made; that is to say, the higher the β value the lower the log-dispersion. Similarly, since η represents the product strength, then the higher the η value the better the product strength, implying that if its value is lower compared with the previous period, no improvement was made.

On the other hand, for failures generated by human errors, by considering that human errors do not depend on time, their reliability was monitored by monitoring the λ parameter of Equation (2). Consequently, because λ is the inverse of the mean time to failure, if its value is lower compared with the previous period, then this implies that no improvement was made.

We started from the identification of the critical problem (FMEA analysis), as in the case of the cutting tool. Referring to the unidentified wear of the edge, we propose that an analysis of the life–strength relationship be carried out and a model be generated as a way of solving the problem. A specific material, a cutting tool, and a process/machine that were identified as critical were considered. From Table 14, the critical wear failure of the edge of the cutting tool was considered, whose NPR value is 30. Thus, to determine the parameters β and η, the minimum and maximum forces generated by the machine were measured. Then, by using the method proposed in [32], the Weibull shape (β) and scale (η) parameters were determined based only on the observed maximal (σ_1) and minimal (σ_2) applied effort. The method's efficiency was based on the following facts:

(1) The square root of σ_1/σ_2 represents the base life on which the Weibull lifetimes are estimated;
(2) The mean of the logarithms of the expected lifetimes ($g(x)$) is completely determined by the determinant of the analyzed stress matrix;
(3) The Weibull distribution is a circle centered on the arithmetic mean (μ), and it covers the entire span of the principal stresses;
(4) σ_1/σ_2 and $g(x)$ completely determine the σ_{1i} and σ_{2i} values, which correspond to any lifetime in the Weibull analysis; and
(5) σ_1/σ_2 and η completely determine the minimal and maximal lifetime, which correspond to any σ_{1i} and σ_{2i} values. Additionally, the β and η parameters are used when the stress is either constant or variable.

Table 14. Machine efforts were recorded for roughing with the cutting tool according to the number of cycles per part.

C	9	18	27	36	45	54	63	72	81	90	99	108	117
X	7–16	7–21	7–19	4–19	6–17	5–17	4–13	4–17	5–10	4–14	4–19	4–12	2–19
Y	5–10	5–12	4–11	4–9	3–10	3–10	2–10	4–9	2–10	4–13	3–10	3–10	2–10

C, cycles per piece; X, Effort S (%); Y, Effort S (%).

Life–effort relationship analysis as a method for improving the reliability of the man–machine system in the machining area. The data in Table 14 aim to generate a model of the life–effort relationship between the cutting tool used in the manufacturing process of different steel parts in the CNC machining center. The model was built as follows. The % strength represents the excess effort to which the machine is subjected in order to carry out the operation on the part for the x and y axes.

Table 15 shows the values captured from the screen of the machining center board referring to the effort generated in the z axis for drilling the pieces with a 3/8" drill according to the number of accumulated cycles.

Table 15. Machine efforts recorded for the z axis, for drilling, according to the number of cycles per part.

C	9	18	27	36	45	54	63	72	81	90	99	108	117
Z	27–29	27–29	26–28	26–28	27–29	26–28	26–28	27–29	22–28	27–29	27–29	27–29	27–29

C, cycles per piece; Z, Effort S (%).

Table 16 shows the values captured from the screen of the machining center board referring to the efforts generated in the z axis for drilling the pieces with a 1/4" center drill according to the number of accumulated cycles.

Table 16. Machine efforts recorded for the z axis, for drilling, according to the number of cycles per part.

C	1	2	3	4	5	6	7	8	9	10	11	12	13
Z	23–25	23–25	24–26	23–25	23–25	24–26	23–24	25–28	23–28	25–29	25–29	24–29	22–25

C, cycles per piece; Z, Effort S (%).

A comparison among Tables 14–16 shows that efforts presenting the most variation occur when roughing the part on the x axis, which range from 4% to 21%. Second, there are efforts in roughing the part on the y axis, which range from 2% to 13%. The efforts exerted on the z axis to drill the pieces, with two types of drill bits, range from 22% to 29%.

Considering the most significant variation in the effort values, corresponding to that exerted on the x axis (minimum 4% and maximum 21%), these values were used to calculate the parameters β and η given in Table 17 according to [28]. Additionally, the 95% confidence interval was constructed in order to monitor the maximal machine resistance. Tables 5–13 are given In Table 17.

Table 17. Calculation of the β and η parameters using the minimum and maximum values of resistance.

i	Y	R(t)	toi	Applied Effort	Minimal Resistance
1	−3.4034833	0.9673	0.06751	0.619	135.764
2	−2.491662	0.9206	0.13899	1.274	65.942
3	−2.0034632	0.8738	0.20460	1.875	44.796
4	−1.6616459	0.8271	0.26821	2.458	34.172
5	−1.3943983	0.7804	0.33143	3.038	27.653
6	−1.1720537	0.7336	0.39525	3.622	23.189
		0.704	0.43644	4.000	21.000
7	−0.9793812	0.6869	0.46040	4.220	19.907
8	−0.8074473	0.6402	0.52756	4.835	17.373
9	−0.6504921	0.5935	0.59739	5.475	15.342
10	−0.5045088	0.5467	0.67061	6.146	13.667
11	−0.3665129	0.5	0.74806	6.856	12.252
12	−0.2341223	0.4533	0.83075	7.614	11.032
13	−0.1052851	0.4065	0.92000	8.432	9.962
	0	0.3679	1.00000	9.165	9.165
14	0.0219284	0.3598	1.01752	9.326	9.007
15	0.1495258	0.3131	1.12572	10.317	8.142
16	0.279845	0.2664	1.24811	11.439	7.343
17	0.4159621	0.2196	1.39018	12.741	6.593
18	0.562502	0.1729	1.56126	14.309	5.870
19	0.7276158	0.1262	1.77937	16.308	5.151
20	0.9293107	0.0794	2.08757	19.133	4.390
21	1.2296598	0.0327	2.64818	24.271	3.461

Side calculations:

Max=	21
Min=	4
Beta=	1.26264954
Eta=	9.16515139

From the maximum range of values:

3.146	Standard deviation
1.4353	Variation

22.435	Upper limit for resistance
19.565	Lower limit for resistance

The above-mentioned analysis was performed by using the Weibull parameters fitted from the data as

$$\beta = \frac{-4\mu y}{0.0954 \ln(\sigma_1 / \sigma_2)} \tag{5}$$

$$\eta = \sqrt{\sigma_1 * \sigma_2} \tag{6}$$

where σ_1 and σ_2 are the maximum and minimum efforts at which the machine is performing, and μy is the mean of the median rank approach estimated from the elements of the Y vector determined by using the sample size from Piña-Monarrez et al., [32] given as

$$\mathbf{n} = \frac{-1}{ln(R(t))} \tag{7}$$

The n elements of the Y vector used to estimate μy were determined based on the median rank approach as

$$Y_i = \ln\left(-\ln\left(1 - \frac{(i - 0.3)}{(n + 0.4)}\right)\right) \tag{8}$$

From these Y elements, the corresponding standardized base lifetime t_{0i} values that allowed us to determine both the applied effort and its corresponding minimal resistance (see Table 17) are given as

$$t_{0i} = \exp\left\{\frac{Y_i}{\beta}\right\} \tag{9}$$

Therefore, from Equations (5), (6), and (9) the applied effort and the minimal resistance are given as

$$Applied\ effort = \eta * t_{0i} \tag{10}$$

$$Minimal\ resistance = \eta / t_{0i} \tag{11}$$

Finally, the mean and standard deviation for the maximum values of resistance and the corresponding 95% confidence interval are given as

$$\mu_{max} = \sum_{i=1}^{n} \frac{max\ values\ of\ resistance}{n} \tag{12a}$$

$$\sigma_{max} = \sqrt{\sum_{i=1}^{n}\left(\frac{maxvalue\ of\ resistance - \mu_{max}}{n-1}\right)^2} \tag{12b}$$

$$CI = \mu_{max} \pm \left(\frac{1.96}{\sqrt{n}}\right) \tag{13}$$

As shown Table 17, the corresponding reliability indicator was established for each machine resistance. For a machine that has a minimum resistance of 21% over its nominal value, it has a reliability of 0.704 (the minimum resistance that the machine must have to have this reliability). When the effort is 9.1651 (η) and the process has a resistance of 21, the reliability is 0.7040. When the resistance is higher, the reliability is higher too. This method consists of monitoring the critical value of 21% through the confidence interval, hoping that the maximum value of the machine effort does not exceed the upper limit of 22.4353. It is also possible to evaluate whether the maximum efforts produced by the machine exceed the standard efforts considered in the confidence interval.

4. Results

Phase 1: Operational Context. The importance of the participation of different functions and activities developed by operators and technicians in the machine tool, the responsibility they have when performing their functions, as well as the possession of sufficient knowledge, skills, and abilities to efficiently and effectively develop the use of conventional and semi-automated CNC machines and tools, without forgetting the fundamental aspects of safety, formed the fundamental basis for establishing the production system of the machine tool. Regarding the Production System of the machine tool, Figure 3 helped to identify the relationships between functions and activities involved in the production system and made it possible to identify the human–machine interactions within the production system, which were considered when establishing the human–machine system.

Phase 2: Human–Machine System of the machine tool. The component–subsystem relationship is the basis of the model shown in Figure 4. Admission Profile and Institutional Conditions/Attitude were present in every one of the six subsystems. Maintenance/Setup, Supplier, Calibration, and Information/Communication only influenced some blocks. The serial–parallel configuration of this system helped us analyze and assess the system's reliability.

Phase 3: Reliability evaluation models. With the evaluation of the reliability indices for the component–subsystem relationships, comparing Tables 10 and 11, the component–subsystem relationships with the lowest reliability values were 3,1 and 1,1, referring to Process/Operation Maintenance/Setup and Machine Maintenance/Setup, respectively, being Maintenance/Setup the common component affecting the reliability. Regarding the evaluation of the reliability for subsystems, Table 12 shows that in the 7-h period after the start of the shift, in subsystem 1 (machine), the probability that there were no failures for scrap was 5.24%, while in subsystem 2 (tool) it was 47.86% and in subsystem 3 (process/operation) it was 4.64%. The lowest reliability was obtained for subsystem 3. Table 13 shows that in the 7-h period after the start of the shift, in subsystem 1 (machine), the probability that there were no failures for scrap was 0.57%, while in subsystem 2 (tool) it was 13.46% and in subsystem 3 (process/operation) it was 0.16%. The lowest reliability was obtained for subsystem 3. Subsystem 3 (process/operation) was the most common subsystem affecting the reliability. Regarding the reliability of the man–machine system, Table 12 shows a probability of 52.89% that no scrap failures would occur in the 7-h period after the shift started. Regarding the reliability of the man–machine system, Table 13 shows a probability of 14.09% that no scrap failures would occur in the 7-h period after the shift started, with the second period having the lowest human–machine system reliability.

Phase 4: Improvement. Table 14 (FMEA) presents the corresponding activity for assembling, welding, and polishing, whose defect is the presentation of porous pieces due to contaminated tungsten, which presented the highest NPR value. A comparison between Tables 15–17 shows that the efforts presenting the most variation occurred when roughing the part on the x axis, which range from 4% to 21%. Second, there are efforts in roughing the part on the y axis, which range from 2% to 13%. The efforts exerted on the z axis to drill the pieces, with two types of drill bits, range from 22% to 29%. Considering the most significant variation in the effort values, corresponding to that exerted on the x axis (minimum 4% and maximum 21%), the corresponding reliability indicator was established for each machine resistance as shown in Table 17. For a machine that has a minimum resistance of 21% over its nominal value, it has a reliability of 0.704 (the minimum resistance that the machine must have to have this reliability). When the effort is 9.1651 (η) and the process has a resistance of 21, the reliability is 0.7040. When the resistance is higher, the reliability is higher too. The proposed method consists of monitoring the critical value of 21% through the confidence interval, hoping that the maximum value of the machine effort does not exceed the upper limit of 22.4353. Table 17 shows the maximum resistance that the machine must have to obtain a certain degree of reliability. It is also possible to evaluate whether the maximum efforts produced by the machine exceed the standard efforts considered in the confidence interval.

5. Discussion

The results obtained were congruent with the hypotheses formulated. From them, one hypothesis was raised: the man–machine system of the machining area is defined from the component–subsystem relationship if the non-conforming parts present in the machine tool have been considered, analyzed, and classified for a specific period. Its validation was supported by the structured analysis of the collected information, which allowed us to obtain the conditions to be analyzed with the series–parallel system approach. Another hypothesis raised was: the reliability of the man–machine system can be accurately evaluated from the component–subsystem relationship if the used statistical model considers the subsystems and the component–subsystem relationship that have a series–parallel system configuration. Monitoring the reliability of components and subsystems and probability distribution parameters improves the reliability of the man–machine system of the machining area if it is complemented with tools for the identification of failure modes, such as the FMEA, and actions are taken based on it.

In our case, we presented a methodology to evaluate and improve the reliability of the man–machine system in a machine tool workshop by considering the system characteristics presented in workshops of Chihuahua city. Our proposed methodology can be used to conveniently assess the component–subsystem relationship, the subsystems, and the system to detect the lowest reliability. These allow us to reduce the scrap rate and to improve the system's reliability. Regarding the evaluation of the reliability of a man–machine system, the tools used by other authors do not consider the establishment of an evaluation model based on a man–machine system designed with the characteristics of a specific machine tool area that aims to reduce scrap. Additionally, the tools proposed to improve the man–machine system are based on the assumptions of the Weibull and exponential distributions for the analysis of failures of mechanical and human-error origin as they were the most common kinds of origin.

Other authors have applied the Weibull and exponential distributions in different contexts, methodologies, and tools from the perspective of reliability. Their research work is explained below in order to highlight their different contributions that contrast with the contribution of this research work, both in the context of the application and in the methodology used, but having in common the process reliability perspective. Zhang, Xie, and Tang [33] presented a study of efficiency using robust regression methods over the ordinary least-squares regression method based on a Weibull probability plot. The emphasis was on the estimation of the shape parameter of the two-parameter Weibull distribution. Both the case of small data sets with outliers and the case of data sets with multiple-censoring were considered. Maximum-likelihood estimation was also compared with linear regression methods. Simulation results showed that robust regression is an effective method for reducing bias and performs well in most cases.

Flores, Torres, and Alcaraz [34] analyzed three technological alternatives to produce electricity and compress gas in offshore crude oil processing facilities to be installed in Mexico. The comparison of alternatives was performed based on system reliability estimations by using the "reliability block diagram" method. The fundamental concepts of systems reliability theory were pointed out, and the reliability model was defined as a parallel arrangement with redundancy in passive reserve and without maintenance for any component.

Assuming that Weibull time-to-fail distributions cannot be correctly estimated from field data when manufacturing populations from different vintages have different failure modes, Rand and McLinn [35] investigated the pitfalls of ongoing Weibull parameter estimation and analyzed two cases based upon real events. Assessment of the mixed population at each month of calendar time resulted in an increasing Weibull shape parameter estimate at each assessment. When the two populations were separated and estimated properly, a better fit with more accurate estimates of Weibull shape parameters resulted.

Pascual and Zhang [36] proposed control charts for monitoring changes in the Weibull shape parameter β. These charts were based on the range of a random sample from the smallest extreme value distribution. The control chart limits depended only on the sample size, the desired stable average run length (ARL), and the stable value of β. They derived control limits for both one- and two-sided control charts. They were unbiased with respect to the ARL. Pascual and Zhang discuss sample size requirements when estimating the stable value of β from past data. The proposed method was applied to data on the breaking strengths of carbon fibers. For this case, the authors recommended one-sided charts for detecting specific changes in β because they were expected to signal out-of-control sooner than the two-sided charts.

Guevara, Valera-Cárdenas, and Gómez-Camperos [37] proposed a methodology to assess the reliability factor in the management of industrial equipment designing. The complexity of several industrial procedures and the equipment required establishes that not all of an asset's failure patterns may be easily handled through maintenance service activities done after its manufacture and use. To avoid all kinds of high-impact failures when using the product, there must be a stage of elimination by removing some maintenance needs, and it should be done considering their own foundations from the very first moment when the asset is designed and produced. These failures might emerge in production, quality, safety, the environment, and costs, among others, which are hard to identify and control. All this requires different concepts and tools in industrial maintenance and service and the engineering of reliability during the design phase.

Piña-Monárrez [38] proposed conditional Weibull control charts using multiple linear regression. Because, in Weibull analysis, the key variable to be monitored is the lower reliability index ($R(t)$), and because the $R(t)$ index is completely determined by both the lower scale parameter (η) and the lower shape parameter (β), a pair of control charts to monitor a Weibull process were proposed based on the direct relationships between η and β with the log-mean (μx) and the log-standard deviation (σx) of the analyzed lifetime data. Moreover, because of the fact that, in Weibull analysis, right-censored data are common, and because they introduce uncertainty into the estimated Weibull parameters, in the proposed charts, μx and σx are estimated based on the conditional expected times of the related Weibull family. After that, both μx and σx are used to monitor the Weibull process. In particular, μx was set as the lower control limit to monitor η, and σx was set as the upper control limit to monitor β. Numerical applications were used to show how the charts work.

Ferreira and Silva [39] analyzed the parameter estimation for the Weibull distribution with right-censored data using the EM algorithm. Maximum-likelihood estimation (MLE) is a method for estimating the parameters of a statistical model for given data set. This method allowed the authors to estimate the unknown parameters of a statistical model. These parameters were obtained by maximizing the likelihood function of the model in question. The estimation of the parameters of the Weibull distribution by the maximum-likelihood method based on information from a historical record with right-censored data showed this difficulty. The solution used the Expectation Maximization (EM) algorithm.

Although the Weibull distribution for B = 1 mimics the exponential distribution, both distributions are different and behave differently. This fact is due to the Weibull distribution being generated by a non-homogeneous Poisson process in which the risk depends on time, and the exponential distribution is based on a Poisson process in which the risk does not depend on time (a lack of memory property). Based on this fact, we used the Weibull distribution for components and subsystems whose failure mode is mechanical. Due to the mechanical stress that accumulates damage over time, the failure risk depends on time as in the Weibull distribution (or the non-homogeneous Poisson process). On the other hand, by considering that human errors do not depend on time, we used the exponential distribution for components and subsystems whose failure mode is human error (the Poisson process).

This is an innovative contribution because it is based on the participation of the failures in different machining areas. The man–machine system proposed was designed based on the causal relationship between the failures occurring in different machining areas, where the components represent the sub-causes of failures and the subsystems represent the places where failures occur. The series–parallel configuration given to this system allowed us to find statistical models to evaluate the reliability of the component–subsystem relationship (Weibull and exponential distributions), of the subsystems (serial configuration), and of the man–machine system (parallel configuration). No man–machine systems were found to have been configured in a serial–parallel manner based on the participation of the failures.

The Weibull and Exponential distributions were used due to the fact that, at the beginning of the research, it was assumed that the random variables involved in the failures within the man–machine system had a Weibull distribution and an exponential distribution. As the study developed, it was found that failures of mechanical and human-error origin were the most frequent occurrences in the machining area, proving these assumptions to be correct. The Weibull distribution is suitable for representing failures of mechanical origin because they are time-dependent, and the exponential distribution represents failures of human-error origin, which are not time-dependent.

To monitor two sequential periods that have an exponential distribution in one of their component–subsystem relationships, it is only a matter of checking whether their reliability has improved or not and to act on the affected component. This is not the case with the Weibull distribution, where its parameters must be monitored to check whether the reliability has improved with the actions taken. Some proposed improvement actions, such as the use of the life–effort model, may help when the failure is caused by cutting tool wear.

Because, in Weibull analysis, the shape parameter β characterizes the aging of the system, then in the Weibull reliability analysis the used β value plays a key role in accurately determining the reliability $R(t)$ of the analyzed product. Moreover, because in practice β cannot be accurately estimated, then it is generally selected from historical data or from a baseline (a similar product). The estimation of the Weibull scale parameter η depends on both the estimated β value and on the desired $R(t)$ index. The η parameter of the stress distribution is estimated by using $R(t) = 0.9535$ with n as an integer.

6. Conclusions

The failure analysis of a man–machine system is flexible because it allows us to include any quantity of components, systems, and component–subsystem relationships for the analyzed period. Once the man–machine system had been designed, using the series–parallel system approach proposed to assess the reliability of the man–machine system turned out to be efficient. The man–machine system and the statistical models can be adapted to any company in the metal-mechanic sector with low-volume and high-mix production whose operations are carried out on conventional and CNC machine equipment that utilizes the characteristics of batch production. The man–machine system reliability indicators provide useful information that can be monitored over time. Due to the unique relationship between the Weibull parameters and the reliability index, monitoring the Weibull parameters is an effective way to improve the reliability of the system, the components, and the subsystems.

Limitations of the proposed methodology include the lack of records related to failures and, therefore, the analysis of data. In some cases, the same failure was registered with a different name (standardization) and the data collection process was not continuous (systematization). The information in this paper can be used as the basis for the development of a process to improve the reliability of machine shops considering the lower reliability indices.

Considering that Industry 4.0 works together with machines in innovative and highly productive ways, and due to the dependence of the reliability evaluation on the ability of the operator, the technician, and the programmer to determine the causes and sub-causes of failure, to strengthen this research it would be convenient to automate the information captured in the databases as well as the monitoring of improvement actions. Industry 4.0 is an area of great entrepreneurship, where technologies are used to promote new models. Industry 4.0 is also a space of new opportunities, creativity, and innovation. In response to the problems presented by the machining workshops in Chihuahua City, the digitalization of connected Industry 4.0 is a good option because it means lower production costs, faster lead times, and greater responsiveness to customer demand. Customer expectations have changed due to the advent of connected devices and platforms. The needs in production volume have changed from mass production to customized production, which is happening at an accelerated pace. The introduction of digital technologies in manufacturing processes implies the need for systems to operate and manage broadband information and IT infrastructure, as well as buildings and traffic systems. Digital transformation of the industrial sector with automation, data sharing, uploading to the cloud, Big Data, artificial intelligence, the Internet of Things, and technological techniques can be used to achieve smart industrial and manufacturing goals by interacting with people, new technologies, and innovation. The basis of our proposal is the analysis of the failures that occur in the machine tool, which is obtained from a database containing Big Data. Then, automation of the database with the manufacturing processes of the company, the use of artificial intelligence (algorithms) during development for task performance, and on-demand manufacturing for customized prototypes and parts in short-run productions (smart factories) must be considered.

Innovative applications in Smart Cities. Innovation is a driver of the development of current and future Smart Cities. Innovation, understood as newness, improvement, and spread, is often promoted by Information and Communication Technologies (ICTs) that make it possible to automate, accelerate, and change the perspective of the way that economic and "social good" challenges can be addressed. In economics, innovation is generally considered to be the result of a process that brings together various novel ideas to affect society and increase competitiveness. In this sense, the economic competitiveness of future Smart City societies is defined as an increase in consumers' satisfaction given by the right product price/quality ratio. Therefore, it is necessary to design production workflows that maximize the resources used to produce products and services of an appropriate quality. Companies' competitiveness refers to their capacity to produce goods and services efficiently (decreasing prices and increasing quality), making their products attractive in global markets. Thus, it is necessary to achieve high productivity levels that increase profitability and generate revenue. Beyond the importance of stable macroeconomic environments that can promote confidence and attract capital and technology, a necessary condition to build competitive societies is to create virtuous creativity circles that can propose smart and disruptive applications and services that can spread across different social sectors and strata. Smart Cities are willing to create technology-supported environments to make urban, social, and industrial spaces friendly, competitive, and productive contexts in which natural and material resources can be accessible to people and citizens can develop their potential skills in the best conditions possible. Since countries in different geographic locations and with different natural, cultural, and industrial ecosystems must adapt their strategies to these conditions, Smart City solutions are materialized differently. This study shows an example of an experience where industrial planning, urban planning, and health and sanitary problems are addressed with technology, leading to disruptive-data- and artificial-intelligence-centered applications. By sharing research experiences and results that, to date, have mostly been applied in Latin American countries, the authors and editors show how they have contributed to making cities and new societies smart through scientific development and innovation.

Author Contributions: Conceptualization, R.M.A.-T. and J.S.-L.; methodology, R.M.R.-M.; software, R.M.A.-T.; validation, M.R.P.-M. and J.d.l.R.-R.; formal analysis, K.C.A.-S.; investigation, E.R.P.-O.; resources, R.M.A.-T.; data curation, J.S.-L.; writing—original draft preparation, M.R.P.-M.; writing—review and editing, K.C.A.-S.; visualization, E.R.P.-O.; supervision, J.d.l.R.-R.; project administration, R.M.R.-M.; funding acquisition, R.M.A.-T. All authors have read and agreed to the published version of the manuscript.

Funding: This research was funded by TecNM Campus Chihuahua II and PRODEP, grant document number 511-6/2019-7987, 5 July 2019.

Institutional Review Board Statement: The study was conducted according to the guidelines of the Declaration of Helsinki and approved by the Institutional Review Board (or Ethics Committee) of TecNM Campus Ciudad Juárez (document number M00/1121/2018, 9 July 2018).

Informed Consent Statement: Not applicable.

Data Availability Statement: The data can be accessed at: https://docs.google.com/spreadsheets/d/1SibI hTb6NnC1HuQLJwAPb8nUmo5I6ikatPcWpkHsC2c/edit?usp=sharing (accessed on 30 December 2021) https://docs.google.com/spreadsheets/d/1RuCvT-iTUQtMaOUiFcg49Es6F-QASz37/edit?usp=sharing&ouid=115575865940574164949&rtpof=true&sd=true (accessed on 30 December 2021).

Acknowledgments: The authors would like to thank TecNM for providing a commission grant to the principal investigator Rosa Ma Amaya-Toral, the participating machine shops for facilitating access for failure identification and analysis, and CENALTEC Chihuahua for providing the facilities for the development of the life–strength model.

Conflicts of Interest: The authors declare no conflict of interest.

References

1. Takasaki, H.; Kikkawa, K.; Chiba, H.; Handa, Y.; Sesé-abad, A.; Fernández-domínguez, J.C. Cross-cultural adaptation of the health sciences evidence-based practice questionnaire into Japanese and its test–Retest reliability in undergraduate students. *Prog. Rehabil. Med.* **2021**, *6*, 1–21. [CrossRef]
2. Sharma, B. A focus on reliability in developmental research through Cronbach's Alpha among medical, dental and paramedical professionals. *Asian Pac. J. Health Sci.* **2016**, *3*, 271–278. [CrossRef]
3. Haruna, H.; Tshuma, N.; Hu, X. Health information needs and reliability of sources among nondegree health sciences students: A prerequisite for designing eHealth literacy. *Ann. Glob. Health* **2017**, *83*, 369–379. [CrossRef] [PubMed]
4. Weber, D.; Nasim, M.; Mitchell, L.; Falzon, L. A method to evaluate the reliability of social media data for social network analysis. In Proceedings of the 2020 IEEE/ACM International Conference on Advances in Social Networks Analysis and Mining (ASONAM), Hague, The Netherlands, 7–10 December 2020; pp. 317–321.
5. Jager, N.W.; Newig, J.; Challies, E.; Kochskämper, E.; von Wehrden, H. Case study meta-analysis in the social sciences. Insights on data quality and reliability from a large-N case survey. *Res. Synth. Methods* **2022**, *13*, 12–27. [CrossRef] [PubMed]
6. Dhillon, B.S. *Engineering Systems Reliability, Safety, and Maintenance: An Integrated Approach*; CRC Press: New York, NY, USA, 2017.
7. Silva, I.; Guedes, L.A.; Portugal, P.; Vasques, F. Reliability and availability evaluation of wireless sensor networks for industrial applications. *Sensors* **2012**, *12*, 806–838. [CrossRef]
8. Jia, G.; Tabandeh, A.; Gardoni, P. Life-Cycle Analysis of Engineering Systems: Modeling Deterioration, Instantaneous Reliability, and Resilience. In *Risk and Reliability Analysis: Theory and Applications: In Honor of Prof. Armen Der Kiureghian*; Gardoni, P., Ed.; Springer International Publishing: Cham, Switzerland, 2017; pp. 465–494.
9. Vojtov, V.; Berezchnaja, N.; Kravcov, A.; Volkova, T. Evaluation of the Reliability of Transport Service of Logistics Chains. *Int. J. Eng. Technol.* **2018**, *7*, 270–274. [CrossRef]
10. Duan, C.; Deng, C.; Li, N. Reliability assessment for CNC equipment based on degradation data. *Int. J. Adv. Manuf. Technol.* **2019**, *100*, 421–434. [CrossRef]
11. Li, H.; Deng, Z.-M.; Golilarz, N.A.; Soares, C.G. Reliability analysis of the main drive system of a CNC machine tool including early failures. *Reliab. Eng. Syst. Saf.* **2021**, *215*, 107846. [CrossRef]
12. Li, H.; Soares, C.G.; Huang, H.-Z. Reliability analysis of a floating offshore wind turbine using Bayesian Networks. *Ocean. Eng.* **2020**, *217*, 107827. [CrossRef]
13. Bobbio, A.; Portinale, L.; Minichino, M.; Ciancamerla, E. Improving the analysis of dependable systems by mapping fault trees into Bayesian networks. *Reliab. Eng. Syst. Saf.* **2001**, *71*, 249–260. [CrossRef]
14. Guo, J.; Li, Y.-F.; Zheng, B.; Huang, H.-Z. Bayesian degradation assessment of CNC machine tools considering unit non-homogeneity. *J. Mech. Sci. Technol.* **2018**, *32*, 2479–2485. [CrossRef]
15. Langseth, H.; Portinale, L. Bayesian networks in reliability. *Reliab. Eng. Syst. Saf.* **2007**, *92*, 92–108. [CrossRef]
16. Mi, J.; Li, Y.; Li, H.; Peng, W.; Huang, H.-Z. Reliability analysis of CNC hydraulic system based on fuzzy fault tree. In Proceedings of the 2011 International Conference on Quality, Reliability, Risk, Maintenance, and Safety Engineering, Xi'an, China, 17–19 June 2011; pp. 208–212.
17. Lo, H.-W.; Liou, J.J.H.; Huang, C.-N.; Chuang, Y.-C. A novel failure mode and effect analysis model for machine tool risk analysis. *Reliab. Eng. Syst. Saf.* **2019**, *183*, 173–183. [CrossRef]
18. Adamyan, A.; He, D. Analysis of sequential failures for assessment of reliability and safety of manufacturing systems. *Reliab. Eng. Syst. Saf.* **2002**, *76*, 227–236. [CrossRef]
19. Bai, S.; Huang, H.; Li, Y.; Yu, A.; Deng, Z. A modified damage accumulation model for life prediction of aero-engine materials under combined high and low cycle fatigue loading. *Fatigue. Fract. Eng. Mater. Struct.* **2021**, *44*, 3121–3134. [CrossRef]
20. Dundulis, G.; Žutautaitė, I.; Janulionis, R.; Ušpuras, E.; Rimkevičius, S.; Eid, M. Integrated failure probability estimation based on structural integrity analysis and failure data: Natural gas pipeline case. *Reliab. Eng. Syst. Saf.* **2016**, *156*, 195–202. [CrossRef]
21. Eryilmaz, S.; Kan, C. Reliability based modeling and analysis for a wind power system integrated by two wind farms considering wind speed dependence. *Reliab. Eng. Syst. Saf.* **2020**, *203*, 107077. [CrossRef]
22. Fan, J.W.; Zhou, Z.Y.; Wang, Z.L.; Miao, W. Research on the Evaluation of Small Sample Reliability for CNC Grinding Machine Tools Based on Bayes Theory. *Adv. Mater. Res.* **2014**, *971–973*, 688–692. [CrossRef]
23. Li, J. Research on big data acquisition bus technology of NC machine tool. *J. Phys. Conf. Ser.* **2021**, *2029*, 012104. [CrossRef]

24. Li, H.; Huang, H.-Z.; Li, Y.-F.; Zhou, J.; Mi, J. Physics of failure-based reliability prediction of turbine blades using multi-source information fusion. *Appl. Soft Comput.* **2018**, *72*, 624–635. [CrossRef]
25. Mi, J.; Li, Y.-F.; Peng, W.; Huang, H.-Z. Reliability analysis of complex multi-state system with common cause failure based on evidential networks. *Reliab. Eng. Syst. Saf.* **2018**, *174*, 71–81. [CrossRef]
26. Vineyard, M.; Amoako-Gyampah, K.; Meredith, J.R. Failure rate distributions for flexible manufacturing systems: An empirical study. *Eur. J. Oper. Res.* **1999**, *116*, 139–155. [CrossRef]
27. Soltani, R.; Sadjadi, S.J.; Tofigh, A.A. A model to enhance the reliability of the serial parallel systems with component mixing. *Appl. Math. Model.* **2014**, *38*, 1064–1076. [CrossRef]
28. Realyvásquez-Vargas, A.; Arredondo-Soto, K.C.; Carrillo-Gutiérrez, T.; Ravelo, G. Applying the plan-do-check-act (PDCA) cycle to reduce the defects in the manufacturing industry. A Case Study. *Appl. Sci.* **2018**, *8*, 2181. [CrossRef]
29. Arredondo-Soto, K.C.; Blanco-Fernández, J.; Miranda-Ackerman, M.A.; Solís-Quinteros, M.M.; Realyvásquez-Vargas, A.; García-Alcaraz, J.L. A plan-do-check-act based process improvement intervention for quality improvement. *IEEE Access* **2021**, *9*, 132779–132790. [CrossRef]
30. Realyvásquez-Vargas, A.; Arredondo-Soto, K.C.; Blanco-Fernandez, J.; Sandoval-Quintanilla, J.D.; Jiménez-Macías, E.; García-Alcaraz, J.L. Work standardization and anthropometric workstation design as an integrated approach to sustainable workplaces in the manufacturing industry. *Sustainability* **2020**, *12*, 3728. [CrossRef]
31. Realyvásquez-Vargas A, Arredondo-Soto KC, García-Alcaraz JL, Macías EJ Improving a manufacturing process using the 8ds method. *A Case Study A Manuf. Co. Appl. Sci.* **2020**, *10*, 2433.
32. Piña-Monarrez, M.R. Weibull stress distribution for static mechanical stress and its stress/strength analysis. *Qual. Reliab. Eng. Int.* **2018**, *34*, 229–244. [CrossRef]
33. Zhang, L.-F.; Xie, M.; Tang, L.-C. Robust regression using probability plots for estimating the Weibull shape parameter. *Qual. Reliab. Eng. Int.* **2006**, *22*, 905–917. [CrossRef]
34. Flores, M.P.; Torres, J.G.; Rodríguez, J.H.; Alcaraz, A.M. Confiabilidad operativa de sistemas para compresión de gas y generación eléctrica en complejos petroleros. *Inf. Tecnol.* **2010**, *21*, 13–25. [CrossRef]
35. Rand, D.R.; McLinn, J.A. Analysis of field failure data when modeled Weibull slope increases with time. *Qual. Reliab. Eng. Int.* **2011**, *27*, 99–105. [CrossRef]
36. Pascual, F.; Zhang, H. Monitoring the Weibull shape parameter by control charts for the sample range. *Qual. Reliab. Eng. Int.* **2011**, *27*, 15–26. [CrossRef]
37. Guevara, W.; Cárdenas, A.V.; Camperos, J.A.G. Metodología para evaluar el factor confiabilidad en la gestión de proyectos de diseño de equipos industriales. *Tecnura* **2015**, *19*, 129–141.
38. Piña-Monarrez, M.R. Conditional weibull control charts using multiple linear regression. *Qual. Reliab. Eng. Int.* **2017**, *33*, 785–791. [CrossRef]
39. Ferreira, L.A.; Silva, J.L. Parameter estimation for Weibull distribution with right censored data using EM algorithm. *Eksploat. I Niezawodn. Maint. Reliab.* **2017**, *19*, 310–315. [CrossRef]

*applied
sciences*

Article

Characterization and Design for Last Mile Logistics: A Review of the State of the Art and Future Directions

Hyeong Suk Na [1], Sang Jin Kweon [2] and Kijung Park [3,*]

[1] Department of Industrial Engineering, South Dakota School of Mines & Technology, Rapid City, SD 57701, USA; hyeongsuk.na@sdsmt.edu

[2] Department of Industrial Engineering, Ulsan National Institute of Science and Technology, Ulsan 44919, Korea; sjkweon@unist.ac.kr

[3] Department of Industrial and Management Engineering, Incheon National University, Incheon 22012, Korea

* Correspondence: kjpark@inu.ac.kr

Abstract: One of the most challenging problems in last mile logistics (LML) has been the strategic delivery due to various market risks and opportunities. This paper provides a systematic review of LML-related studies to find current issues and future opportunities for the LML service industry. To that end, 169 works were selected as target studies for in-depth analysis of recent LML advances. First, text mining analysis was performed to effectively understand the underlying LML themes in the target studies. Then, the novel definition and typology of LML delivery services were suggested. Finally, this paper proposed the next generation of LML research through advanced delivery technique-based LML services, environmentally sustainable LML systems, improvement of LML operations in real industries, effective management of uncertainties in LML, and LML delivery services for decentralized manufacturing services. We believe that this systematic literature review can serve as a useful tool for LML decision makers and stakeholders.

Keywords: last mile logistics; systematic literature review; last mile logistics innovation; logistics system design; logistics service characterization

Citation: Na, H.S.; Kweon, S.J.; Park, K. Characterization and Design for Last Mile Logistics: A Review of the State of the Art and Future Directions. *Appl. Sci.* **2022**, *12*, 118. https://doi.org/10.3390/app12010118

Academic Editor: Paolo Delle Site

Received: 8 November 2021
Accepted: 19 December 2021
Published: 23 December 2021

Publisher's Note: MDPI stays neutral with regard to jurisdictional claims in published maps and institutional affiliations.

1. Introduction

The evolution of e-commerce markets has drawn great attention to various delivery service issues [1,2] which are intrinsically related to last mile problems. As last mile delivery services require various logistics elements to satisfy faster and more reliable delivery conditions [3], last mile problems have been treated as more inefficient, more expensive, and less environmentally-friendly issues than any other problems in supply chain networks [4]. Indeed, the last segment of supply chains has been increasingly fragmented and has caused additional delivery costs due to the distributed collection/destination locations of customers using e-commerce services [5]. The route from the shop to the courier or shipping agency that will handle the delivery is known as the first mile. In contrast, the transit of products from the shipping agency or transportation hub to the intended final destination is referred to as the last mile. The final delivery between the local parcel depot and the customer who has purchased goods traditionally accounts for up to 28% of the total shipping cost but the lack of economies of scale that makes the last mile costly and ineffective necessitates new last mile strategies and solutions [4,6]. In particular, over the last few years, the stakeholders of last mile deliveries have faced many challenges in providing strategic deliveries due to a wide variety of market risks (e.g., COVID-19 pandemic, urban population growth, densification, traffic congestion and safety, customers' behavior change, and greenhouse gas emissions) and opportunities (e.g., globalization, advanced transport systems, innovative information and communication technologies, Internet of Things, and Industry 4.0) [1,4,7–12]. Accordingly, last mile logistics (LML) that inevitably involve complex operations of last mile deliveries have been more critical in the modern logistics industry.

Issues in LML can be viewed from various geographical dimensions of logistics. Many consumers in urban areas require same-day and on-demand delivery services that involve LML for groceries, prepared meals, and retail purchases as LML can provide customers with convenience and flexibility [13]. Furthermore, LML suppliers in urban areas have experienced many challenges such as traffic congestion, limited delivery time, service regulations, inefficient delivery vehicle routing, and inappropriate supply chain network design [4,14,15]. Consequently, urban logistics should be addressed to enhance the economic, environmental, and social sustainability of LML [11].

Many reports have presented that economic growth, improvement of environmental sustainability, resolution of urban unemployment, safe urban freight transport, and regional economy development are particularly related to LML [10]. As such, an appropriate design for city logistics is considered essential to improve welfare in society [16]. For instance, the first mile in a city and the last mile to the final consignee can be operated by conventional trucks and environmentally friendly city freighters or technology-advanced delivery cars, respectively [17]. Furthermore, the high demand of LML delivery services has increased complexity in design of urban logistics [14,18]. In 2014, 54% of the world population already lived in cities or urban areas, and by 2050 a projected two thirds will be urban [19]. This means that more than six billion people globally will live in urban areas, which leads to great pressure on all aspects of urban planning, including LML [20]. The necessity of LML clearly shows that the overarching objective of LML efforts is to elevate the prosperity of city logistics while alleviating their emerging negative consequences [10,12]. In this regard, existing LML-related studies have provided a wide variety of approaches to solve last mile problems in urban areas.

This paper aims to provide a comprehensive and structured review of recent studies relevant to LML and focuses on operational as well as technological aspects. Our state-of-the-art review identifies various potentials and opportunities associated with LML-related research. Although individual studies have successively achieved the ultimate goal of LML, they tend to provide redundant and ambiguous concepts across critical LML terminology defined by different contexts and properties in supply chain management. In other words, there has been no widely agreed upon framework to characterize LML and its innovations in a supply chain, and only a few studies have investigated the conceptualization of LML to contemplate a comprehensive framework for practical LML services. In order to improve the operational efficiency of last mile applications and offer practical guidelines for future research, it is necessary to first develop consensus on the underlying characteristics and design for LML. Along with previous LML issues and approaches that have been primarily addressed in existing studies, major opportunities for more innovative design and development are suggested to motivate various future applications in this paper.

The rest of this paper is organized as follows. In Section 2, we introduce the methodology behind our systematic literature review. We employ a text mining of an extensive set of existing LML-related studies to effectively understand major themes in the literature as a basis for a comprehensive literature review. In Section 3, existing definitions and concepts of LML in the literature are examined to provide a standard definition of LML. Section 4 discusses four main issues (i.e., sharing economy, proximity stations/points and hubs, environmentally sustainable LML, and delivery technology innovation), which are identified from LML themes through text mining, are further reviewed in order to characterize current issues and approaches for LML. New opportunities for design and development of innovative LML are proposed in Section 5. Finally, Section 6 offers guidelines for practice and offers opportunities for future research.

2. Systematic Literature Review Methodology and Analysis

A literature review for LML was performed based on a systematic literature review procedure widely adopted in other domains [10,21]. First, the objective and scope of the literature review on LML were established. In this study, the main objective was to provide a systematic review of recent LML-related studies to investigate the definition and typology of LML, main issues and concerns in LML, and new opportunities for the next generation of LML. Based on a preliminary search of highly cited LML studies, significant keywords that include "last mile logistics", "urban logistics", "city logistics", "typology of last mile logistics", "last mile logistics innovation", "collaborative last mile logistics", "cooperative last mile logistics", "sustainable last mile logistics", and "robust last mile logistics" were identified to better search target research articles. Different combinations of these keywords using Boolean AND and OR operators were used to retrieve initial articles published between 2001 and the end of August 2021 in scientific platforms—EBSCO, Elsevier, Google Scholar, JSTOR, Science Direct, Springer, Web of Science, and Wiley Online Library. Examples of the keyword combinations for the literature search are ("last mile" AND "logistics"), ("last mile logistics" AND "urban logistics"), ("last mile logistics" AND city logistics"), ("last mile logistics" AND "innovation"), and ("collaborative last mile logistics" OR "cooperative last mile logistics"). Various research works including these keywords or their combination in the title, keyword list, or abstract were disclosed. Consequently, more than 400 research works written in English were collected as the initial article set for review. Next, the following search criteria were applied to refine the initial article set: (i) inclusion of research areas related to LML, and (ii) inclusion of peer-reviewed journal articles having meaningful keywords in their abstract. Then, 169 research articles from 84 unique journals were identified as LML studies that contributed to recent LML advances, but 15 journals were associated with more than three articles each (Figure 1). This indicates that a small number of journals dominate LML studies over most other journals. Finally, the 96 research articles which had more than 10 citations were selected as target articles. The literature demonstrates that LML is an emerging research area with rapid growth (Figure 2a). The black solid line and the green dashed line represent the total number of and the average number of publications, respectively, while the red line means the number of publications for each year. Moreover, Figure 2b shows the top eight countries producing the most LML research based on the affiliation of the first author. Furthermore, the top five articles in terms of the number of citations are [22] (476), Ref. [23] (331), Ref. [24] (243), Ref. [25] (236), and [26] (233). An overview of the literature review method used for this study is presented in Figure 3.

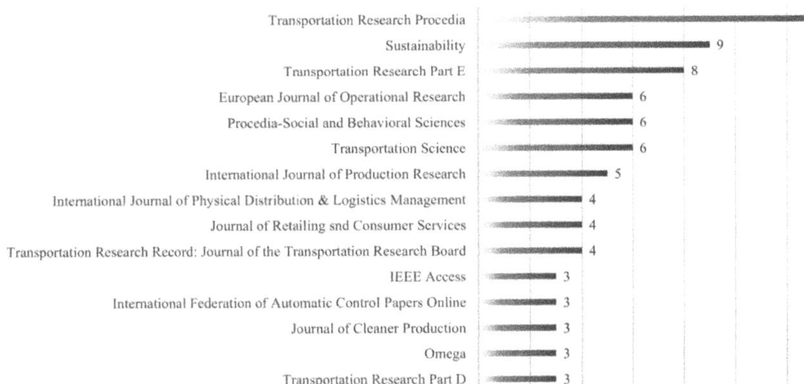

Figure 1. Top 15 journals according to the number of papers.

To better understand the underlying themes addressed in the target research article set, text mining analysis was performed using the following procedure. First, text data were collected from the title, abstract, and keywords of each article and separately saved to a text file. The set of text data was processed by the *tm* package for the *R* statistical program language [27] to standardize the text data; stop-words, whitespaces, numbers, and punctuation were removed from the original text dataset, and all letters were transformed to lower case. In addition, each word in the set of text data was stemmed by the *SnowballC* package [28] to regard word variants originated from the same root as the same term (e.g., "destin" for "destination" and "destinations"). As a result of the above pre-processing, an initial document-term matrix (*F*) was produced consisting of 96 target articles and 2616 terms in Equation (1). The individual frequency of 2616 terms observed from the target research articles was represented by the following ($m \times n$) sparse matrix:

$$F = \begin{bmatrix} f(a_1,t_1) & f(a_2,t_2) & \cdots & f(a_1,t_n) \\ f(a_2,t_1) & f(a_2,t_2) & \cdots & f(a_2,t_n) \\ \vdots & \vdots & \vdots & \vdots \\ f(a_m,t_1) & f(a_m,t_2) & \cdots & f(a_m,t_n) \end{bmatrix}, \tag{1}$$

where a_m refers to the *m*th target research article for all $m \in M = \{1,2,\cdots,96\}$, t_n is the *n*th term appearing in the set of articles a_m for all $n \in N = \{1,2,\cdots,2616\}$ and for all $m \in M$, and $f(a_m,t_n)$ is the frequency of term t_n in article a_m for all $n \in N$ and for all $m \in M$. The above procedure was performed by the application provided in [29].

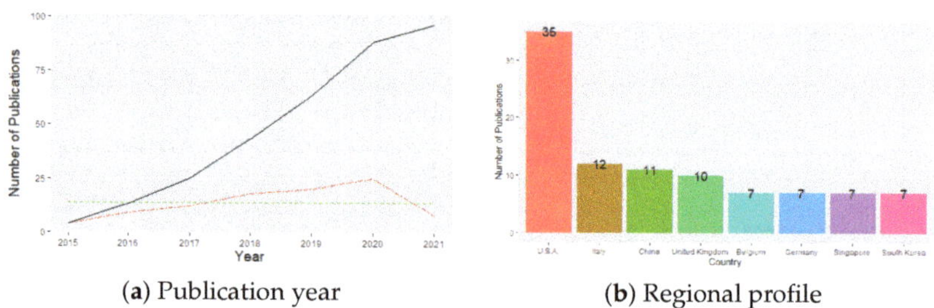

(**a**) Publication year   (**b**) Regional profile

Figure 2. Systematic review analysis of target research articles.

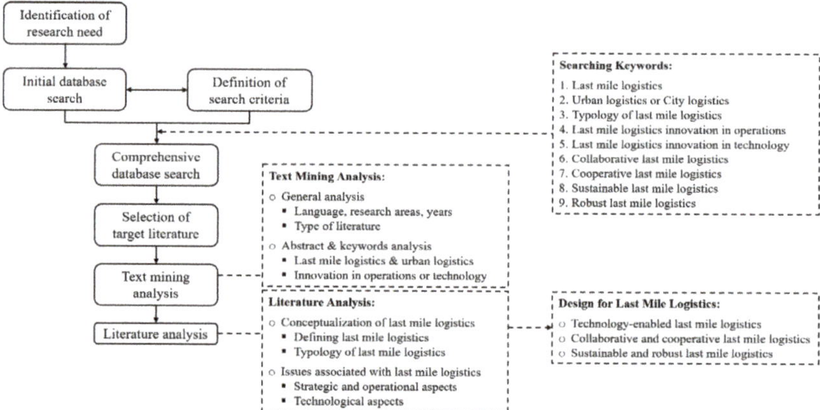

Figure 3. Framework for systematic literature review.

As common terms (e.g., last mile and logistics) that frequently appeared across the target research articles did not help explain specific features of the different works, the frequency $f(a_m,t_n)$ for all $m \in M$ and for all $n \in N$ was transformed into the Term Frequency-Inverse Document Frequency (TF-IDF) [30]. The TF-IDF transformation for a document-term matrix treats common terms frequently appearing in a specific set of articles as more important than common terms widely observed in most articles. First, TF-IDF values of each term for all the articles are calculated using Equation (2) and then averaged. Herein, the relative term frequency ($= t_n / \sum t_n$ of each article) was used to avoid a situation where a lengthy article has a high TF-IDF. Next, terms having a mean TF-IDF average were sorted to finalize the document-term matrix (*F*) using the equation in Equation (2):

$$f^*(a_m, t_{n*}) = f(a_m, t_{n*}) \times \log(|M|/|M_{t_n}|), \tag{2}$$

where t_{n*} is the relative term frequency, $|M|$ is the cardinality of the target research article set M (i.e., $|M| = 96$), and $|M_{t_n}|$ is the cardinality of the set of articles, including the specific term t_n for all $n \in N = \{1, 2, \cdots, 2616\}$. The term selection based on TF-IDF was performed based on the R-codes provided by [31]. As a result, a new document-term matrix consisting of 96 target articles and 827 terms was generated for the term-frequency analysis of the target articles.

Figure 4 shows the 50 most frequent terms that describe significant features of the target research articles. Overall, the frequently appearing terms across the LML studies indicate that the 169 target research articles have focused on strategic and operational improvements in LML; for example, "hub", "omnichannel", "crowdsourc (e.g., crowdsource)", "platform", "collabor (e.g., collaborative)", "crowdship (e.g., crowdshipping)", "consolid (e.g., consolidation)", "alloc (e.g., allocation)", "energi (e.g., energy)", "stochast (e.g., stochastic)", "price", "global", and "profit". Furthermore, the frequency of terms shows that technological advances in LML are another main focus in the literature; for example, "smart", "autonom (e.g., autonomous)", "intellig (e.g., intelligent)", "electr (e.g., electric)", "drone", and "bike". This paper thus focuses on (i) strategic and operational aspects and (ii) technological aspects of LML to highlight research issues and identify future opportunities.

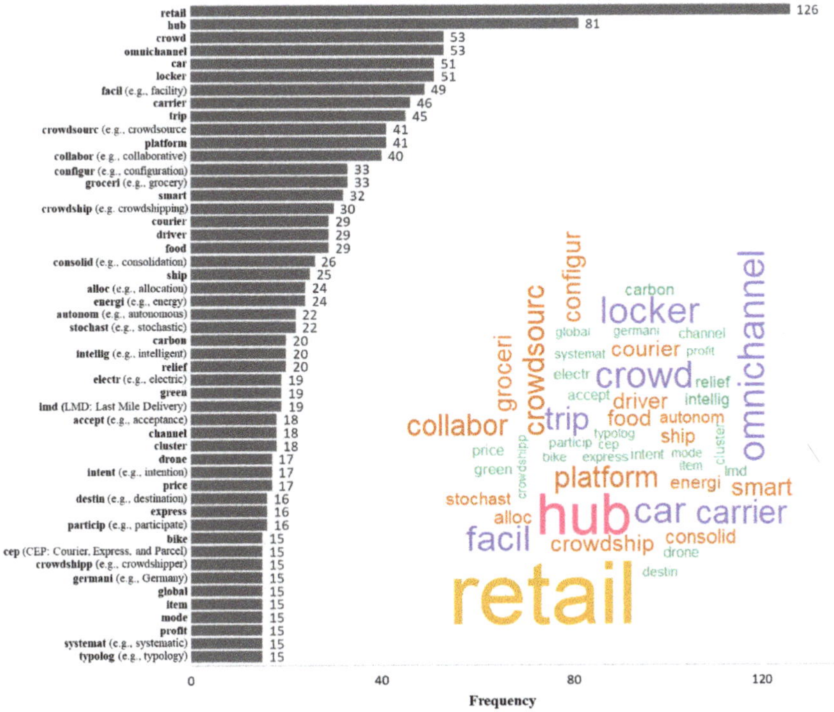

Figure 4. The 50 most frequent terms identified from the target research articles.

3. Conceptualization of Last Mile Logistics

The term last mile was first coined in the telecommunications industry [14] and was originally used to indicate the final leg of a delivery system for traditional deliveries from brick-and-mortar retailers [32]. Early applications of LML in the literature narrowly extended supply chains directly to the end consumer, which simply represented a home delivery service for consumers [33,34] and the last part of a delivery process [13]. In this regard, traditional LML was conceived of as deliveries for the last mile supply chain, last mile, final mile, home delivery, B2C distribution, or grocery delivery.

However, the traditional definitions have not been sufficient to describe newly emerging LML features driven by recent e-commerce changes that inherently involve new uncertainties [35,36], exclusion of in-store order fulfilment processes [23], and non-specific final locations [33,37]. Despite a steady increase in the number of contributions to LML research, LML has been conceptualized inconsistently across studies without consensus around a single definition of LML [11]. As various LML approaches have been available, a consistent and robust definition of LML is necessary to be a basis for LML design frameworks. Table 1 shows the various LML-related definitions used in our target research articles.

Table 1. Definitions of LML-related terminologies.

Terminology	Reference	Definition
Last mile	[32]	Final leg in a B2C delivery process in which the parcels are delivered to the destination, either at the recipient's place or at a collection point
	[38]	Last part of a delivery process of physical goods from a last transit point to a final drop point
	[39]	Distance from the main traffic station, such as rail transit, to the destination
	[40]	Last segment of distribution for a delivery with the specific distance
	[41]	Transport of goods from a local contact place to a point of consumption
Last mile delivery	[14]	Final leg of transport of goods in the supply chain to their consumption point
	[42]	Delivery of purchased items to the doors of customers
	[43]	Delivery of goods to the home in the last link of the supply chain
	[44]	Delivery of parcels to their destination in a city
	[45]	Last segment of a delivery process that involves all required activities and processes of the delivery chain
	[46]	Delivery from the last upstream transit point to the last recipient
	[47]	Transport from the retailer's local point to the final recipient's place
	[48]	Delivery of items to their final recipient's point within a city
Last mile distribution	[49]	Last part of the supply chain delivery process, including necessary activities from the last transit point to
Last mile parcel distribution	[50]	Delivery of parcels from distribution centers or substations to individual addresses
Last mile logistics	[51]	Movement of goods from a distribution center to the last recipient's doorstep
	[13]	Last stretch of a B2C consignment delivery process
	[11]	Last stretch of the logistics system from the last distribution point to the recipient's preferred final drop point
	[1]	Last stage of a delivery from a distribution center to a customer's place
	[52]	Last stretch of a B2C parcel delivery process of goods from a penetration point to the final consignee's point

An important insight from the target research articles is the functional scope of three different logistics domains: city logistics, urban logistics, and LML. The three logistics domains can be clarified with different perspectives [14]. City logistics is a critical field in urban areas and mainly focuses on stakeholders' interrelationships from the perspective of macro-level logistics. However, urban logistics refers to how parcels can be effectively transported in urban areas at a meso level. On the other hand, LML is related to delivery processes at a micro level. Considering these three different functional scopes, in this paper, a working definition for LML is proposed to synthesize all the above LML-related definitions as follows:

LML is the final branch of parcel delivery services that involves a point of delivery to a final consignee's preferred, predefined collection point or destination location after order placement.

Based on the above LML definition, the scope of interest in this paper covers a research stream from leaving warehouses of the supplier or logistics provider to arriving at the collection point designated by a final consignee. The underlying characteristics of LML concepts under the LML definition have been presented in various types of LML-related research. The operational and technological aspects of LML identified in Section 2 show that the nature of LML-related research can be categorized into four LML main research areas in Section 4 (see Table 2).

Table 2. Typology of LML-related research.

Research Area	Main Issue
Sharing economy	Impact of sharing economy in LML to employment market Operations of sharing economy in LML Environmental impact of sharing economy in LML
Proximity stations/points and hubs	Integrating proximity stations/points into the existing LML seamlessly Searching for potential locations for these stations/points Location-routing problem for LML Vehicle routing problem for LML Assessment of distributed network strategies for LML
Environmentally sustainable LML	Multi-criteria decision making for sustainable LML Environmental impact assessment and sustainable strategies for e-commerce LML Integration of environmental sustainability into new LML approaches
Delivery technology innovation	Limitations of traditional truck-or van-based last mile delivery services Transition to innovative and environment-friendly last mile delivery services using advanced vehicle technologies

4. Current Issues of Last Mile Logistics

The keywords frequently appearing in existing LML studies address that the main issues of LML in the literature are associated with sharing economy, proximity stations/points and hubs, environmental sustainability, and delivery technology innovation. The following subsections show extant LML studies that discussed each issue through various approaches.

4.1. Sharing Economy

The advancement of information and communications technologies has accelerated the introduction of sharing economy in LML. Sharing economy in LML involves a way of parcel deliveries by individual, self-contracted contractors joined at a crowdsourcing platform. New technologies that help find optimal routing and parking spots in real time have been lowering the entry barriers of crowdsourcing so that more logistics service providers can join a crowdsourcing platform for last mile parcel deliveries. Crowdsourcing for last mile deliveries is likely to spread further as the digital economy grows, which may be one of the viable options for tackling rising unemployment [53]. Using a crowdsourcing platform, passenger cars in urban areas can be utilized for home delivery services. Crowd workers in LML can be characterized by open-loop car routes, drivers' wage-response behavior, interplay with the ride-share market, and service zone sizes for fulfilling last mile deliveries from shared logistics. Although sharing economy in LML is not as scalable as a traditional truck-only system, this transition to the sharing economy paradigm for LML has a potential to create economic benefits by reducing original truck fleet sizes and exploiting additional operational flexibilities [54]. Specifically, from the logistics company's perspective, using a personal car to carry delivery parcels by an individual contractor can keep costs low, as it does not require additional fleet capability even if more parcels need to be carried. Instead, the company apply surge pricing to resolve a demand-supply imbalance for the last mile parcel deliveries. This allows the logistics company to be more flexible to the logistics-related market. From the self-contracted worker's perspective, their idle capacity can be utilized for carrying parcels. Many people have idle capacity in the course of their day, and using small slivers of time to carry parcels can be one of the ways to effectively use such an idle capacity. Furthermore, this provides an avenue toward the gig economy and shows options for resolving unemployment issues. From the final customer's perspective, the reduction in high overhead business intermediaries with a low-cost technology platform can reduce the delivery cost to the customer's site.

Table 3 summarizes the major issues and findings of LML studies related to sharing economy. Ref. [55] presented a mixed integer programming (MIP) model that solves a vehicle routing problem (VRP) in which vehicles are operated by crowdsourced occasional drivers. Ref. [56] extended the work of [55] to an MIP model considering multiple parcel deliveries for each crowdsourced driver. Ref. [42] performed a simulation analysis based on logistic regression and an agent-based transportation simulator to identify the benefits of LML for retail store order pickups using a social network of customers. The results showed that employing friends in a social network for LML can decrease transportation emissions and delivery costs while maintaining delivery speed and reliability. Ref. [18] performed a discrete event simulation analysis based on a conceptual framework with five principles to facilitate the integration of crowdsourced delivery into a conventional delivery network. Ref. [57] identified the effectiveness of crowdsourcing last mile delivery through a simulation model for same-day delivery using crowdsourced vehicles. They found that crowdsourced fleets can be effective to maximize the total number of deliveries when late delivery penalties are not severe. Ref. [25] proposed a route planning problem to model a dynamic crowdsourced delivery platform that automatically matches delivery tasks, ad-hoc drivers, and dedicated backup vehicles for deliveries uncovered by ad-hoc drivers. They found a total cost reduction through ad-hoc drivers, along with backup vehicles and an increase in the cost-efficiency of the system depending on the drivers' stop willingness. Ref. [58] presented a model to identify factors that affect the acceptability and preferences of crowdshipping attributes between consignors and consignees for package deliveries. They found that preferences can be described distinctively depending on shipment distance. Ref. [50] proposed a last mile parcel delivery system that employs ridesharing strategy based on internet of vehicles intelligence for deliveries in a smart city.

Table 3. LML-related issues and research findings: Sharing economy.

LML-Related Issue	Research Finding	Reference
Social impact	Crowdsourcing delivery to decrease unemployment	[53]
Economic impact	MIP model to address the VRP for crowdsourced drivers	[55]
	Total cost reduction by ad-hoc drivers with backup vehicles	[25]
	Effectiveness of crowdsourcing last mile delivery	[57]
	Delivery cost reduction by a dual-channel logistics system	[18]
	Multiple parcel deliveries for each crowdsourced driver	[56]
	Acceptability and preferences of crowdshipping attributes characterized depending on shipment distance	[58]
	Potential of crowdsourcing last mile delivery	[54]
Environmental impact	Negative impacts of crowdsourcing deliveries on overall environmental performance	[59] [54]
	Reduced transportation emissions and delivery costs due to the use of social network	[42]
	Positive effect of shared mobility on greenhouse gas emissions	[60]
	Reduced carbon emissions and delivery distance in urban and suburban areas by a social network-enabled package pickup	[61]

4.2. Proximity Stations/Points and Hubs

Using proximity stations or proximity points is a new way to improve the efficiency of LML. This is useful for small- to medium-sized delivery parcels when customers are not at home. Deliverers store parcels at a depot station near a customer's address when home deliveries fail, and customers pick up their products later. In addition, delivery time can be reduced by visiting a proximity point during the night when traffic volume is low [62].

The applications of proximity stations and/or proximity points to LML are aimed, not only to integrate proximity points into existing LML, but also to identify locations for proximity points (see Table 4). Ref. [63] proposed a modular bento-box system to store delivery parcels until customers pick up their deliveries. Shopping malls, central squares, and residential districts were suggested as examples for the possible location of proximity points. Ref. [64] proposed an effective mobile crowd-tasking model that handles many citizens as crowd workers to perform LML. They formulated a proximity station problem in LML as a minimal-cost network flow problem, which minimizes the additional cost to assign all delivery parcels in pick-own-parcel stations to the most convenient crowd workers. Similarly, Ref. [26] investigated the utilization of parcel locker pickups for LML in the Polish InPost Company system and found that the proper location of parcel lockers was one of the most important factors for delivery efficiency.

Table 4. LML-related issues and research findings: Proximity stations/points and hubs.

LML-Related Issue	Research Finding Integrating Proximity Stations/Points	Reference
Integrating proximity stations/points into the existing LML	Improvement of delivery time, increase in average travel speed, and reduction in greenhouse gas emissions by using proximity stations/points	[62]
	Modular bento-box system for customer pickup	[63]
Identifying potential locations for proximity stations/points	Network min-cost flow problem with pick-own-parcel stations to maximize resources using a collaborative approach	[64]
	Evaluation of using parcel lockers in the Polish InPost Company system	[26]
Routing problem for last mile delivery hubs	Location-routing model to determine the placement of last mile delivery hubs	[65,66]
	Development of a hybrid genetic algorithm to efficiently solve the computational complexity issue of the location-routing problem with large-size instances for LML hubs	[67]
	Development of a two-stage stochastic travel time model to solve a delivery VRP to the set of final customer's homes and the set of hub locations for pickup stores	[68]

Hub location problems can be also applied to properly address the location of proximity stations and/or points and to consolidate last mile deliveries across urban areas that can reduce traffic and cut greenhouse gas emissions (see Table 4). Depending on the type of demand allocation for final customers, a last mile hub location problem can be partitioned into single and multiple allocation problems. The single allocation problem assigns every final customer to a single last mile delivery hub, while the multiple allocation problem assigns every final customer to more than one last mile delivery hub. Setting up efficient hub locations for LML can lead to great economic/temporal benefits for both consignors and consignees in delivery costs, quality of delivery services, and environmental impacts [69].

The location of last mile delivery hubs affects both the fixed cost of hub placement and the variable cost of delivery parcels to final customers. Refs. [65,66] respectively proposed a location-routing model to determine multiple last mile delivery hubs based on the total location and delivery vehicle routing costs for each customer's pickup and delivery needs. Ref. [67] further proposed a hybrid genetic algorithm that efficiently handles the computational complexity issue of existing LML location-routing models. Ref. [68] proposed a two-stage stochastic travel time model to solve a delivery VRP for the sets of both final customers and hub locations of pickup stores.

With a number of studies focusing on the potential locations for last mile delivery hubs, last mile delivery hubs have been established in practice. For instance, the City of London Corporation determined its first last mile delivery hub for the purpose of removing large numbers of delivery vehicles from city streets. This last mile delivery hub is expected to accelerate the use of e-cargo bikes and people on foot for the final leg of parcel deliveries and to take up to 23,000 fewer vehicle journeys in central London every year [70].

4.3. Environmentally Sustainable LML

As the environmental impact of operational activities in industries has become an urgent global issue, environmental sustainability is now essential in the management and operations of the entire supply chain [71]. In particular, LML involves various externalities such as gas emissions, air pollution, noise, and congestion [11]. The significant role of LML in the environmental sustainability of a supply chain has urged the transformation of the conventional LML into a more environmentally sustainable system. Accordingly, various studies have adopted environmental factors in modeling LML problems along with traditional time and cost factors through proposed LML approaches (Table 5).

Table 5. LML-related issues and research findings: Environmental impact.

LML-Related Issue	Research Finding	Reference
Multi-criteria decision making for sustainable LML	Conceptual framework to evaluate LML from economic, social, and environmental aspects	[72]
	Bi-criteria auction process of last mile delivery orders that maximizes both economic and environmental sustainability	[73]
	A distributed network based on crowd logistics as the most sustainable LML strategy	[74]
Environmental impact assessment and sustainable strategies for e-commerce LML	Lower carbon footprints in last mile deliveries through e-commerce than conventional brick-and-mortar stores	[43]
	Effective reduction in greenhouse gas emissions through local collection and delivery points for failed home deliveries	[75]
	Stochastic last mile model to calculate probabilistic estimates of traveling distances and break-even point at which last mile delivery causes less carbon emissions than customer pickup	[76]
	Reduction in light goods vehicle traffic and associated environmental impacts through LML	[77]
	A framework to reduce CO_2 emissions in e-commerce LML	[78]
	Principles for sustainable LML	[79]
Integration of environmental sustainability into new LML approaches	Reduction in carbon emissions and delivery distances through a social network enabled package pickup	[61]
	Importance of local authorities to promote cargo cycles	[24]
	Reduction in CO_2 emissions per person through shared mobility	[60]
	Emissions and cost savings by using a social network in LML for retail store order pickups	[42]
	Slightly more emissions in minimizing operating costs for the last mile delivery system than minimizing emissions for the system	[54]
	Negative impact of crowdsourcing in LML on the environment of the road	[59]
	Crowd logistics that can be environmentally-friendly only if it is optimized for existing delivery trips	[80]
	Both environmental and economic benefits obtained by crowd-shipping through public transportation in urban areas	[81]

The importance of environmental sustainability in LML led to multi-criteria decisions to simultaneously consider both economic and environmental factors that have trade-offs in LML. Ref. [72] provided a conceptual framework to evaluate last mile options from economic, social, and environmental aspects that involve design criteria and performance attributes for industrial, institutional, and consumer group stakeholders. Ref. [73] proposed a bi-criteria auction process to assign last mile delivery orders to trucks that maximizes both the economic and environmental sustainability of LML. Ref. [74] evaluated three LML strategies (i.e., centralized distribution network, decentralized distribution network through home-delivery, and decentralized distribution network based on crowd logistics) through a system dynamics simulation and its multi-criteria decision analysis using the Preference Ranking Organization METHod for Enrichment of Evaluations (PROMETHEE) on economic, technological, environmental, and societal criteria. Their case study of local food LML showed that a distributed network based on crowd logistics is the most sustainable LML strategy.

The sustainability role of LML in e-commerce was also widely discussed to suggest new sustainable LML strategies in the literature. Ref. [43] addressed that last mile deliveries through e-commerce resulted in lower carbon footprints than conventional brick-and-mortar stores. Ref. [75] addressed that local collection and delivery points for failed home deliveries of e-commerce can effectively reduce greenhouse gas emissions against traditional shipping methods. Ref. [76] developed a stochastic LML model to calculate probabilistic estimates of traveling distances, and they compared carbon emissions generated by conventional customer pickup and last mile delivery through e-commerce. Ref. [77]

examined LML for light goods from e-commerce retailers in London, and they concluded that last mile deliveries for e-commerce can reduce light-goods vehicle traffic and associated environmental impacts. Ref. [78] developed a framework to reduce carbon dioxide (CO_2) emissions in e-commerce LML and suggested potential solutions (e.g., re-allocation of vehicles, use of electric vehicles (EVs), reduction in delivery failures, proper vehicle planning, alternative fuels, and urban consolidation centers) to reduce CO_2 emissions in LML. Ref. [79] examined the urban delivery performance of LML for business-to-consumer and business-to-business cases in the United Kingdom.

Recent studies have paid more attention to environmental sustainability of new LML operations and approaches that are expected to provide both economic and environmental benefits in urban logistics. Ref. [61] compared last mile delivery distances and greenhouse gas emissions in last mile delivery systems (i.e., door-to-door delivery, designated package pickup, and socially networked package pickup) for online purchases, which were modeled through a genetic algorithm. Their results showed that a social-network-enabled package pickup can significantly reduce carbon emissions, as well as delivery distances in both urban and suburban scenarios. Ref. [24] proposed a sustainable urban logistics framework to boost the potential use of cargo cycles based on empirical case studies in the United Kingdom. They claimed that local authorities play a critical role in promoting cargo cycles to make them major urban freight transport. Ref. [54] proposed a novel logistics planning model for last mile deliveries based on shared movability, which can characterize open-loop car routes, wage-response properties, and competition in the ride share and delivery services. Refs. [54,59] found that last mile delivery through shared mobility may increase greenhouse gas emissions due to an increase in vehicle traveling distance. Ref. [60] conducted a survey from 363 car sharing respondents in the Netherlands to analyze the effect of shared mobility on car use and CO_2 emissions, and they concluded that shared mobility positively affected the environment in Belgium. Ref. [42] claimed that LML for retail store order pickups using a social network of customers can reduce emissions from ground logistics as well as delivery costs while maintaining delivery speed and reliability. Ref. [80] suggested crowd logistics platforms, and compared the performance their proposed platform with traditional logistics counterparts in terms of the unit delivery cost and the environmental impact of crowd logistics. Their multi-actor multi-criteria analysis to evaluate the delivery scenarios showed that crowd logistics can be environmentally-friendly only if they are optimized for existing delivery trips. Ref. [81] evaluated the environmental and economic impacts of crowdshipping through public transportation in the city of Rome. The scenario analysis based on a discrete choice model in [81] showed that the crowdshipping platform can have both environmental and economic benefits.

4.4. Delivery Technology Innovation

Along with the studies addressing operational challenges, LML studies related to overcoming technical issues were also abundant in the literature. Newly emerging delivery technologies such as unmanned aerial vehicles (UAVs) or drones, connected autonomous vehicles (CAVs), EVs, automated lockers, real-time data transmission systems, dynamic route planning systems, fleet management solutions, tracking devices, and identification means and devices [4,12,82] were discussed to improve LML performance (see Table 6).

Table 6. LML-related issues and research findings: Delivery technology innovation.

LML-Related Issue	Research Finding	Reference
Limitations of traditional truck- or van-based last mile delivery services	Service transition from truck- or van-based goods delivery to drone-assisted delivery in urban areas	[82]
	Social costs of home delivery due to increased delivery traffic flows in residential areas	[83]
Transition to innovative and environment-friendly last mile delivery services using advanced vehicle technologies	Possibility that UAVs or drones can carry heavier goods while hovering	[22]
	Challenges of drones used for LML services in urban areas	[1]
	Importance of using CAVs for future LML delivery services in urban areas	[84]

An innovative transition of delivery services from truck- or van-based parcels deliveries to drone-assisted deliveries in urban areas has received attention in logistics [4,82]. According to the 2016 Material Handling Industry (MHI) Annual Industry Report [85], 59% of survey respondents recognized that emerging technologies, including UAVs or drones, CAVs, EVs, or robots, were having influence on logistics. The report also claimed that adoption rates for such technologies are expected to grow to higher than 50% over the next decade. Indeed, drones have been extensively utilized in the logistics field, deployed for delivery services in LML [86,87]. However, challenges in the legal restrictions and restricted service areas, labor and logistics costs, battery safety, and cost efficiency of drone applications for LML prevented drones from being prevalently used for urban LML deliveries [1]. For more reliable deliveries by drones in LML, operational challenges for drone-assisted delivery or truck–drone hybrid delivery will need to be tackled [88–91]. With improved adoption processes, drones can be an essential strategic delivery method to innovate conventional LML delivery services if they can carry heavier goods while hovering [22].

Traffic flows in urban areas have frequently fluctuated by uncertain traffic factors on the road. Accordingly, home delivery services impose many social costs due to a sudden increase in delivery service vehicles in residential areas [83]. The use of CAVs for future LML delivery services in urban areas has been the kernel of recent studies as traditional logistics delivery services have caused intrinsic drawbacks [84]. Furthermore, autonomous trucks or vans can be integrated with drones or robots, which may significantly improve LML delivery services in urban areas [92]. EVs have also received attention in LML as an alternative to internal combustion engine vehicles, with increasing interest in new forms of delivery vehicles which are more economically and environmentally sustainable.

5. New Opportunities for Design and Development of Innovative Last Mile Logistics

The current interests and focuses of LML delivery service studies should be more closely scrutinized to find new potentials in innovative LML fields. The following subsections show new opportunities for design and development of innovative LML delivery services to improve the current LML-related issues.

5.1. Advanced LML Services with New Delivery Techniques

As we discussed in Section 4.4, LML delivery services in urban areas will accelerate technological innovation due to the extensive use of unmanned vehicles, CAVs, UAVs or drones, EVs, automated lockers, real-time data transmission systems, dynamic route planning systems, tracking devices, and identification means or devices. These can provide more opportunities for application of new delivery techniques in LML including the following areas:

- Smart scheduling and urban consolidation through new delivery technologies: The innovative and advanced delivery technologies will lead to new methods of boosting effective LML delivery services. Several Industry 4.0 solutions for LML delivery services can be actualized from advanced techniques for enabling smart scheduling and developing real-time stochastic optimization models [1]. Furthermore, the construction of consolidation centers in urban areas can improve urban LML delivery services with micro consolidation and distribution centers [84]. These overarching trends are expected to continue into the future of urban LML delivery systems.

- Improvement in operations of new delivery techniques: UAVs or drones will dominate LML delivery services in a few years. For instance, many global logistics powerhouses have already initiated drone-assisted delivery services for food and industrial products. Consumer goods represent finished products that are sold to and consumed by people, in general, whereas industrial products are materials used in the production of other commodities. Industrial products are purchased and used for both industrial and commercial purposes. They consist of machinery, manufacturing plants, raw materials, and any other commodity or component that is used by industries or businesses. Similarly, the surge in adoption of new delivery techniques, including drone taxis and small/large drones, will prominently attract a lot of attention in LML-related markets in the near future. Another promising opportunity to enhance urban LML delivery services will be realized by diversifying LML delivery channels with EVs, which is also an environmentally-friendly delivery service. However, potential issues associated with using EVs are their short drive distance, which is generally less than 150–200 miles per charge, and long recharging time of batteries compared to their internal combustion engine counterparts. Along with these issues, the LML industry also needs to improve the performance of solid-state batteries (e.g., lithium-ion and sodium-ion solid-state batteries) and recharging technologies with lower installation costs [93].

- Development of optimization models for last mile delivery operations using new advanced techniques: Owing to the commercialization of new advanced techniques and the need for innovative last mile delivery services, there are many delivery routing problems which require resolution within a short time period [94], e.g., modified VRPs or extended traveling salesman problems in LML using drones [91,95–98], autonomous vehicles [99], robots [9,92], and drone-robot integration [100]. Ref. [95] proposed a truck–drone delivery optimization model using mixed integer linear programming (MILP) and its solution approach that reflects drone energy consumption and restricted flying zones. Ref. [97] developed an MILP model to minimize the total completion time of LML delivery services using autonomous drones and delivery trucks. Ref. [91] presented an MILP to minimize the customer waiting time when using a single truck and multiple drones for delivery. Yu (2018) proposed a mixed integer, non-linear programming model to decide optimal delivery schedules (i.e., allocation, routing and battery charging) of autonomous vehicles, not only to minimize driving distance, but also to maximize renewable energy utilization. Furthermore, the delivery scheduling problems in LML need to be addressed for adopting autonomous robots [9,101]. This significant stream of research for last mile delivery operations using advanced technologies can stimulate the LML markets' innovation, as well as create steady demand for LML in the future.

5.2. Innovative LML Applications Using New Technologies and Systems for Environmental Sustainability

Although many extant studies addressed the significant effects of new LML delivery technologies on environmental sustainability (see Section 4.3), adverse effects of new technologies on the environment should be also investigated. For example, using crowdsourcing may have a negative impact on the environment in the road owing to an increase in traffic flow [59]. In particular, greenhouse gas emissions may increase due to open-loop routes of delivery vehicles, and a resultant low per-kilometer emission rate of crowdsourced vehicles can be offset in that the aggregated car trip distance is over 35% longer than that of trucks due to cars' much smaller capacity [54]. In this regard, the role of future LML delivery services in terms of environmental sustainability should be investigated in more detail to facilitate a transition to environmentally sustainable LML as follows:

- Management of negative impacts on the environment through technology enabled LML services: From the perspective of environmental sustainability in LML systems, a major challenge is to significantly decrease greenhouse gas emissions that affect climate changes. Replacing existing light delivery vehicles that are powered by an internal combustion engine to EVs for LML services can be more sustainable than traditional delivery trucks or vans [1,12,102]. New technologies to achieve sustainable LML can be performed with innovative management strategies for LML deliveries such as LML delivery services during the night avoiding high traffic congestion [103] and using EVs [78] for CO_2 emission reduction with fuel saving.

- Sustainable LML system design in urban areas: Collaborative urban logistics services should be considered for better consolidation of existing infrastructure and resources to leverage the environmental sustainability of LML deliveries in urban areas. An environmentally-friendly LML system can be operated by fewer delivery vehicles and light-weight vehicles to lessen emissions [104]. Moreover, the innovative digitalization of LML systems is required for the successful development of a sustainable smart city. Future LML delivery services are expected to integrate automation technologies and digitization into operational strategies to facilitate real-time decisions [88,105]. Thus, an intelligent system that can monitor and analyze environmental impacts of deliveries in LML in real-time will be necessary to enhance environmental sustainability in urban areas. Similarly, integrating drone- and/or robot-assisted last mile deliveries should be evaluated from both cost and environmental perspectives to suggest sustainable LML operations. For LML innovations through new technologies, policies and regulations for different stakeholders can affect successful implementation of LML [10]. Development of novel methodologies from different perspectives and that of tools to assess the viability of sustainable LML should be also addressed in future LML studies.

5.3. Effective Management of Uncertainties in LML

Supply chain systems inevitably involve structural and operational complexity as there are a wide variety of stakeholders, information, and materials that should be managed with their associated uncertainties [106,107]. Supply chain complexity becomes more critical in LML in which highly distributed supply channels and end-user locations exist with multiple delivery methods; more uncertain elements should be carefully controlled to achieve the effectiveness of LML in practice. For this, the following issues should be discussed more in depth to properly handle an increase in uncertainties while operating LML:

- Conceptualization and measurement of LML complexity: Complexity in supply chains has been variously defined depending on aspects and domains in the existing literature relevant to traditional supply chains [108]. However, the concepts and measures of complexity in LML have not been widely addressed in LML studies to date. Uncertainties that are caused by the variety, size, and dynamic operations of LML may be distinct from those in conventional logistics. Therefore, complexity in LML needs changes in definitions and measures that were originally used for conventional logistics, although basic concepts for supply chain complexity may be partially applicable to LML cases. Along with efforts to define and develop complexity measures for LML, the impact of complexity on LML performance should be investigated to understand underlying complexity dynamics in an LML system.

- Modeling LML with dynamic factors: The dynamic nature of a city's logistics system should be reflected on LML models with new indicators and aspects that properly describe uncertainties in LML. Multiple indicators to comprehensively represent traffic congestion conditions and parking space availability, which can affect delivery productivity, should be considered as sources of complexity in LML [109]. In addition, various types of uncertainty such as demand volatility, infrastructure accessibility, conflicting objectives among stakeholders [10] are required to be incorporated into LML models. Indeed, real-time data, fleet management and dynamic route planning, and tracking devices will be important resources to develop algorithms and optimization techniques in LML to reflect real-world problems.
- Mitigation of uncertainties in LML: The complexity of LML would exponentially increase as crowd-sourcing and new delivery technologies such as drones and UAVs are actively employed in logistics operations. In this regard, studies that discuss how the impact of new uncertainties in advanced LML approaches is effectively controlled are needed. For example, operational strategies that enable a highly distributed delivery network to be simplified are required to avoid complex routes [18]. Autonomous robots that assist truck-based LML can be effective in making deliveries in relatively small areas with multiple delivery stops [92].
- Developing solution approaches: Addressing VRP or location routing problems for LML generally makes associated mathematical models more complicated. This leads to high computational complexity due to its combinatorial structure in searching for optimality. As integrating new LML concepts and approaches into existing LML operations usually increases computational complexity, well-designed algorithms and/or heuristics should be developed to solve a large-scaled problem in a polynomial time. Refs. [110,111] classified multiple VRP variants for urban freight transportation and the associated algorithms solving the various LML problem.

5.4. LML for Decentralized Manufacturing Services

The growth of 3D printing technologies and services has accelerated a new paradigm of manufacturing. 3D printing provides a new opportunity that can decentralize manufacturing by producing parts near consumption, which can lead to shortened supply chains and reduced inventory [112]. Under decentralized manufacturing, LML would be more important as a large set of low-volume and lightweight final products made by decentralized manufacturing may need to be effectively delivered to highly distributed consumer locations through local transports [113]. The 3D printing industry can provide new opportunities in logistics operations, and more insights into the effects of 3D printing on LML should be scrutinized to prepare for the transition to this new paradigm of LML. In this regard, the following topics should be further discussed:

- LML for local fabshop-based 3D printing: Local 3D printing fabshops can provide a new business opportunity for logistics companies if they are able to offer 3D printing services as well as delivery services at the same time through their local 3D printing shops [114]. This type of local 3D printing is effective for orders that are required to fabricate customized design shapes and high-quality products [113]. As customers simply need to order what they want to manufacture through an online system, the role of LML for the deliveries of 3D printed products to individual customers will become more important. A dynamic local logistics network that is assisted by new mobility options (e.g., drones or UAVs) will be helpful to achieve operational efficiency if a few local fabshops should handle individual low-volume orders requested from highly distributed end-customers.

- LML for consumer-based (home) 3D printing: Consumer-based (home) 3D printing is more suitable for low-end and common products [113]. The acceleration of 3D printing at home would make material flows from material suppliers to end-users more critical [82,112], and conventional last mile delivery for final products may need a significant transition to last-mile delivery for raw materials. Therefore, rapid replenishment of a wide variety of low-volume materials for customers would be an important decision-making problem in LML. In addition, reverse logistics to reproduce raw materials from disused 3D printed products will emerge as a new competitive edge in LML.

- Concurrent manufacturing logistics (mobile manufacturing): Amazon Technologies, Inc. recently filed patents for mobile manufacturing that enables carriers equipped with 3D printers to take and produce orders at the same time during delivery [115,116]. This new form of logistics will eliminate the current strict separation between manufacturing and logistics and will provide a great deal of flexibility and competitiveness for offering final products to customers [117]. Mobile manufacturing is expected to open a new way to optimize order lead-time by valuably using delivery time, which has traditionally been considered as cost addition for final delivery. This new approach will be applicable not only to mobile manufacturing of low-variety products within small-sized vehicles (e.g., cars and small trucks) but also to mobile manufacturing of high-variety products within large-sized vehicles (e.g., large trucks and aircraft). For both cases, product quality and autonomous manufacturing in a moving vehicle should be technically and operationally supported for successful LML operations of mobile manufacturing.

6. Conclusions

LML delivery services have been a critical part of the supply chain management. Our findings in this study further indicate that additional research is required to enhance environmental, operational, and technical sustainability of LML delivery services. In this paper, we identified four new opportunities for LML: (i) Advanced LML services with new delivery techniques, (ii) innovative LML applications using new technologies and systems for environmental sustainability, (iii) effective management of uncertainties in LML, and (iv) LML for decentralized manufacturing services. The identified opportunities indicate that additional efforts in the LML research field should be given to enhance the operational, technological, and environmental sustainability of LML services.

This paper presents a literature review of LML. The main goal of this review study was to provide a systematic review of LML-related studies by investigating the definition and the typology of LML, exploring issues and concerns in LML, and discussing new opportunities and future directions for the next generations of LML. In this literature review, more than 400 works written in English were initially identified, and 169 target research articles were eventually found to meet the study criteria and identified as complete LML studies that contributed to recent LML advances. Furthermore, to better understand the underlying themes addressed in the target research, text mining analysis was performed.

Considering all reference papers comprehensively, a novel definition of LML and a typology of LML-related research are proposed in this paper. The challenges constraining the innovation of LML services are discussed with various current issues: (i) the development of information and communications technology has recently accelerated the introduction of the sharing economy era; (ii) using proximity stations or proximity points is a new way to improve the efficiency of LML, and the hub location problem has also has attention re-paid to consolidate last mile deliveries across urban areas to reduce traffic and cut greenhouse gas emissions; (iii) the innovation of techniques in LML services may be helpful to design environmentally sustainable LML; and (iv) various advanced delivery technologies can be contemplated (e.g., UAVs or drones, robots, CAVs, and EVs) to improve performance in LML services.

To improve LML services, more efforts and suggestions should be addressed for various applications that are associated with new transportation modes for LML operations and their environmental sustainability, digital transformation, and novel methodology to improve LML operations of real industries, managing uncertainties in LML, and LML for decentralized manufacturing services. In this paper, such major opportunities for design and development of innovative LML services were discussed from technological and operational perspectives.

We believe that this systematic literature review can serve as a useful tool for LML decision makers and stakeholders to model an efficient delivery system. In addition, the future directions and suggestions proposed in this paper can be leveraged for smart city designs with smart logistics systems considering sudden increases in last mile deliveries. Our study would leverage interdisciplinary collaboration with various logistics researchers and urban planners to develop a system to meet their needs.

Author Contributions: Conceptualization, H.S.N.; methodology, H.S.N.; software, K.P.; investigation, H.S.N., S.J.K., and K.P.; writing—original draft preparation, H.S.N., S.J.K., and K.P.; writing—review and editing, H.S.N., S.J.K., and K.P.; visualization, K.P.; and funding acquisition, K.P. All authors have read and agreed to the published version of the manuscript.

Funding: This work was supported by Incheon National University Research Grant in 2020 for Kijung Park.

Conflicts of Interest: The authors declare no conflict of interest.

References

1. Bosona, T. Urban Freight Last Mile Logistics-Challenges and Opportunities to Improve Sustainability: A Literature Review. *Sustainability* **2020**, *12*, 8769. [CrossRef]
2. Ewedairo, K.; Chhetri, P.; Jie, F. Estimating transportation network impedance to last-mile delivery. *Int. J. Logist. Manag.* **2018**, *29*, 110–130. [CrossRef]
3. Le, T.V.; Stathopoulos, A.; Van Woensel, T.; Ukkusuri, S.V. Supply, demand, operations, and management of crowd-shipping services: A review and empirical evidence. *Transp. Res. Part C Emerg. Technol.* **2019**, *103*, 83–103. [CrossRef]
4. Ranieri, L.; Digiesi, S.; Silvestri, B.; Roccotelli, M. A review of last mile logistics innovations in an externalities cost reduction vision. *Sustainability* **2018**, *10*, 782. [CrossRef]
5. Kin, B.; Spoor, J.; Verlinde, S.; Macharis, C.; Van Woensel, T. Modelling alternative distribution set-ups for fragmented last mile transport: Towards more efficient and sustainable urban freight transport. *Case Stud. Transp. Policy* **2018**, *6*, 125–132. [CrossRef]
6. Deutsch, Y.; Golany, B. A parcel locker network as a solution to the logistics last mile problem. *Int. J. Prod. Res.* **2018**, *56*, 251–261. [CrossRef]
7. Aljohani, K.; Thompson, R.G. Optimizing the Establishment of a Central City Transshipment Facility to Ameliorate Last-Mile Delivery: A Case Study in Melbourne CBD. In *City Logistics 3: Towards Sustainable and Liveable Cities*; Wiley Online Library: Hoboken, NJ, USA, 2018; pp. 23–46.
8. Buldeo Rai, H.; Verlinde, S.; Macharis, C.; Schoutteet, P.; Vanhaverbeke, L. Logistics outsourcing in omnichannel retail: State of practice and service recommendations. *Int. J. Phys. Distrib. Logist. Manag.* **2019**, *49*, 267–286. [CrossRef]
9. Chen, C.; Demir, E.; Huang, Y.; Qiu, R. The adoption of self-driving delivery robots in last mile logistics. *Transp. Res. Part E Logist. Transp. Rev.* **2021**, *146*, 102214. [CrossRef]
10. Dolati Neghabadi, P.; Evrard Samuel, K.; Espinouse, M.L. Systematic literature review on city logistics: Overview, classification and analysis. *Int. J. Prod. Res.* **2019**, *57*, 865–887. [CrossRef]
11. Olsson, J.; Hellström, D.; Pålsson, H. Framework of last mile logistics research: A systematic review of the literature. *Sustainability* **2019**, *11*, 7131. [CrossRef]
12. Viu-Roig, M.; Alvarez-Palau, E.J. The Impact of E-Commerce-Related Last-Mile Logistics on Cities: A Systematic Literature Review. *Sustainability* **2020**, *12*, 6492. [CrossRef]
13. Lim, S.F.W.; Jin, X.; Srai, J.S. Consumer-driven e-commerce: A literature review, design framework, and research agenda on last-mile logistics models. *Int. J. Phys. Distrib. Logist. Manag.* **2018**, *48*, 308–332. [CrossRef]
14. Cardenas, I.; Borbon-Galvez, Y.; Verlinden, T.; Van de Voorde, E.; Vanelslander, T.; Dewulf, W. City logistics, urban goods distribution and last mile delivery and collection. *Compet. Regul. Netw. Ind.* **2017**, *18*, 22–43. [CrossRef]
15. Janjevic, M.; Winkenbach, M. Characterizing urban last-mile distribution strategies in mature and emerging e-commerce markets. *Transp. Res. Part A Policy Pract.* **2020**, *133*, 164–196. [CrossRef]
16. Rao, C.; Goh, M.; Zhao, Y.; Zheng, J. Location selection of city logistics centers under sustainability. *Transp. Res. Part D Transp. Environ.* **2015**, *36*, 29–44. [CrossRef]

17. Cleophas, C.; Cottrill, C.; Ehmke, J.F.; Tierney, K. Collaborative urban transportation: Recent advances in theory and practice. *Eur. J. Oper. Res.* **2019**, *273*, 801–816. [CrossRef]
18. Guo, X.; Jaramillo, Y.J.L.; Bloemhof-Ruwaard, J.; Claassen, G. On integrating crowdsourced delivery in last-mile logistics: A simulation study to quantify its feasibility. *J. Clean. Prod.* **2019**, *241*, 118365. [CrossRef]
19. United Nations. 68% of the World Population Projected to Live in Urban Areas by 2050. United Nations (UN). Available online: https://www.un.org/development/desa/en/news/population/2018-revision-of-world-urbanization-prospects.html (accessed on 9 December 2021)
20. Agussurja, L.; Cheng, S.F.; Lau, H.C. A State Aggregation Approach for Stochastic Multiperiod Last-Mile Ride-Sharing Problems. *Transp. Sci.* **2019**, *53*, 148–166. [CrossRef]
21. Scheidegger, A.P.G.; Pereira, T.F.; de Oliveira, M.L.M.; Banerjee, A.; Montevechi, J.A.B. An introductory guide for hybrid simulation modelers on the primary simulation methods in industrial engineering identified through a systematic review of the literature. *Comput. Ind. Eng.* **2018**, *124*, 474–492. [CrossRef]
22. Savelsbergh, M.; Van Woensel, T. 50th anniversary invited article—City logistics: Challenges and opportunities. *Transp. Sci.* **2016**, *50*, 579–590. [CrossRef]
23. Hübner, A.H.; Kuhn, H.; Wollenburg, J.; Towers, N.; Kotzab, H. Last mile fulfilment and distribution in omni-channel grocery retailing: A strategic planning framework. *Int. J. Retail. Distrib. Manag.* **2016**, *44*. [CrossRef]
24. Schliwa, G.; Armitage, R.; Aziz, S.; Evans, J.; Rhoades, J. Sustainable city logistics—Making cargo cycles viable for urban freight transport. *Res. Transp. Bus. Manag.* **2015**, *15*, 50–57. [CrossRef]
25. Arslan, A.M.; Agatz, N.; Kroon, L.; Zuidwijk, R. Crowdsourced delivery—A dynamic pickup and delivery problem with ad hoc drivers. *Transp. Sci.* **2019**, *53*, 222–235. [CrossRef]
26. Iwan, S.; Kijewska, K.; Lemke, J. Analysis of parcel lockers' efficiency as the last mile delivery solution—The results of the research in Poland. *Transp. Res. Procedia* **2016**, *12*, 644–655. [CrossRef]
27. Feinerer, I.; Hornik, K. tm: Text Mining Package. Available online: https://cran.r-project.org/web/packages/tm/index.html (accessed on 9 December 2021)
28. Bouchet-Valat, M. SnowballC: Snowball stemmers based on the Clibstemmer UTF-8 library. Available online: https://cran.r-project.org/web/packages/SnowballC/index.html (accessed on 9 December 2021)
29. Williams, G. Hands-on data science with R text mining. Available online: https://www.scribd.com/doc/252462619/Hands-on-Data-Science-with-R-Text-Mining (accessed on 9 December 2021)
30. Salton, G.; Buckley, C. Term-weighting approaches in automatic text retrieval. *Inf. Process. Manag.* **1988**, *24*, 513–523. [CrossRef]
31. Hornik, K.; Grün, B. topicmodels: An R package for fitting topic models. *J. Stat. Softw.* **2011**, *40*, 1–30.
32. Gevaers, R.; Van de Voorde, E.; Vanelslander, T. Characteristics and typology of last-mile logistics from an innovation perspective in an urban context. In *City Distribution and Urban Freight Transport: Multiple Perspectives, Edward Elgar Publishing*; Edward Elgar Publishing: Cheltenham, UK, 2011; pp. 56–71.
33. Kull, T.J.; Boyer, K.; Calantone, R. Last-mile supply chain efficiency: An analysis of learning curves in online ordering. *Int. J. Oper. Prod. Manag.* **2007**, *27*, 409–434. [CrossRef]
34. Punakivi, M.; Yrjölä, H.; Holmström, J. Solving the last mile issue: Reception box or delivery box? *Int. J. Phys. Distrib. Logist. Manag.* **2001**, *31*, 427–439. [CrossRef]
35. Dablanc, L.; Giuliano, G.; Holliday, K.; O'Brien, T. Best practices in urban freight management: Lessons from an international survey. *Transp. Res. Rec.* **2013**, *2379*, 29–38. [CrossRef]
36. Ehmke, J.F.; Mattfeld, D.C. Vehicle routing for attended home delivery in city logistics. *Procedia-Soc. Behav. Sci.* **2012**, *39*, 622–632. [CrossRef]
37. Esper, T.L.; Jensen, T.D.; Turnipseed, F.L.; Burton, S. The last mile: An examination of effects of online retail delivery strategies on consumers. *J. Bus. Logist.* **2003**, *24*, 177–203. [CrossRef]
38. Aized, T.; Srai, J.S. Hierarchical modelling of Last Mile logistic distribution system. *Int. J. Adv. Manuf. Technol.* **2014**, *70*, 1053–1061. [CrossRef]
39. Li, Q.; Wang, Y.; Li, K.; Chen, L.; Wei, Z. Evolutionary dynamics of the last mile travel choice. *Phys. Stat. Mech. Its Appl.* **2019**, *536*, 122555. [CrossRef]
40. Swanson, D. A simulation-based process model for managing drone deployment to minimize total delivery time. *IEEE Eng. Manag. Rev.* **2019**, *47*, 154–167. [CrossRef]
41. Halldórsson, Á.; Wehner, J. Last-mile logistics fulfilment: A framework for energy efficiency. *Res. Transp. Bus. Manag.* **2020**, *37*, 100481. [CrossRef]
42. Devari, A.; Nikolaev, A.G.; He, Q. Crowdsourcing the last mile delivery of online orders by exploiting the social networks of retail store customers. *Transp. Res. Part Logist. Transp. Rev.* **2017**, *105*, 105–122. [CrossRef]
43. Edwards, J.B.; McKinnon, A.C.; Cullinane, S.L. Comparative analysis of the carbon footprints of conventional and online retailing: A "last mile" perspective. *Int. J. Phys. Distrib. Logist. Manag.* **2010**, *40*, 103–123. [CrossRef]
44. Giret, A.; Carrascosa, C.; Julian, V.; Rebollo, M.; Botti, V. A crowdsourcing approach for sustainable last mile delivery. *Sustainability* **2018**, *10*, 4563. [CrossRef]
45. Yuen, K.F.; Wang, X.; Ng, L.T.W.; Wong, Y.D. An investigation of customers' intention to use self-collection services for last-mile delivery. *Transp. Policy* **2018**, *66*, 1–8. [CrossRef]

46. Vakulenko, Y.; Shams, P.; Hellström, D.; Hjort, K. Service innovation in e-commerce last mile delivery: Mapping the e-customer journey. *J. Bus. Res.* **2019**, *101*, 461–468. [CrossRef]
47. McLeod, F.; Cherrett, T.; Bektas, T.; Allen, J.; Martinez-Sykora, A.; Lamas-Fernandez, C.; Bates, O.; Cheliotis, K.; Friday, A.; Piecyk, M.; et al. Quantifying environmental and financial benefits of using porters and cycle couriers for last-mile parcel delivery. *Transp. Res. Part Transp. Environ.* **2020**, *82*, 102311. [CrossRef]
48. Palanca, J.; Terrasa, A.; Rodriguez, S.; Carrascosa, C.; Julian, V. An agent-based simulation framework for the study of urban delivery. *Neurocomputing* **2020**, *423*, 679–688. [CrossRef]
49. Hsiao, Y.H.; Chen, M.C.; Lu, K.Y.; Chin, C.L. Last-mile distribution planning for fruit-and-vegetable cold chains. *Int. J. Logist. Manag.* **2018**, *29*, 862–886. [CrossRef]
50. Wang, F.; Wang, F.; Ma, X.; Liu, J. Demystifying the crowd intelligence in last mile parcel delivery for smart cities. *IEEE Netw.* **2019**, *33*, 23–29. [CrossRef]
51. Amonde, T.M.; Ajagunna, I.; Iyare, N.F. Last mile logistics and tourist destinations in the Caribbean. *Worldw. Hosp. Tour. Themes* **2017**, *9*, 17–30. [CrossRef]
52. Risher, J.J.; Harrison, D.E.; LeMay, S.A. Last mile non-delivery: Consumer investment in last mile infrastructure. *J. Mark. Theory Pract.* **2020**, *28*, 484–496. [CrossRef]
53. Thacker, S. How Crowdworking is Making Headways as a Booming Ecosystem In 2018. Available online: https://www. entrepreneur.com/article/307017 (accessed on 9 December 2021)
54. Qi, W.; Li, L.; Liu, S.; Shen, Z.J.M. Shared mobility for last-mile delivery: Design, operational prescriptions, and environmental impact. *Manuf. Serv. Oper. Manag.* **2018**, *20*, 737–751. [CrossRef]
55. Archetti, C.; Savelsbergh, M.; Speranza, M.G. The vehicle routing problem with occasional drivers. *Eur. J. Oper. Res.* **2016**, *254*, 472–480. [CrossRef]
56. Macrina, G.; Pugliese, L.D.P.; Guerriero, F.; Laganà, D. The vehicle routing problem with occasional drivers and time windows. In *International Conference on Optimization and Decision Science*; Springer: Berlin/Heidelberg, Germany, 2017; pp. 577–587.
57. Castillo, V.E.; Bell, J.E.; Rose, W.J.; Rodrigues, A.M. Crowdsourcing last mile delivery: Strategic implications and future research directions. *J. Bus. Logist.* **2018**, *39*, 7–25. [CrossRef]
58. Punel, A.; Stathopoulos, A. Modeling the acceptability of crowdsourced goods deliveries: Role of context and experience effects. *Transp. Res. Part E Logist. Transp. Rev.* **2017**, *105*, 18–38. [CrossRef]
59. Bellos, I.; Ferguson, M.; Toktay, L.B. The car sharing economy: Interaction of business model choice and product line design. *Manuf. Serv. Oper. Manag.* **2017**, *19*, 185–201. [CrossRef]
60. Nijland, H.; van Meerkerk, J. Mobility and environmental impacts of car sharing in the Netherlands. *Environ. Innov. Soc. Transitions* **2017**, *23*, 84–91. [CrossRef]
61. Suh, K.; Smith, T.; Linhoff, M. Leveraging socially networked mobile ICT platforms for the last-mile delivery problem. *Environ. Sci. Technol.* **2012**, *46*, 9481–9490. [CrossRef] [PubMed]
62. Edwards, J.; McKinnon, A.; Cherrett, T.; McLeod, F.; Song, L. The impact of failed home deliveries on carbon emissions: Are collection/delivery points environmentally-friendly alternatives. In Proceedings of the 14th Annual Logistics Research Network Conference, Citeseer, UK, 9–11 September 2009; p. M117.
63. Dell'Amico, M.; Hadjidimitriou, S. Innovative logistics model and containers solution for efficient last mile delivery. *Procedia-Soc. Behav. Sci.* **2012**, *48*, 1505–1514. [CrossRef]
64. Wang, Y.; Zhang, D.; Liu, Q.; Shen, F.; Lee, L.H. Towards enhancing the last-mile delivery: An effective crowd-tasking model with scalable solutions. *Transp. Res. Part E Logist. Transp. Rev.* **2016**, *93*, 279–293. [CrossRef]
65. Çetiner, S.; Sepil, C.; Süral, H. Hubbing and routing in postal delivery systems. *Ann. Oper. Res.* **2010**, *181*, 109–124. [CrossRef]
66. Karaoglan, I.; Altiparmak, F.; Kara, I.; Dengiz, B. The location-routing problem with simultaneous pickup and delivery: Formulations and a heuristic approach. *Omega* **2012**, *40*, 465–477. [CrossRef]
67. Moon, I.; Salhi, S.; Feng, X. The location-routing problem with multi-compartment and multi-trip: formulation and heuristic approaches. *Transp. A Transp. Sci.* **2020**, *16*, 501–528. [CrossRef]
68. Zhou, F.; He, Y.; Zhou, L. Last mile delivery with stochastic travel times considering dual services. *IEEE Access* **2019**, *7*, 159013–159021. [CrossRef]
69. Freeman, O. London's Last Mile Logistic Hub For Sustainable Practices. Available online: https://www.www.supplychaindigital. com/logistics-1/londons-last-mile-logistic-hub-sustainable-practices (accessed on 9 December 2021)
70. Brown, B. Last Mile Logistics Hub to Consolidate Deliveries Across City of London. Available online: https://www.citymatters. london/last-mile-logistics-hub-to-consolidate-deliveries-across-city-of-london/ (accessed on 9 December 2021)
71. Seuring, S.; Müller, M. From a literature review to a conceptual framework for sustainable supply chain management. *J. Clean. Prod.* **2008**, *16*, 1699–1710. [CrossRef]
72. Harrington, T.S.; Singh Srai, J.; Kumar, M.; Wohlrab, J. Identifying design criteria for urban system 'last-mile'solutions–a multi-stakeholder perspective. *Prod. Plan. Control* **2016**, *27*, 456–476. [CrossRef]
73. Handoko, S.D.; Lau, H.C.; Cheng, S.F. Achieving economic and environmental sustainabilities in urban consolidation center with bicriteria auction. *IEEE Trans. Autom. Sci. Eng.* **2016**, *13*, 1471–1479. [CrossRef]
74. Melkonyan, A.; Gruchmann, T.; Lohmar, F.; Kamath, V.; Spinler, S. Sustainability assessment of last-mile logistics and distribution strategies: The case of local food networks. *Int. J. Prod. Econ.* **2020**, *228*, 107746. [CrossRef]

75. Song, L.; Guan, W.; Cherrett, T.; Li, B. Quantifying the greenhouse gas emissions of local collection-and-delivery points for last-mile deliveries. *Transp. Res. Rec.* **2013**, *2340*, 66–73. [CrossRef]

76. Brown, J.R.; Guiffrida, A.L. Carbon emissions comparison of last mile delivery versus customer pickup. *Int. J. Logist. Res. Appl.* **2014**, *17*, 503–521. [CrossRef]

77. Allen, J.; Piecyk, M.; Piotrowska, M.; McLeod, F.; Cherrett, T.; Ghali, K.; Nguyen, T.; Bektas, T.; Bates, O.; Friday, A.; et al. Understanding the impact of e-commerce on last-mile light goods vehicle activity in urban areas: The case of London. *Transp. Res. Part D Transp. Environ.* **2018**, *61*, 325–338. [CrossRef]

78. Awwad, M.; Shekhar, A.; Iyer, A. Sustainable Last-Mile Logistics Operation in the Era of Ecommerce. In Proceedings of the International Conference on Industrial Engineering and Operations Management, Washington, DC, USA, 6–8 March 2018; pp. 27–29.

79. Bates, O.; Friday, A.; Allen, J.; Cherrett, T.; McLeod, F.; Bektas, T.; Nguyen, T.; Piecyk, M.; Piotrowska, M.; Wise, S.; et al. Transforming last-mile logistics: Opportunities for more sustainable deliveries. In Proceedings of the 2018 CHI Conference on Human Factors in Computing Systems, Montreal, QC, Canada, 21–26 April 2018, pp. 1–14.

80. Rai, H.B.; Verlinde, S.; Macharis, C. Shipping outside the box. Environmental impact and stakeholder analysis of a crowd logistics platform in Belgium. *J. Clean. Prod.* **2018**, *202*, 806–816.

81. Gatta, V.; Marcucci, E.; Nigro, M.; Patella, S.M.; Serafini, S. Public transport-based crowdshipping for sustainable city logistics: Assessing economic and environmental impacts. *Sustainability* **2019**, *11*, 145. [CrossRef]

82. Mckinnon, A.C. The possible impact of 3D printing and drones on last-mile logistics: An exploratory study. *Built Environ.* **2016**, *42*, 617–629. [CrossRef]

83. Kapser, S.; Abdelrahman, M. Acceptance of autonomous delivery vehicles for last-mile delivery in Germany–Extending UTAUT2 with risk perceptions. *Transp. Res. Part C Emerg. Technol.* **2020**, *111*, 210–225. [CrossRef]

84. Kassai, E.T.; Azmat, M.; Kummer, S. Scope of Using Autonomous Trucks and Lorries for Parcel Deliveries in Urban Settings. *Logistics* **2020**, *4*, 17. [CrossRef]

85. MHI. Accelerating Change: How Innovation is Driving Digital, Always-on Supply Chains. The 2016 MHI Annual Industry Report, 2016, pp. 1–52. Available online: https://www.mhi.org/publications/report (accessed on 9 December 2021)

86. Aurambout, J.P.; Gkoumas, K.; Ciuffo, B. Last mile delivery by drones: An estimation of viable market potential and access to citizens across European cities. *Eur. Transp. Res. Rev.* **2019**, *11*, 30. [CrossRef]

87. Sanders, N.R.; Boone, T.; Ganeshan, R.; Wood, J.D. Sustainable supply chains in the age of AI and digitization: Research challenges and opportunities. *J. Bus. Logist.* **2019**, *40*, 229–240. [CrossRef]

88. Bányai, T. Real-time decision making in first mile and last mile logistics: How smart scheduling affects energy efficiency of hyperconnected supply chain solutions. *Energies* **2018**, *11*, 1833. [CrossRef]

89. Boysen, N.; Briskorn, D.; Fedtke, S.; Schwerdfeger, S. Drone delivery from trucks: Drone scheduling for given truck routes. *Networks* **2018**, *72*, 506–527. [CrossRef]

90. Han, Y.q.; Li, J.q.; Liu, Z.; Liu, C.; Tian, J. Metaheuristic algorithm for solving the multi-objective vehicle routing problem with time window and drones. *Int. J. Adv. Robot. Syst.* **2020**, *17*, 1729881420920031. [CrossRef]

91. Moshref-Javadi, M.; Hemmati, A.; Winkenbach, M. A truck and drones model for last-mile delivery: A mathematical model and heuristic approach. *Appl. Math. Model.* **2020**, *80*, 290–318. [CrossRef]

92. Simoni, M.D.; Kutanoglu, E.; Claudel, C.G. Optimization and analysis of a robot-assisted last mile delivery system. *Transp. Res. Part E Logist. Transp. Rev.* **2020**, *142*, 102049. [CrossRef]

93. Gold, A. Toyota's Solid-State Battery Prototype Could Be an EV Game Changer. Available online: https://www.motortrend.com/news/toyota-solid-state-battery-ev-2021/ (accessed on 9 December 2021)

94. Gayialis, S.P.; Konstantakopoulos, G.D.; Tatsiopoulos, I.P. Vehicle routing problem for urban freight transportation: A review of the recent literature. In *Operational Research in the Digital era–ICT Challenges*; Springer: Berlin/Heidelberg, Germany, 2019; pp. 89–104.

95. Jeong, H.Y.; Song, B.D.; Lee, S. Truck-drone hybrid delivery routing: Payload-energy dependency and No-Fly zones. *Int. J. Prod. Econ.* **2019**, *214*, 220–233. [CrossRef]

96. Kitjacharoenchai, P.; Min, B.C.; Lee, S. Two echelon vehicle routing problem with drones in last mile delivery. *Int. J. Prod. Econ.* **2020**, *225*, 107598. [CrossRef]

97. Kitjacharoenchai, P.; Ventresca, M.; Moshref-Javadi, M.; Lee, S.; Tanchoco, J.M.; Brunese, P.A. Multiple traveling salesman problem with drones: Mathematical model and heuristic approach. *Comput. Ind. Eng.* **2019**, *129*, 14–30. [CrossRef]

98. Salama, M.; Srinivas, S. Joint optimization of customer location clustering and drone-based routing for last-mile deliveries. *Transp. Res. Part C Emerg. Technol.* **2020**, *114*, 620–642. [CrossRef]

99. Yu, S.; Puchinger, J.; Sun, S. Two-echelon urban deliveries using autonomous vehicles. *Transp. Res. Part E Logist. Transp. Rev.* **2020**, *141*, 102018. [CrossRef]

100. Lemardelé, C.; Estrada, M.; Pagès, L.; Bachofner, M. Potentialities of drones and ground autonomous delivery devices for last-mile logistics. *Transp. Res. Part E Logist. Transp. Rev.* **2021**, *149*, 102325. [CrossRef]

101. Boysen, N.; Schwerdfeger, S.; Weidinger, F. Scheduling last-mile deliveries with truck-based autonomous robots. *Eur. J. Oper. Res.* **2018**, *271*, 1085–1099. [CrossRef]

102. Oliveira, C.M.d.; Albergaria De Mello Bandeira, R.; Vasconcelos Goes, G.; Schmitz Gonçalves, D.N.; D'Agosto, M.D.A. Sustainable vehicles-based alternatives in last mile distribution of urban freight transport: A systematic literature review. *Sustainability* **2017**, *9*, 1324. [CrossRef]
103. Sadhu, S.S.; Tiwari, G.; Jain, H. Impact of cycle rickshaw trolley (CRT) as non-motorised freight transport in Delhi. *Transp. Policy* **2014**, *35*, 64–70. [CrossRef]
104. Patella, S.M.; Grazieschi, G.; Gatta, V.; Marcucci, E.; Carrese, S. The Adoption of Green Vehicles in Last Mile Logistics: A Systematic Review. *Sustainability* **2021**, *13*, 6. [CrossRef]
105. de Souza, R.; Goh, M.; Lau, H.C.; Ng, W.S.; Tan, P.S. Collaborative urban logistics–synchronizing the last mile a Singapore research perspective. *Procedia-Soc. Behav. Sci.* **2014**, *125*, 422–431. [CrossRef]
106. Birkie, S.E.; Trucco, P.; Campos, P.F. Effectiveness of resilience capabilities in mitigating disruptions: leveraging on supply chain structural complexity. *Supply Chain. Manag.* **2017**, *22*, 506–521. [CrossRef]
107. Christopher, M. The agile supply chain: Competing in volatile markets. *Ind. Mark. Manag.* **2000**, *29*, 37–44. [CrossRef]
108. Turner, N.; Aitken, J.; Bozarth, C. A framework for understanding managerial responses to supply chain complexity. *Int. J. Oper. Prod. Manag.* **2018**, *38*, 1433–1466. [CrossRef]
109. Amaral, J.C.; Cunha, C.B. An exploratory evaluation of urban street networks for last mile distribution. *Cities* **2020**, *107*, 102916. [CrossRef]
110. Cattaruzza, D.; Absi, N.; Feillet, D.; González-Feliu, J. Vehicle routing problems for city logistics. *EURO J. Transp. Logist.* **2017**, *6*, 51–79. [CrossRef]
111. Konstantakopoulos, G.D.; Gayialis, S.P.; Kechagias, E.P. Vehicle routing problem and related algorithms for logistics distribution: A literature review and classification. *Oper. Res.* **2020**, 1–30. [CrossRef]
112. Fan, Z.; Meixner, L. *3D Printing: A Guide for Decision-Makers*; World Economic Forum: Geneva, Switzerland, 2020; Volume 1, p. 012039.
113. Boon, W.; Van Wee, B. Influence of 3D printing on transport: A theory and experts judgment based conceptual model. *Transp. Rev.* **2018**, *38*, 556–575. [CrossRef]
114. Wieczorek, A. Impact of 3D printing on logistics. *Res. Logist. Prod.* **2017**, *7*, 443–450. [CrossRef]
115. Apsley, L.K.; Bodell, C.I.; Danton, J.C.; Hayden, S.R.; Kapila, S.; Lessard, E.; Uhl, R.B. Providing Services Related to Item Delivery via 3D Manufacturing on Demand. U.S. Patent 9,898,776, 20 February 2018.
116. Apsley, L.K.; Bodell, C.I.; Danton, J.C.; Reyes-Guerrero, E.; Hayden, S.R.; Kapila, S.; Lessard, E.; Uhl, R.B. Vendor Interface for Item Delivery via 3D Manufacturing on Demand. U.S. Patent 9,858,604, 2 January 2018.
117. Ryan, M.J.; Eyers, D.R.; Potter, A.T.; Purvis, L.; Gosling, J. 3D printing the future: Scenarios for supply chains reviewed. *Int. J. Phys. Distrib. Logist. Manag.* **2017**, *47*, 992–1014. [CrossRef]

Review

A Taxonomy of Idea Management Tools for Supporting Front-End Innovation

Di Zhu [1,2], Abdullah Al Mahmud [1,*] and Wei Liu [2,*]

[1] School of Design and Architecture, Swinburne University of Technology, Melbourne 3122, Australia
[2] Faculty of Psychology, Beijing Normal University, Beijing 100875, China
* Correspondence: aalmahmud@swin.edu.au (A.A.M.); wei.liu@bnu.edu.cn (W.L.)

Abstract: Idea management is a crucial pillar of corporate management. Organizations may save research expenses, influence future development, and maintain distinctive competency by controlling front-end ideas. To date, several idea management tools have been developed. However, it is unknown to what extent they support the idea management process. Therefore, this scoping review aims to understand the classification of idea management tools and their effectiveness through an overview of the academic literature. Electronic databases (Scopus, ACM Digital Library, Web of Science Core Index, Elsevier ScienceDirect, and SpringerLink) were searched, and a total of 38 journal papers ($n = 38$) from 2010 to 2020 were retrieved. We identified 30 different types of idea management tools categorized as digital tools ($n = 21$), guidelines ($n = 5$), and frameworks ($n = 4$), and these tools have been utilized by software designers, hardware designers, and stakeholders. The identified tools may support various stages of idea management, such as capturing, generating, implementing, monitoring, refinement, retrieving, selection, and sharing. However, most tools only support a single stage (either capture or generate), and they cannot track the life cycle of the ideas, which may lead to misunderstanding. Therefore, it is essential to develop tools for managing ideas that would allow end users, designers, and other stakeholders to minimize bias in selecting and prioritizing ideas.

Keywords: information system; fuzzy front end; engineering design; idea management; idea generation

Citation: Zhu, D.; Al Mahmud, A.; Liu, W. A Taxonomy of Idea Management Tools for Supporting Front-End Innovation. *Appl. Sci.* **2023**, *13*, 3570. https://doi.org/10.3390/app13063570

Academic Editor: Panagiotis Tsarouhas

Received: 31 December 2022
Revised: 6 March 2023
Accepted: 7 March 2023
Published: 10 March 2023

1. Introduction

Idea management, which is at the forefront of innovation management, is a crucial pillar of corporate management [1], and idea management can be viewed as a subsidiary innovation management process with the objectives of efficient and effective idea production, assessment, and selection [2]. Therefore, a company must innovate to become more competitive considering globalization. Researchers and practitioners proposed different solutions to managing improvement, such as idea management described in innovation and information technology [3]. Companies aim to achieve breakthroughs of innovation. However, their ideas are the real issue. In recent decades, front-end innovation has been recognized for its potential to strengthen innovation capability [4]. The early stage of new product development (NPD) has broadly recognized the same meaning as the fuzzy front-end phase [5]. Researchers believe that the fuzzy front end significantly influences the whole process and the outcome (input–output process) since it substantially affects the innovation's design and total costs [6].

Therefore, the given field has paid more attention to fuzzy front-end innovation. Ideation is a vital phase of ambiguous front-end innovation and an integral part of the design [7] critical to business success [8]. Particularly conceptual ideas are the foundation of a product that continuously determines the design solutions' success and has the most significant impact on the product development cost [9]. From a business perspective, improved design ideas can shift buying decision criteria, benefiting companies' competitiveness [10]. The idea represents the potential starting point of innovation that strengthens individual and innovative company capability [4]. The research chooses to pursue a higher quality of ideas since quality ideas heavily impact success [11,12]. How can we improve the effectiveness and appropriateness of idea management systems, applications, tools, or methods?

Idea management is an approach for capturing, organizing, and developing ideas. Several researchers have investigated the idea management area to improve idea quality [5,10,13]. Various challenges to proposing creative ideas arose because of too many collected ideas and limited relevant information [2]. Additionally, by managing ideas, companies can save on research costs [14], extend users' coverage [15], impact future development [16], maintain unique competency [10], etc. Many companies established research and development (R&D) and user experience (UX) departments [13]. Idea management in the R&D lab managed idea flow as information processing [17]. Researchers have discussed the process of idea management from insights to validation [18,19].

Moreover, they have discovered external idea generation, in which community innovation requires collecting ideas through the idea management system [13,20]. Using tools to track and record idea development, researchers can further discuss, reflect, and offer the ideation [21]. The essential requirements for designers are capturing, managing, and utilizing design ideas [22]. Furthermore, because management is not single-use, research in different contexts may require various idea management tools. Tools help to illuminate the apparent path for the growth of ideas and all evidence contexts [12]. Idea management tools allow one to capture, generate, and save ideas from stakeholders. As such, researchers can obtain new knowledge to iterate current products, propose new products [23], and expand the problem space [8].

Nevertheless, previous research only investigated the quality of ideas and did not include UX of the idea management system. Researchers discuss the process model [24], life cycle [18], success factor [12], and quality of ideas [25]. Still, research should address the interaction of people and idea management systems, which would help showcase how different concepts are managed in other contexts [4]. There is a gap between an ideal proposal and implementation in the real world. It is challenging to connect product planning, design, and engineering teams with trust and certainty [12]. For example, UX designers reflect on formal ideas [26]. They need to pay more attention to the importance of a methodical and sustainable process since the idea management tools should build on a specific conceptual framework [3,27]. Only a few studies mentioned the design of idea management tools from the perspective of UX. For example, Inie and Dalsgaard investigated the cognitive and social processes of users [22]. Moon and Han [28] developed a creative idea generation methodology for the fuzzy front end (FFE) of radical innovation from the perspective of UX.

It is imperative to understand user behavior, especially interaction when using the idea management system to suggest a more holistic and efficient idea management system. It is used to conduct market research. Therefore, this scoping review aimed to understand the criteria, potentials, tools, and issues emerging in the idea management area from 2010 to 2020 through an overview of the academic literature. The specific objectives were to examine (1) the overview of the idea management tools and (2) the evaluation methods and criteria of these tools.

2. Materials and Methods

A scoping review represents a practical way to map fields of study in which the material range could be more straightforward. It summarizes and disseminates current research findings from a broader range of views [29]. This study follows the JBI reviewer's manual [30]. We follow the PRISMA-ScR protocol standards [31]. This scoping review aims to understand what idea management tools, systems, or applications were utilized and how those were evaluated.

2.1. Inclusion and Exclusion Criteria

Studies were included, highlighting idea management applications and defined outcomes to assist designers over the last ten years (from 2010 to 2020). Studies published in the English language were considered, including studies that reported one of the eligible outcomes, including the tool, strategy, application, product, system, or platform. However, the research team excluded studies that were review articles or systematic reviews and had objectives only as an iteration of current tools instead of developing new tools or entailing research commentary without initial results [32]. The included studies should clearly describe the process and results of idea management tools. Analysis without users' data was excluded. The studies where idea management was not the prime focus of the study (e.g., only mentioned in the discussion or references) were excluded. The study must include information to assess the methodological quality. Full papers that could not be retrieved were excluded.

2.2. Search Strategy

To reflect the position of all uses of the idea management tools in the design context, keywords about idea management tools were used (see Table A1 in Appendix A). Literature search strategies were developed using idea management titles and outcomes related to devices, applications, platforms, software, design, and systems. Since an "idea" may have many combinations, this study looked for the exact phrases using database settings. The electronic database searches were supplemented by combining Scopus, ACM Digital Library, Web of Science Core Index, Elsevier ScienceDirect, and SpringerLink. Only journals were considered. The literature searches were limited to English, with a range covering the last ten years (2010–2020). To ensure literature saturation, the research team scanned the reference lists of the included studies for additional sources and authors' files to guarantee that all relevant material had been identified.

2.3. Screening

The database search resulted in 82 titles, with 78 identified from a database search and four from a Google Scholar search. As per Figure 1, it was a three-level screening. The review authors independently screened the titles. Then, abstracts were yielded by the investigation against the inclusion criteria. The first screen level entailed rapid title screening, followed by the reviewing of author pairs, filtering of full-text reports, and deciding whether these met the inclusion criteria. The team sought additional information from study authors to resolve eligibility questions, and disagreements were resolved through discussion. The next level was full-text screening that screened the titles and abstracts. Thirty-seven full-text articles were assessed for eligibility. Commentary without actual results was excluded since idea management was not the focus of the study (e.g., only mentioned in the discussion or references). Reports that required more information to assess the methodological quality or the complete reports that could not be retrieved were excluded. Ten articles were identified through reference scanning, and finally, 38 articles were included in the review.

2.4. Data Extraction and Analysis Method

The results, including focus, findings, target users, and tools, are presented in diagrams and tables. Two analysis parts summarize articles based on a standardized form and thematic synthesis. Using a standardized format (as Table A1 in Appendix A shows) and a detailed instruction manual that informs specific tailoring of the NVivo 12 file, two reviewers extracted data independently and duplicated them from each eligible study. To guarantee consistency across reviewers, the team conducted calibration exercises before starting the review. The data abstracts included demographic information, methodology, design details, and all reported essential outcomes. The reviewers resolved disagreements through discussion, and one of the two arbitrators adjudicated unresolved disputes. The data extraction covered basic information about the study, research methods, and findings.

The research team utilized thematic synthesis to identify trends and opportunities for developing idea management tools. The thematic synthesis had two stages, including coding text and developing descriptive themes. To structure and gather idea management outcomes, researchers selected texts that reflected the perception and used the context of tools, applications, and strategies.

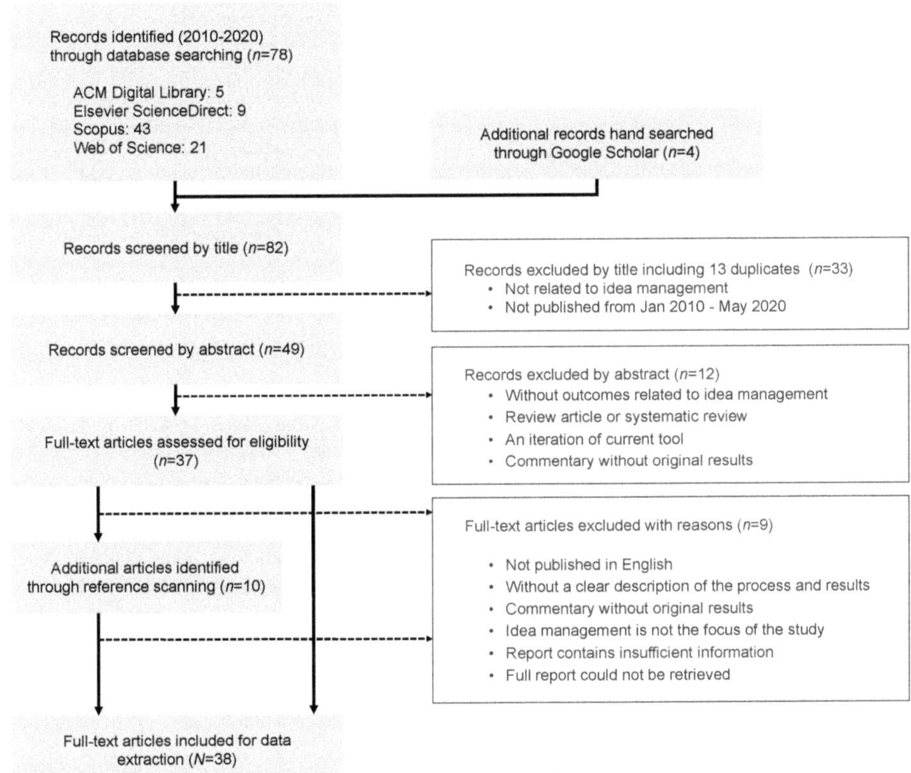

Figure 1. Article screening process of the scoping review.

At first, the team sorted publications per the tools of idea management, with selected texts reflecting the views and descriptions of idea management tools. The team then organized them in a matrix where texts were clustered and coded [33]. The general characteristics of the literature were organized into descriptive themes. The research team identified and described codes reflecting how the concept was described in the literature. Thus, the encoding was to bring out the essence of the texts that illustrated the concept of idea management tools.

3. Results

3.1. Overview of the Tools

As Table 1 shows, we classified 38 tools based on outcome classification, theories to support idea management work, type of user, and design task.

3.1.1. Classification of Tools

Three categories exist within idea management research: digital tools, guidelines, and frameworks.

- Digital tools (21 papers, 70%). This category contains idea management tools that support idea management work including local work and online collaboration. For example, these tools may zoom in on idea generation and can display images and texts with different inputs that inspire and promote idea management, especially idea generation. Furthermore, these applications tend to focus on a single context. Most applications are designed for computer terminals, with only one being a mobile app [21]. These applications support the entire process of ideas, for instance, product planning [16], idea generation [2,7], and ideation [34,35]. Some applications provide vital structure and suggestions [23] to express explicit knowledge [34]. Other applications rely on the company's internal system [36]. One research paper evaluates the classroom climate by using the application to stimulate the generation of ideas and originality [37]. Some of the digital tools are more complete, supporting more general context with users [5]. It can test more users in natural settings following their diverse experiences [5,38,39]. The purpose of systems is to increase collaboration [40], transparency [38], and quality [41,42]. Three research studies involve more users via the Internet, such as forums based on information and communication technology, developing a distributed-innovation capability [36]. As such, this research focuses on fostering the habit of posting ideas on the system [38] and increasing the ideas' quality. The idea management system concentrates more on anonymous online collaboration, attracting users to contribute.
- Guidelines (five papers, 16.6%). Guidelines are instructions or printed materials to support overall idea management work. Users can follow the instruction of strategies and guidelines to work on idea management effectiveness. Idea management strategies are designed more toward professional practitioners such as design-driven entrepreneurs [10] and experienced engineering teams [43]. Guidelines explain how designers used product characteristics for concept identification and how previous concepts were turned into new solutions by adjusting their characteristics [9]. These strategies are summarized by experienced designers [43] or those proposing high-quality ideas [25]. Furthermore, cards are a way to display the context and explanation [43]. Researchers identify a conceptual framework by understanding the development of new ideas for familiar problems or daily design activities [22]. Guidelines are derived from users' design activities, including the activities of designers, engineers, and entrepreneurs. It means research adopts more user research methods than other format outcomes.
- Framework (four papers, 13.3%). A framework helps explain the process, structures, and relationships of idea management processes. It usually includes the holistic and contextual views. The framework shows the metrics [15], life cycle [19], business model [44], and process [45] regarding idea management. These frameworks were developed following a more apparent scope and purpose than the other three outcome formats. For example, the metrics introduce the annotation of ideas with a domain-independent taxonomy that describes concepts, gathering creative ideas from large groups [15]. It presents a relationship between different stages and elements from a new perspective, leading to more innovative ideas.

In conclusion, more studies focus on digital tools to support idea capturing, generation, refinement, and selection. Guidelines assist professional designers who have the expertise to conduct idea management. The framework shows a visual overview of the idea management process.

3.1.2. Use of Theory to Design the Idea Management Tool

The designed idea management tools used different theories to develop their key features. Below we summarize the theories mentioned in the 35 studies.

Cognitive psychology (10 papers, 40%). This perspective is focused on individual studies. Based on cognitive psychology theory, participants have a distinct way of experiencing the world. From the standpoint of cognitive psychology, idea management tools related to emotion and memory search converge and diverge. To improve idea management activities, consumers' emotions using sentiment analysis can be studied [13]. From a memory search perspective, researchers propose a cognitive model [46]. Treating idea management as a creative activity is conducted according to the reflective practice theory of designing [47]. The focus is on the example design by inspiration and fixation influence [48] and fragment ideas using morphological analysis [10]. Specific studies are revised from the former design behavior framework [21] or complementary basic schemata [35]. Stimuli discussion is the most complicated topic given the many types of stimuli, such as visual and verbal, abstract and concrete, emotional, and cognition. The purpose is to assist designers in producing creative ideas [2]. The analogy applies the structured knowledge from a familiar field to one less familiar. One tool uses analogical reasoning and ontology to provoke participants to generate more creative ideas [7]. The study highlights that the heartbeat has an impact on the creativity of participants. Therefore, using heartbeat variations to create high-quality ideas during brainstorming can be helpful [11]. From the perspective of the cognitive psychology, ideas are influenced by various types of ideations, intuitive and logical [49], type of design [8], and inspirational resources [18]. Cognitive strategies can broaden the analysis perspective regarding further exploration [44]. Furthermore, because of inherent design knowledge, there are cognition differences between participants, such as experienced designers and novice designers with idea generation reference behavior [49], and engineering performs better in individual brainstorming sessions [9]. Triggers promote a higher quality of ideas [50] for individual designers or design teams [13]. Studies show that stimuli significantly impact problem-solving styles [47], with words, images, and videos as potential triggers [21]. There are different types of incentives, low levels (concrete) and high levels (abstract) [48], tests, and visual [46]. Sharing diverse ideas can stimulate other ideas [25].

- Information management technology (eight papers, 32%). Idea management is a sub-category of information management. Information management technology can foster collaboration and transparency among participants with diverse experiences [5], possibly leading to a new service idea [44]. Three studies target crowds that gain more advantages thanks to information technology than others since it helps manage a large amount of information [38]. Public service, aging, and intelligent space fields require diverse user feedback [15]. The other existing tool analogy recommends an approach that helps organizations to improve how they generate new ideas [42]. Ideas could be considered knowledge, specifically explicit knowledge [41]. Users have a hold on the consumers' domain-specific knowledge [51]. These two studies aim to express the evident expertise of participants.

Table 1. Overview of the tools. * Denotes the characteristics of the tool.

Author	Purpose of the Tool	Outcome Type			Use of Theory (to Support Idea Management)			Participant Categories			Design Task	
		Digital Tools	Guidelines	Frameworks	Cognitive Psychology	Information Management Technology	Social Psychology	Software Designers	Hardware Designers	Stakeholders	Individual	Group
Cheng, 2016 [21]	Support the idea development	*			*			*			*	
Munemori et al., 2018 [41]	Enable the expression of explicit knowledge	*				*		*				*
Riedel et al., 2010 [34]	Promote associative ideation	*					*			*		
Žavbi et al., 2013 [35]	Assist engineering designers in generating concept designs	*			*				*		*	
Ardaiz-Villanueva et al., 2011 [37]	Encourage creativity and analyze classroom atmosphere	*					*		*		*	
Benbya et al., 2018 [36]	Develop a distributed-innovation capability	*				*				*		
Murah et al., 2013 [50]	Provides the structure and the platform to contribute ideas	*								*		
Howard et al., 2011 [8]	Use internally sourced stimuli	*							*		*	
Fiorineschi et al., 2018 [23]	Make recommendations for new uses for products	*						*			*	
Han et al., 2018 [2]	Assist designers to produce creative ideas	*			*			*			*	
Parjanen et al., 2012 [5]	Investigate how brokerage works in a virtual setting	*				*				*	*	
Xie et al., 2010 [42]	Manage the whole process of idea and support team creation	*								*	*	
El et al., 2017 [43]	Organizations to improve their ways of generating new ideas	*				*						

Table 1. *Cont.*

Author	Purpose of the Tool	Outcome Type			Use of Theory (to Support Idea Management)			Participant Categories			Design Task	
		Digital Tools	Guidelines	Frameworks	Cognitive Psychology	Information Management Technology	Social Psychology	Software Designers	Hardware Designers	Stakeholders	Individual	Group
Gonçalves et al., 2014 [49]	Report preferences for inspirational approaches	*						*			*	
Han et al., 2018 [7]	Create ontologies that facilitate reasoning	*			*					*	*	
Yu et al., 2013 [40]	Increases the creativity of ideas across generations	*					*			*		
Alessi et al., 2015 [38]	Stay in line with the needs of society	*				*				*		*
Sadriev et al., 2014 [52]	Build up direct purposefully the innovation development processes	*										
Kokogawa et al., 2013 [53]	Provide photographs are used to support idea generation	*						*			*	
Bacciotti et al., 2016 [16]	Support product planning in ideation processes	*							*		*	
Munemori et al., 2020 [11]	Use heartbeat variations for creating high quality ideas	*			*							
Bayus, 2012 [25]	Maintaining an ongoing supply of quality ideas from the crowd		*				*			*		*
Daly et al., 2012 [9]	Define concepts using product characteristics		*				*		*		*	
Tanyavutti et al., 2018 [10]	An idea generation method for the concept		*		*					*		*
Yilmaz et al., 2013 [43]	Suggest how to develop new ideas for familiar problems		*				*		*			*
Inie et al., 2020 [22]	Identify a conceptual framework of ten strategies		*					*				
Westerski et al., 2013 [15]	Collect ideas for innovation from large communities			*		*				*		

Table 1. *Cont.*

Author	Purpose of the Tool	Outcome Type			Use of Theory (to Support Idea Management)			Participant Categories			Design Task	
		Digital Tools	Guidelines	Frameworks	Cognitive Psychology	Information Management Technology	Social Psychology	Software Designers	Hardware Designers	Stakeholders	Individual	Group
Jeong et al., 2016 [44]	Lead to a new service idea			*		*				*		*
Westerski et al., 2011 [19]	Aid gather, organize, choose, and manage the creative ideas			*						*		*
Börekçi et al., 2015 [45]	Provide insights for generating ideas in design thinking			*				*				
López-Mesa et al., 2011 [47]	Effects of additional stimuli on the design process				*				*			
Schlecht et al., 2014 [54]	Influence of resource constraints on idea generation								*			
Luo et al., 2015 [51]	Enhance consumer performance in online idea generation platforms					*				*		
Vasconcelos et al., 2017 [48]	Compare the inspiration effects from these two types of stimuli				*			*			*	
Liikkanen et al., 2010 [46]	Experimentation in conceptual product design				*				*			
Oldschmidt et al., 2011 [55]	Present stimuli along with a design problem							*			*	
Sozo et al., 2019 [13]	Emotional stimuli for creativity				*			*			*	
Yang et al., 2021 [56]	Understand social interaction									*		*

- Social psychology (six papers, 24%). Idea management is commonly considered group work, especially regarding idea generation and selection. Enhancing interaction can lead to higher-quality idea management by considering the participants as a group. Heuristics, peer interaction, and social context are the primary theories adopted by studies. A heuristic is a potential shortcut to reducing psychological consumption. It is a simple rule allowing one to make complex decisions or inferences quickly and effortlessly, providing design heuristics to designers, and helping engineers to facilitate concept generation [9]. Thus, the study recruits experienced designers to summarize their idea's development [43]. The other perspective is activating interaction between participants by establishing a social context [37] and sharing ideas with group members [34], especially with peers. During group collaboration, including face-to-face and remote modes, participants are more likely to generate higher-quality ideas and increase their creativity [39]. One of the most practical problems is tracking the concept over time to identify the challenges [25]. It would be easier for the right question to lead to more creative ideas. A straightforward design solution space that involves known elements is needed [9]. It includes specific dimensions, contexts, constraints, and goals [36]. Regulations can change the design solution. These constraints could be time awareness [54], paper-based or computer-based time design activity, time on searching online [21], and detailed or less detailed design briefs [45].

3.1.3. Tools to Support Individual and Group Work

Unique tools are included, supporting personal work and group tools that aid team collaboration.

- Individual work (16 papers, 61.5% total). Table 2 shows these tools are designed for individuals and most likely apply to university students or individual designers. These studies aimed to generate unique ideas and provide different triggers [48] and tools [9] to test better ways of generating more creative ideas that uphold specific constraints. Some of these tools have another hypothesis. For example, stimuli presented to student designers through texts and design problems would enhance the quality of their design solutions [55]. Moreover, tools can be tested in a lab environment. One of the tools is based on computer-aided design (CAD), designed by Autodesk. Bacciotti's research aids NPD initiatives' product preparation and ideation processes [14]. Individual work only has a single user and relies on the quality of ideas. Therefore, these individual tools focus more on promoting the unique potential of creative design. The differences of the tools depend on the design briefs users may work on; if users have limited scope of the design topics, the tools provide more concrete support such as CAD to support product design [16]. If users have broader design aims, the tools offer more abstract guidance to support various design direction, such as design steps [40].

- Group work (10 papers, 38.5% total). As Table 3 shows, creative design activities consistently involve a group of participants as part of the concept design. These participants have different roles, including designers [46], entrepreneurs [10], service providers [38], mobile carriers [44], and crowdsourcing users [25]. The main goal when asking a group of participants is to improve the quality of ideas during the idea generation stage. These tools aim to collect diverse users' ideas, suggestions, and feedback. Following the generation of better ideas is the first goal; the second goal is selection [46], refinement [41], and sharing [36]. Furthermore, some researchers recruit stakeholders to develop ideas of higher feasibility, especially regarding public service, providing service to diverse users [38]. The idea management collaboration consistently occurs in a professional environment with many employees [36], design teams [46], and engineering teams [43]. The different focus is linked to the relationship between the users and company, such as employees [43], informants [41], and co-creator [25]. Besides inter-role collaborations in the professional group work, such as idea management in UI design teams [13], others may involve the interdisciplinary group members [40]. The interdisciplinary collaboration may balance the different knowledge reserve and different level of involvement and accelerate the collaboration. Therefore, understanding the relationship between users' social interactions and innovation contributions may promote team idea management.

Table 2. List of individual-based tools.

Author	Name of the Tool	Purpose of the Tool
Bacciotti et al., 2016 [16]	CAD	Support product planning in ideation processes of new product development (NPD) initiatives
Han et al., 2018 [7]	The Retriever	Create ontologies that facilitate reasoning over real-world datasets that are sufficiently deep and comprehensive to inspire original thought
Parjanen et al., 2012 [5]	Not mentioned	Investigate how brokerage works in a virtual setting where experts from various fields and perspectives engage
Daly et al., 2012 [9]	Design Heuristics	Support designers' defined concepts using product characteristics and modify them to create new solutions
Cheng, 2016 [21]	AGCI interface prototype	Support the idea development of individual designers and greatly affect idea communication
Žavbi et al., 2013 [35]	Computer tools	Assist engineering designers in generating concept designs
Ardaiz-Villanueva et al., 2011 [37]	Wikideas and Creativity Connector tools	Encourage creativity and analyze classroom atmosphere
Xie et al., 2010 [40]	IMS	Manage the whole process of idea and support team creation
Vasconcelos et al., 2017 [48]	Not mentioned	Compare the inspiration effects from these two types of stimuli (abstract and concrete)
Oldschmidt et al., 2011 [55]	Not mentioned	Present stimuli along with a design problem, which would improve the quality of their design solution
Howard et al., 2011 [8]	Sweeper	Use internally sourced stimuli
Sozo et al., 2019 [13]	Emotriggers	Emotional stimuli for creativity
Fiorineschi et al., 2018 [23]	A nine-step method	Make recommendations for finding potential new uses for current products and/or technologies
Han et al., 2018 [2]	The Combinator	Assist designers to produce creative ideas and be beneficial in expanding the design space
Kokogawa et al., 2013 [53]	GUNGEN-PHOTO	Provide photographs that are used to support idea generation
Gonçalves et al., 2014 [49]	Not mentioned	Report preferences for inspirational approaches

Table 3. List of group-based tools.

Author	Name of the Tool	Purpose of the Tool
Tanyavutti et al., 2018 [8]	Not mentioned	An idea generation method for the concept
Alessi et al., 2015 [38]	Sentiment analysis tool and gamification	Stay in line with the needs of society and foster collaboration and transparency
López-Mesa et al., 2011 [47]	SCAMPER	Effects of additional stimuli on the design process and on the creativity of the outcomes
Jeong et al., 2016 [44]	To-Be curve	Lead to a new service idea and new business models for smart spaces with adequate technology and market feasibility
Benbya et al., 2018 [36]	Not mentioned	Develop a distributed-innovation capability
Munemori et al., 2018 [41]	GUNGEN-Web II	Enable the expression of explicit knowledge

Table 3. *Cont.*

Author	Name of the Tool	Purpose of the Tool
Westerski et al., 2011 [19]	Not mentioned	Aid in gathering, organizing, choosing, and managing the creative ideas offered by the communities established around businesses or organizations
Yang et al., 2021 [56]	Not mentioned	Understand the relationship between users' social interactions and innovation contributions
Bayus, 2012 [25]	Dell's IdeaStorm	Challenges in maintaining an ongoing supply of quality ideas from the crowd over time
Yilmaz et al., 2013 [44]	Design Heuristics	Suggest how experienced designers develop new ideas for familiar problems

3.1.4. Users of Idea Management Tools

We identified three types of users, software designers, hardware designers, and stakeholders, who use idea management tools.

- Software designers (11 papers, 31.4% total). Design is the activity of conceiving and planning what does not yet exist. Software designers design interfaces, features, and processes, including interaction designers, graphic designers, and UX designers. They are mentioned in the highest frequency alongside a variety of types, including novice designers [51], student designers [55], and professional designers [49]. Some research discusses the impacts of remaining stimuli, such as abstract and concrete stimuli [51] and emotional triggers [13]. Designers must capture inspiration in daily life. Therefore, some tools considered this use scenario. Furthermore, different types of designers may have diverse needs, such as interaction designers [22]. All the research highlights the importance of creativity and proposed tools to help designers improve their creations.
- Hardware designers (nine papers, 25.7% total). Hardware designers design products, structures, and environments and include industrial and engineering designers. Usually, they need support in managing their ideas, especially engineering designers. Throughout these studies, engineering students are equal to industrial [9] and novice designers [47]. Instead of designers, they have limited space for innovation since the tangible prototype is limited by material and structure. Furthermore, some of them may already have patents protected. As such, studies aim to understand how experienced designers transform ideas into solutions [9] and then follow the cognitive process, assisting engineering designers in generating concept designs [35]. Design challenges can focus on a smaller scope, such as product line improvement [43]—an engineer's idea management concentrates on improving the idea generation of tangible solutions.
- Stakeholders (15 papers, 42.9% total). The remaining roles in idea management research are within the professional field, considered as stakeholders. The functions include entrepreneurs [10], service providers [38], mobile carriers [44], many employees [36], and crowdsourcing community users [15]. It switches the focus from ways to improve creation to collaboration. Crowdsourcing attracts users to propose ideas [25], allowing the company to collect more ideas from the idea pool [56]. It is based on quantity leading to quality [1,3]. Additionally, the company can create a community where users can share their ideas and contribute. Peer interaction impacts idea generation [49]. Some researchers recruit users from the community, but design idea management systems for designers come from the future [39]. Crowdsourcing should be traceable over time [25], with the quality and motivation of generating ideas changing.

3.1.5. How Do Tools Support Idea Management Work?

Figure 2 displays how all research considers idea generation since it is the primary consideration of idea management. The phases include capturing, generating, implementing, monitoring, refinement, retrieving, selecting, and sharing.

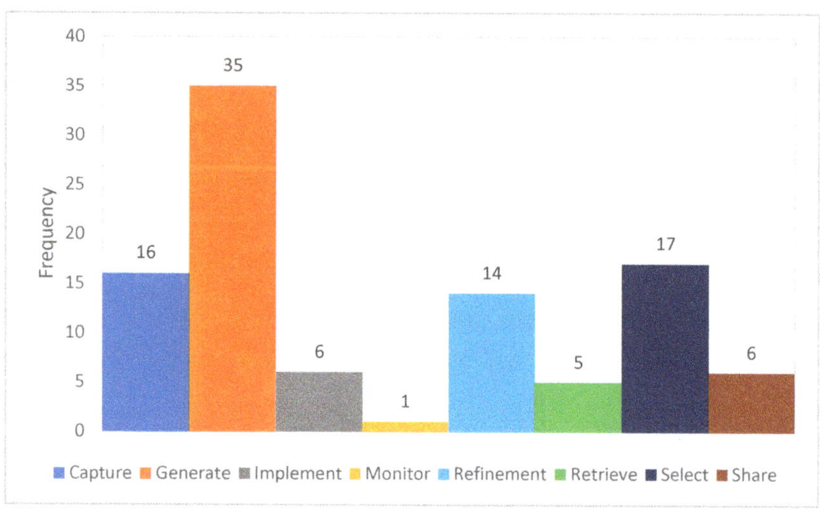

Figure 2. Frequency of idea management research phases.

The idea selection phase was mentioned in seventeen papers. Ideas must be prioritized when crowdsourcing or teams have to provide many ideas. This step focuses on the impact of ideas [19]. Sixteen papers discuss capturing ideas by studying the stimuli, including photographs, abstract, and concrete stimulations [48]. Furthermore, the crowd's innovation has to interact with others' ideas [35,53] through social interaction such as commenting and teamwork [53]. Through refinement, potential ideas are combined to propose a better one. Fourteen papers study idea refinement with creative activity being iterative [18]. Thus, it must combine the former ideas following the current context. Implementation (six papers) is more relevant to teamwork, which must develop ideas [38]. Sharing (six papers) is more about crowdsourcing, which requires sharing ideas in the community. Some designers and engineers need to retrieve ideas (five papers) from the idea pool [45]. Only one study mentions monitoring related to public service [38].

3.2. Effectiveness Evaluation of Idea Management Tools

The section below explains the findings following the idea management tool evaluation.

3.2.1. Evaluation Criteria of Idea Management Tools

According to Table 3, there are various criteria for evaluating idea management tools. Based on the exact definition of criteria, some criteria share the same meaning, such as novelty, obviousness, creativity, and originality, which all represent uniqueness of the ideas; therefore, the research team combined these criteria into a cluster. Most of the standards focus on justifying the idea's quality and impact.

The most frequently mentioned criterion is novelty, meaning the idea should be unusual and unexpected [8], implying originality and creativity. A novel idea commonly expands the design space [46]. Furthermore, quantity is an important criterion. It can be evaluated by the number of views [16], even given a time limit [25]. Quantity should be considered the criterion since quantity leads to quality, which is a mixed criterion. Quality does not have a precise measurement, making researchers think about its adoption intention [51], usefulness, and flexibility [2]. They also invite experts to evaluate the quality of ideas [13]. Variety is seen as a measure of design solution scope. It requires further analysis to define the cluster category [13]. Usability measures how easily a user interacts with the ideas [21] and perceived usefulness [8]. Feasibility means applicable, adaptive task constraints [8], which are both marketable and precise [54]. Satisfaction also requires an evaluation from others, which usually adopts the Likert scale [41]. Improvement equals the added value of recommender systems, which are not considered in older ideas [42].

3.2.2. Testing the Effectiveness of the Idea Management Tool

Applying idea management tools to complete a specific design challenge was the most common tool evaluation method. Fourteen papers utilized experimental comparison, specifically group comparison. For example, a CAD tool supports idea generation, comparing ideation performance between engineering students who use the tool and those without it [16]. They receive the same design brief used for the screwdriver design. Most research defines a transparent object, including a tangible product such as a washing machine [13] and a digital game such as the ultimate game [11].

Digital tools are idea management systems and platforms (five papers) that are created through user developmental testing. An example is the idea management system for team creation. Idea management is a continuous process. The testing lasts years, and a further two months are necessary to evaluate performance by asking to design a die mold [40]. The objects are stable, like the employees and crowds in the community. A survey method is adopted to collect the data.

Studies with interactive designs conduct evaluation experiments, such as application and platform evaluation. Most research adopts experimental comparison as the first step of research with more precise evaluation criteria, and seven papers select students to participate in the experiment, especially design and engineering students. Longitudinal experiments require target users to participate. Therefore, crowds [38,53], employees [36,40], and aging people [5] are considered.

4. Discussion

Numerous idea management tools have been developed to aid designers in generating more innovative ideas and improving their effectiveness.

4.1. How Are Idea Management Tools Classified?

Idea management tools can be divided into three clusters: digital tools (70%), guidelines (16.6%), and framework (13.3%). Following this research, digital tools are the most common outcome of idea management research because these tools could be used to address a specific problem and support idea management work directly. On the other hand, the system is the most complex outcome and should adapt to various challenges. It takes a long time to build prototypes, and it is hard to test them in the field—guidelines and frameworks face the same specific question but with different forms. Strategies use text to explain the details. Frameworks use visual models such as the matrix and life cycle.

Given the complexity of design [52,57], professional designers and students always use computers to solve design-related problems. Computer-based tools are simple to test in a lab environment, with professional designers always conducting idea management on computers. However, mobility is a trend in design research, while mobile phones have a larger screen and a higher-speed processor. Portable devices would help them catch ideas in their everyday life. Thus, we should consider the features on the mobile phone that can support designers anytime and anywhere, such as idea capture and idea selection. Furthermore, these applications focus more on the idea generation phase, providing stimuli to improve idea quality.

There are more opportunities for different phases of idea management. The system manages ideas in a complex process well, such as group work, especially the community's role in crowdsourcing. More stakeholders would be involved in making decisions. Benbya and Leidner emphasized that the system must consider the interaction between users and systems on features, including comments and voting [36]. Guidelines and frameworks differ since they are more theoretical. Strategies provide instruction in idea management, guiding design activities, not through specific tools but with a clear direction. Such a conceptual framework of ten systems supports interaction designers [22]. It summarizes the explicit knowledge of how designers can improve the management of ideas. Tools designed according to the theoretical framework may improve the quality and efficiency of thoughts. Modeling is the other way to organize the results of idea management. It is more detailed than strategies targeting the context, such as emotional metrics [15]. This study argues that methods and framework should be tested in the real world. Therefore, research should explain the tool's theoretical basis and performance.

In addition to various carrier forms, these tools have the following three theoretical origins: cognitive psychology theory (40%), information technology theory (32%), and social psychology theory (24%). Cognitive psychology theory wants to improve individual performance. Information technology theory aims to enhance explicit knowledge transformation, and social psychology theory seeks to provide opportunities for social interaction. Unsurprisingly, cognitive psychology represents wanting to improve idea management performance. The latest research introduces the idea quality forecasting models to measure and forecast the quality [58]. It considers that stimuli would affect users to avoid fixation and provide inspiration [48], such as visual and verbal, abstract and concrete, and emotion and cognition. These triggers define idea quality. Even physical conditions affect them [11]. Information management technology relies on the internet and original data, meaning that the research is more likely to build on the current system. The tools are affected by cognitive models of various job roles. Researchers have mentioned the hierarchical roles of the idea provider, for example, an idea provider with a higher hierarchical rank may have more of an effect of constructive feedback on the idea quality [14]. Users of the idea management tools include software and hardware designers, professionals, and crowds in the community. It has different types of ideations, intuitive and logical [49], types of design [6], and inspirational resources [21]. Still, there needs to be research comparing different cognitive models of idea management. Furthermore, Cheng et al. observed the moderating effects of perceived goal clarity, knowledge self-efficacy, and cognitive demand on the correlations between germane cognitive load and idea convergence quality [59].

The social psychology perspective treats idea management work as a group activity. It introduces facilitator comment activities that increase the possibility of views colliding, leading to higher-quality ideas [37]. The social psychology research summarizes strategies from former idea management activities that can support novice designers in improving their effectiveness. Heuristics is a helpful way to enhance peer interaction [9]. However, the group discussion displays more uncertainty. It requires an experienced facilitator or a higher fault tolerance system. If we desire to generalize the findings, the idea management tools should be stable, practical, and effective. Hence, we should investigate current idea management stimulus, technologies, and strategies. Then, we could decide which combination improves the effectiveness of idea management tools. The right design questions can affect the idea management results. The questions always contain timing of awareness [54], paper-based or computer-based time design activity, time performing online searches [21], and detailed or less detailed design briefs [45]. There is increasing talk about measurable elements such as times and types of activity. Regarding the design brief, it is not easy to evaluate its details. Triggers promote higher-quality ideas [50] for individual designers or design teams [13]. It is still necessary to investigate which stimuli trigger creative ideas more effectively, especially the performance between words and visual stimuli. Future research should study low-level (concrete) and high-level (abstract) mixing.

4.2. Tools to Support Individual Work and Group Work

There are more individual tasks (61.5%) since it is more straightforward for researchers to design individual lab studies or empirical studies, providing concrete results and making it easier to recruit users. Personal tools solve design problems by a single designer or engineer, primarily for academic purposes. With 38.5% of researchers having studied group collaboration, most focus on community idea management. Users post and discuss their ideas via the forum. Group tools can support more idea management work than scenario diversity, even for professional purposes.

However, idea management work continuously occurs within teams [22]. With teamwork studies, it is easier for researchers to understand the actual needs of the users. Only some of them consider professional settings, which involve stakeholders such as designers [47], entrepreneurs [8], service providers [38], and mobile carriers [44]. Furthermore, studies geared toward design and engineering students as subjects to test the feasibility may contribute less to professional scenarios. Future research should focus on the interaction between professions with individual users. The single involvement of users may generate good ideas. However, fewer prove feasible and valuable. Practitioners have relied on users' sense of belonging and used the voting system to collect consumer feedback [57].

Yet, if users find their voice does not contribute to any change, they may feel frustrated and stop sharing their opinions. Researchers offer suggestions for designing online studies that respect respondents' time, effort, and dignity following participants' crowdsourcing experiences [60]. If we rely on algorithms, these choices are frequently complicated, necessitating algorithms that consider users' strategic (or occasionally irrational) behavior and how available tasks or people are distributed over time [61]. The involvement of professionals can generate valuable ideas and can also pay attention to the critical needs of users. Researchers have reported that conflict, decreased collaboration, and diminished innovation are more common in teams with individuals in charge [62]. Therefore, we should create opportunities for professionals and end users to discuss their opinion and make decisions together.

4.3. Who Are the Target Users of Idea Management Tools?

There are three types of users: software designers (31.4%), hardware designers (25.7%), and stakeholders (42.9%). Designers can be divided into software designers, who design interfaces, features, and processes, and hardware designers, who create products, structures, and engineering environments. Software designers must be more creative since the cost of a design application is lower than a product, and they face fewer limitations [63]. Therefore, the research concentrates more on the improvement of creativity with tools. Inspiration proves an essential element [48].

Still, hardware designers have more constraints since the product must be usable. It has a lot of mechanical requirements, given a basic structure. Therefore, they investigate effects from different functional parts, using techniques such as sentiment analysis [38]. The design assignment is simplified to a sub-problem. It requires the design problem to be divided into parts easily. Stakeholders exert authority when making decisions. It depends on different departments and teams. Real-world settings impact the final product. However, more papers must explain the diverse needs of stakeholders only by providing a channel that involves them in the idea management process. It is difficult to determine whether stakeholders express more opinions in the discussion.

4.4. How Do Tools Support Idea Management Work?

Idea management tools can support different stages of idea management. As Figure 2 shows, capturing, generating, refinement, and selection are the most popular topics intensely affecting performance. The generating phase ($n = 35$, 100%) has drawn the most attention. All included papers support idea generation. It has been proven that idea generation's quantitative and qualitative components are vital to concept management [64]. Idea capturing ($n = 16$, 45.7%) and idea selection ($n = 17$, 48.6%) were other frequently mentioned phases. Ideas are generated and shared through the extracting process, while ideas are captured and evaluated utilizing the landing process [65]. The premise that organizations producing many ideas would pick excellent ideas has been the focus of most idea selection research [66]. Nevertheless, idea management is a holistic process. It should not consider only a few stages or a single stage. It should be viewed as a journey of ideas.

4.5. What Are the Evaluation Criteria of Idea Management Tools?

How useful do users consider idea management tools? Do the tools help provide a better idea management experience? Nineteen papers evaluating the performance of ideas management tools, reviewing the evaluation design, and testing the results are necessary to consider. Most of the studies are conducted by academic researchers. Fewer are designed by corporations such as Allianz [36]. Only some tools are tested in real contexts with target users [8,13,22,34,36,40]. As we mentioned, they often use students as subjects. Table 4 displays different criteria to evaluate the idea management tools. The research team combined the requirements with the exact definition. Most standards look to justify the idea's quality and impact. Some criteria are measurable, such as quantity, satisfaction, and usability.

Table 4. Frequency of criteria for evaluating ideas in the literature.

Criteria for Evaluating Ideas	Frequency	Reference
Novelty	11	[2,7,10,13,35,37,39,47,48,54,55]
Quantity	10	[2,7,8,10,13,16,25,34,47,48]
Quality	9	[2,5,7,10,13,25,34,47,51,53]
Variety	6	[13,16,35,43,47,48]
Usability	6	[2,7,10,11,21,54]
Satisfaction	2	[3,6,40]
Feasibility	2	[8,54]
Improvement	2	[34,42]

Regarding quality, improvement, and appropriateness, it is necessary to involve experts or users to identify the qualitative performance and turn it into a score [51]. Novelty, quantity, and quality are the most mentioned criteria. However, measures depend on different idea management requirements, such as radical and incremental innovation. Therefore, we should further explore the effectiveness of other criteria combinations, especially in different innovation types.

Researchers adopted an experimental comparison ($n = 14$, 73.7%) to evaluate the effectiveness and performed longitudinal experiments ($n = 5$, 26.3%). It was found that these methods map the classification of tools since the longitudinal investigation requires the mutual idea management system to test over months or even years. Furthermore, to evaluate the performance of tools, researchers need a benchmark against which to compare the performance. Researchers tend to use the usual methods to solve design problems, such as brainstorming and the K-J method. However, there are more techniques to improve the performance of idea management. Mikelsone and Liela found that researchers have created a wide range of potential routes for further study, demonstrating the infinite potential of this subject [20].

Supposing the idea management system relies on crowdsourcing, it is more technology-oriented and capable of managing a large amount of data. Crowdsourcing markets have gained popularity and show promise, facilitating theoretical and empirical research into the creation of algorithms to improve several features of these markets [61]. Still, these data are more quantitative, including satisfaction and feasibility. The convergence of ideas for further examination from a vast pool of candidate ideas of varying quality poses a significant challenge [59]. Therefore, we must decide which technique has the best performance. Future directions are related to the improvement of selecting the more popular benchmark and investigating how to enhance the idea quality in the crowdsourcing system.

5. Conclusions

We identified three types of idea management tools: digital tools, guidelines, and frameworks. Most tools are validated for the research context, but little is known about users and design tasks affecting their implementation. For those tools that collect users' ideas that will gather a large number of different ideas, designers lack channels to further investigate and understand the user suggestions. The idea management tools may involve stakeholders without a rich experience of managing ideas. Without effective communication between the designers and users, designers and other stakeholders may experience misunderstanding. To minimize the bias, we should further investigate the potential barriers of multiple stakeholders managing ideas and recommendations to overcome these barriers. Since idea generation is iterative, most idea management tools support a single stage (either capture or generate). Therefore, they cannot track the life cycle of the ideas in all stages of idea management. Therefore, it is essential to develop tools for managing ideas that would allow end users, designers, and other stakeholders to minimize bias in selecting and prioritizing ideas.

Author Contributions: Methodology, A.A.M.; formal analysis, D.Z.; data curation, W.L.; writing—original draft preparation, D.Z.; writing—review and editing, A.A.M.; supervision, W.L. All authors have read and agreed to the published version of the manuscript.

Funding: This research did not receive any specific grant from funding agencies in the public, commercial, or not-for-profit sectors.

Institutional Review Board Statement: Not applicable.

Informed Consent Statement: Not applicable.

Data Availability Statement: MDPI Research Data Policies at https://www.mdpi.com/ethics (accessed on 6 March 2023).

Conflicts of Interest: The authors declared no potential conflict of interest concerning this article's research, authorship, and/or publication.

Appendix A

Table A1. Articles used in the review.

Author	Purpose of the Tool	Outcome Type	Use of Theory (to Support Idea Management)	Participant Categories	Design Task	Effectiveness of the Design Outcome	Criteria of Effectiveness	Evaluation Type	Name of the Tools ($n = 38$)
Cheng, 2016 [21]	Support the idea development	Digital tools	Cognitive psychology	Software designers	Individual	Experimental comparison	Usability	Lab study	AGCI interface prototype
Munemori et al., 2018 [41]	Enable the expression of explicit knowledge	Digital tools	Information management technology	Software designers	Group	Experimental comparison	Satisfaction	Lab study	GUNGEN-Web II
Riedel et al., 2010 [34]	Promote associative ideation	Digital tools	Social psychology	Stakeholders	Not mentioned	Experimental comparison	Number, quality, and improvements	Field study	Melodie ICT Tool
Žavbi et al., 2013 [35]	Assist engineering designers in generating concept designs	Digital tools	Cognitive psychology	Hardware designers	Individual	Experimental comparison	Variety and better chance to find innovative solutions	Lab study	Computer tools
Ardaiz-Villanueva et al., 2011 [37]	Encourage creativity and analyze classroom atmosphere	Digital tools	Social psychology	Hardware designers	Individual	Experimental comparison	Creativity and affinity	Internet study	Wikideas and Creativity Connector tools
Benbya et al., 2018 [36]	Develop a distributed-innovation capability	Digital tools	Information management technology	Stakeholders	Not mentioned	Longitudinal experiment	Not mentioned	Field study	Not mentioned
Murah et al., 2013 [50]	Provides the structure and the platform to contribute ideas	Digital tools	Not mentioned	Stakeholders	Not mentioned	No evaluation	Not mentioned	Not mentioned	Kacang Cerdik
Howard et al., 2011 [8]	Use internally sourced stimuli	Digital tools	Not mentioned	Hardware designers	Individual	Experimental comparison	Frequency, originality, appropriateness, and unobviousness	Field study	Sweeper

Table A1. *Cont.*

Author	Purpose of the Tool	Outcome Type	Use of Theory (to Support Idea Management)	Participant Categories	Design Task	Effectiveness of the Design Outcome	Criteria of Effectiveness	Evaluation Type	Name of the Tools (n = 38)
Fiorineschi et al., 2018 [23]	Make recommendations for new uses for products	Digital tools	Not mentioned	Software designers	Individual	No evaluation	Not mentioned	Lab study	A nine-step method
Han et al., 2018 [2]	Assist designers to produce creative ideas	Digital tools	Cognitive psychology	Software designers	Individual	Experimental comparison	Originality, usefulness, fluency, and flexibility	Lab study	The Combinator
Parjanen et al., 2012 [5]	Investigate how brokerage works in a virtual setting	Digital tools	Information management technology	Stakeholders	Individual	Longitudinal experiment	Not mentioned	Internet study	Not mentioned
Xie et al., 2010 [42]	Manage the whole process of idea and support team creation	Digital tools	Not mentioned	Stakeholders	Individual	Longitudinal experiment	Satisfaction	Lab study	IMS
El et al., 2017 [43]	Organizations to improve their ways of generating new ideas	Digital tools	Information management technology	Not mentioned	Not mentioned	Not mentioned	Value for actor and organization	Not mentioned	Recommendation system
Gonçalves et al., 2014 [49]	Report preferences for inspirational approaches	Digital tools	Not mentioned	Software designers	Individual	Not mentioned	Not mentioned	Quantitative research	Not mentioned
Han et al., 2018 [7]	Create ontologies that facilitate reasoning	Digital tools	Cognitive psychology	Stakeholders	Individual	Expert evaluation	Not mentioned	Lab study	The Retriever
Yu et al., 2013 [40]	Increases the creativity of ideas across generations	Digital tools	Social psychology	Stakeholders	Not mentioned	Experimental comparison	Not mentioned	Internet study	Internet-scale idea generation systems
Alessi et al., 2015 [38]	Stay in line with the needs of society	Digital tools	Information management technology	Stakeholders	Group	Longitudinal experiment	Not mentioned	Internet study	Sentiment analysis tool and gamification

Table A1. *Cont.*

Author	Purpose of the Tool	Outcome Type	Use of Theory (to Support Idea Management)	Participant Categories	Design Task	Effectiveness of the Design Outcome	Criteria of Effectiveness	Evaluation Type	Name of the Tools ($n = 38$)
Sadriev et al., 2014 [52]	Build up direct purposefully the innovation development processes	Digital tools	Not mentioned	Not mentioned	Not mentioned	Not mentioned	Not mentioned	Not mentioned	Not mentioned
Kokogawa et al., 2013 [53]	Provide photographs that are used to support idea generation	Digital tools	Not mentioned	Software designers	Individual	Experimental comparison	Quality	Lab study	GUNGEN-PHOTO
Bacciotti et al., 2016 [16]	Support product planning in ideation processes	Digital tools	Not mentioned	Hardware designers	Individual	Experimental comparison	Not mentioned	Lab study	CAD
Munemori et al., 2020 [11]	Use heartbeat variations for creating high-quality ideas	Digital tools	Cognitive psychology	Not mentioned	Not mentioned	Experimental comparison	Usefulness	Lab study	GUNGEN-Heartbeat
Bayus, 2012 [25]	Maintaining an ongoing supply of quality ideas from the crowd	Guidelines	Social psychology	Stakeholders	Group	No evaluation	Quality and quantity	Lab study	Dell's IdeaStorm
Daly et al., 2012 [9]	Define concepts using product characteristics	Guidelines	Social psychology	Hardware designers	Individual	No evaluation	Not mentioned	Lab study	Design Heuristics
Tanyavutti et al., 2018 [10]	An idea generation method for the concept	Guidelines	Cognitive psychology	Stakeholders	Group	Experimental comparison	Not mentioned	Lab study	Not mentioned
Yilmaz et al., 2013 [43]	Suggest how to develop new ideas for familiar problems	Guidelines	Social psychology	Hardware designers	Group	No evaluation	Variety	Field study	Design Heuristics

Table A1. *Cont.*

Author	Purpose of the Tool	Outcome Type	Use of Theory (to Support Idea Management)	Participant Categories	Design Task	Effectiveness of the Design Outcome	Criteria of Effectiveness	Evaluation Type	Name of the Tools ($n = 38$)
Inie et al., 2020 [22]	Identify a conceptual framework of ten strategies	Guidelines	Not mentioned	Software designers	Not mentioned	No evaluation	Not mentioned	Field study	Not mentioned
Westerski et al., 2013 [15]	Collect ideas for innovation from large communities	Frameworks	Information management technology	Stakeholders	Not mentioned	Experimental comparison	Not mentioned	Internet study	Idea management systems
Jeong et al., 2016 [44]	Lead to a new service idea	Frameworks	Information management technology	Stakeholders	Group	Not mentioned	Not mentioned	Case study	To-Be curve
Westerski et al., 2011 [19]	Aid in gathering, organizing, choosing, and managing the creative ideas	Frameworks	Not mentioned	Stakeholders	Group	Not mentioned	Not mentioned	Internet study	Not mentioned
Börekçi et al., 2015 [45]	Provide insights for generating ideas in design thinking	Frameworks	Not mentioned	Software designers	Not mentioned	Not mentioned	Not mentioned	Lab study	Not mentioned
López-Mesa et al., 2011 [47]	Effects of additional stimuli on the design process	Not mentioned	Cognitive psychology	Hardware designers		Process-based and outcome evaluation	Novelty, variety, quantity, and quality	Lab study	SCAMPER
Schlecht et al., 2014 [54]	Influence of resource constraints on idea generation	Not mentioned	Not mentioned	Hardware designers	Not mentioned	Not mentioned	Novelty, appropriateness, technical feasibility, marketability, and clarity	Lab study	Not mentioned
Luo et al., 2015 [51]	Enhance consumer performance in online idea generation platforms	Not mentioned	Information management technology	Stakeholders	Not mentioned	Not mentioned	Idea quality: adoption intent	Lab study	Not mentioned

Table A1. *Cont.*

Author	Purpose of the Tool	Outcome Type	Use of Theory (to Support Idea Management)	Participant Categories	Design Task	Effectiveness of the Design Outcome	Criteria of Effectiveness	Evaluation Type	Name of the Tools (*n* = 38)
Vasconcelos et al., 2017 [48]	Compare the inspiration effects from these two types of stimuli	Not mentioned	Cognitive psychology	Software designers	Individual	No evaluation	Fluency, diversity, commonness, and conformity	Lab study	Not mentioned
Liikkanen et al., 2010 [46]	Experimentation in conceptual product design	Not mentioned	Cognitive psychology	Hardware designers	Not mentioned	No evaluation	Not mentioned	Lab study	Not mentioned
Oldschmidt et al., 2011 [55]	Present stimuli along with a design problem	Not mentioned	Not mentioned	Software designers	Individual	No evaluation	Originality and practicality	Lab study	Not mentioned
Sozo et al., 2019 [13]	Emotional stimuli for creativity	Not mentioned	Cognitive psychology	Software designers	Individual	Experimental comparison	Quantity, quality, variety, and novelty	Field study	Emotriggers
Yang et al., 2021 [56]	Understand relationship between users' social interaction and innovation contribution	Not mentioned	Not mentioned	Stakeholders	Group	Longitudinal experiment	Not mentioned	Lab study	Not mentioned

References

1. Gerlach, S.; Brem, A. Idea management revisited: A review of the literature and guide for implementation. *Int. J. Innov. Stud.* **2017**, *1*, 144–161. [CrossRef]
2. Han, J.; Shi, F.; Chen, L.; Childs, P.R. The Combinator—A computer-based tool for creative idea generation based on a simulation approach. *Des. Sci.* **2018**, *4*, e11. [CrossRef]
3. Mikelsone, E.; Spilbergs, A.; Volkova, T.; Liela, E.; Frisfelds, J. Idea Management System Application Type Impact on Idea Quantity. *Eur. Integr. Stud.* **2020**, *4*, 192–206. [CrossRef]
4. Jensen, A.R.V.; Musura, A.; Petrovecki, K.; Ryeong Kim, S.; Gerlach, S.; Brem, A. A literature review of idea management. *Nord. Conf.* **2017**, *1*, 1–9.
5. Parjanen, S.; Hennala, L.; Konsti-Laakso, S. Brokerage functions in a virtual idea generation platform: Possibilities for collective creativity? *Innov. Manag. Policy Pract.* **2012**, *14*, 363–374. [CrossRef]
6. Herstatt, C.; Verworn, B. *The 'Fuzzy Front End' of Innovation. Bringing Technology and Innovation into the Boardroom*; Palgrave Macmillan: London, UK, 2004; pp. 347–372.
7. Han, J.; Shi, F.; Chen, L.; Childs, P.R.N. A computational tool for creative idea generation based on analogical reasoning and ontology. *Artif. Intell. Eng. Des. Anal. Manuf. AIEDAM* **2018**, *32*, 462–477. [CrossRef]
8. Howard, T.J.; Culley, S.; Dekoninck, E.A. Reuse of ideas and concepts for creative stimuli in engineering design. *J. Eng. Des.* **2011**, *22*, 565–581. [CrossRef]
9. Daly, S.R.; Yilmaz, S.; Christian, J.L.; Seifert, C.M.; Gonzalez, R. Design heuristics in engineering concept generation. *J. Eng. Educ.* **2012**, *101*, 601–629. [CrossRef]
10. Tanyavutti, A.; Anuntavoranich, P.; Nuttavuthisit, K. An idea generation tool harnessing cultural heritage for design-driven entrepreneurs. *Acad. Entrep. J.* **2018**, *24*, 1–16.
11. Munemori, J.; Komori, K.; Itou, J. GUNGEN-heartbeat: A support system for high quality idea generation using heartbeat variance. *IEICE Trans. Inf. Syst.* **2020**, *103*, 796–799. [CrossRef]
12. Ryeong Kim, S. Idea Management. Identifying the factors that contribute to uncertainty in idea generation practices within front end NPD. *Des. J.* **2017**, *20* (Suppl. 1), S4398–S4408. [CrossRef]
13. Sozo, V.; Ogliari, A. Stimulating design team creativity based on emotional values: A study on idea generation in the early stages of new product development processes. *Int. J. Ind. Ergon.* **2019**, *70*, 38–50. [CrossRef]
14. Shani, N.; Divyapriya, P. A role of innovative idea management in HRM. *Int. J. Manag.* **2011**, *2*, 69–78.
15. Westerski, A.; Dalamagas, T.; Iglesias, C.A. Classifying and comparing community innovation in Idea Management Systems. *Decis. Support Syst.* **2013**, *54*, 1316–1326. [CrossRef]
16. Bacciotti, D.; Borgianni, Y.; Rotini, F. A CAD tool to support idea generation in the product planning phase. *Comput. Aided Des. Appl.* **2016**, *13*, 490–502. [CrossRef]
17. Alexy, O.; Criscuolo, P.; Salter, A. Managing unsolicited ideas for R&D. *Calif. Manag. Rev.* **2012**, *54*, 116–139.
18. Bassiti, L. Toward an Innovation Management Framework: A Life-Cycle Model with an Idea Management Focus. *Int. J. Innov. Manag. Technol.* **2013**, *4*, 551–559. [CrossRef]
19. Westerski, A.; Iglesias, C.A.; Nagle, T. The road from community ideas to organisational innovation: A life cycle survey of idea management systems. *Int. J. Web Based Communities* **2011**, *7*, 493–506. [CrossRef]
20. Mikelsone, E.; Liela, E. Literature review of idea management: Focuses and gaps. *J. Bus. Manag.* **2015**, *5*, 107–121.
21. Cheng, P.J. Development of a mobile app for generating creative ideas based on exploring designers' on-line resource searching and retrieval behavior. *Des. Stud.* **2016**, *44*, 74–99. [CrossRef]
22. Inie, N.; Dalsgaard, P. How interaction designers use tools to manage ideas. *ACM Trans. Comput. Hum. Interact.* **2020**, *27*, 1–26. [CrossRef]
23. Fiorineschi, L.; Frillici, F.S.; Gregori, G.; Rotini, F. Stimulating idea generation for new product applications. *Int. J. Innov. Sci.* **2018**, *10*, 454–474. [CrossRef]
24. Bank, J.; Raza, A. Collaborative Idea Management: A Driver of Continuous Innovation. *Technol. Innov. Manag. Rev.* **2014**, *4*, 11–16. [CrossRef]
25. Bayus, B.L. Crowdsourcing new product ideas over time: An analysis of the Dell IdeaStorm community. *Manag. Sci.* **2013**, *59*, 226–244. [CrossRef]
26. Inie, N.; Dalsgaard, P.; Dow, S. Designing Idea Management Tools: Three challenges. In Proceedings of the Design as a Catalyst for Change—DRS International Conference 2018, Limerick, Ireland, 25–28 June 2018; p. 3.
27. Krejci, D.; Missonier, S. Idea Management in a Digital World: An Adapted Framework. In Proceedings of the 54th Hawaii International Conference on System Sciences, Kauai, HI, USA, 5 January 2021; Volume 1, pp. 5851–5860.
28. Moon, H.; Han, S.H. A creative idea generation methodology by future envisioning from the user experience perspective. *Int. J. Ind. Ergon.* **2016**, *56*, 84–96. [CrossRef]
29. Arksey, H.; O'Malley, L. Scoping studies: Towards a methodological framework. *Int. J. Soc. Res. Methodol. Theory Pract.* **2005**, *8*, 19–32. [CrossRef]
30. Munn, Z.; Peters, M.D.J.; Stern, C.; Tufanaru, C.; McArthur, A.; Aromataris, E. Systematic review or scoping review? Guidance for authors when choosing between a systematic or scoping review approach. *BMC Med. Res. Methodol.* **2018**, *18*, 143. [CrossRef]

31. Tricco, A.C.; Lillie, E.; Zarin, W.; O'Brien, K.K.; Colquhoun, H.; Levac, D.; Moher, D.; Peters, M.D.J.; Horsley, T.; Weeks, L.; et al. PRISMA extension for scoping reviews (PRISMA-ScR): Checklist and explanation. *Ann. Intern. Med.* **2018**, *169*, 467–473. [CrossRef]

32. Bazzano, A.N.; Martin, J.; Hicks, E.; Faughnan, M.; Murphy, L. Human-centred design in global health: A scoping review of applications and contexts. *PLoS ONE* **2017**, *12*, e0186744. [CrossRef]

33. Anåker, A.; Heylighen, A.; Nordin, S.; Elf, M. Design Quality in the Context of Healthcare Environments: A Scoping Review. *Health Environ. Res. Des. J.* **2017**, *10*, 136–150. [CrossRef]

34. Riedel, J.C.K.H.; Vodicka, M. Evaluating the Melodie ICT tool for supporting idea generation. *IFAC Proc. Vol.* **2010**, *43*, 331–334. [CrossRef]

35. Žavbi, R.; Fain, N.; Rihtaršič, J. Evaluation of a method and a computer tool for generating concept designs. *J. Eng. Des.* **2013**, *24*, 257–271. [CrossRef]

36. Benbya, H.; Leidner, D. How Allianz UK used an idea management platform to harness employee innovation. *MIS Q. Exec.* **2018**, *17*, 141–157.

37. Ardaiz-Villanueva, O.; Nicuesa-Chacón, X.; Brene-Artazcoz, O.; Sanz De Acedo Lizarraga, M.L.; Sanz De Acedo Baquedano, M.T. Evaluation of computer tools for idea generation and team formation in project-based learning. *Comput. Educ.* **2011**, *56*, 700–711. [CrossRef]

38. Alessi, M.; Camilo, A.; Chetta, V.; Giangreco, E.; Soufivand, M.; Storelli, D. Applying Idea Management System (IMS) approach to design and implement a collaborative environment in public service related open innovation processes. *CEUR Workshop Proc.* **2015**, *1367*, 25–32. [CrossRef]

39. Yu, L.; Nickerson, J.V. An internet-scale idea generation system. *ACM Trans. Interact. Intell. Syst.* **2013**, *3*, 1–24. [CrossRef]

40. Xie, L.; Zhang, P. Idea management system for team creation. *J. Softw.* **2010**, *5*, 1187–1194. [CrossRef]

41. Munemori, J.; Sakamoto, H.; Itou, J. Development of idea generation consistent support system that includes suggestive functions for preparing concreteness of idea labels and island names. *IEICE Trans. Inf. Syst.* **2018**, *101*, 838–846. [CrossRef]

42. El Haiba, M.; Elbassiti, L.; Ajhoun, R. Using recommender systems to support idea generation stage. *J. Eng. Appl. Sci.* **2017**, *12*, 9341–9351.

43. Yilmaz, S.; Daly, S.R.; Christian, J.L.; Seifert, C.M.; Gonzalez, R. Can experienced designers learn from new tools? A case study of idea generation in a professional engineering team. *Int. J. Des. Creat. Innov.* **2014**, *2*, 82–96. [CrossRef]

44. Jeong, S.; Jeong, Y.; Lee, K.; Lee, S.; Yoon, B. Technology-based new service idea generation for smart spaces: Application of 5G mobile communication technology. *Sustainability* **2016**, *8*, 1211. [CrossRef]

45. Börekçi, N.A.G.Z. Usage of design thinking tactics and idea generation strategies in a brainstorming session. *METU J. Fac. Archit.* **2015**, *32*, 1–17. [CrossRef]

46. Liikkanen, L.A.; Perttula, M. Inspiring design idea generation: Insights from a memory-search perspective. *J. Eng. Des.* **2010**, *21*, 545–560. [CrossRef]

47. López-Mesa, B.; Mulet, E.; Vidal, R.; Thompson, G. Effects of additional stimuli on idea-finding in design teams. *J. Eng. Des.* **2011**, *22*, 31–54. [CrossRef]

48. Vasconcelos, L.A.; Cardoso, C.C.; Sääksjärvi, M.; Chen, C.C.; Crilly, N. Inspiration and Fixation: The Influences of Example Designs and System Properties in Idea Generation. *J. Mech. Des. Trans. ASME* **2017**, *139*, 031101. [CrossRef]

49. Gonçalves, M.; Cardoso, C.; Badke-Schaub, P. What inspires designers? Preferences on inspirational approaches during idea generation. *Des. Stud.* **2014**, *35*, 29–53. [CrossRef]

50. Murah, M.Z.; Abdullah, Z.; Hassan, R.; Bakar, M.A.; Mohamed, I.; Amin, H.M. Kacang cerdik: A conceptual design of an idea management system. *Int. Educ. Stud.* **2013**, *6*, 178–184. [CrossRef]

51. Luo, L.; Toubia, O. Improving online idea generation platforms and customizing the task structure on the basis of consumers' domain-specific knowledge. *J. Mark.* **2015**, *79*, 100–114. [CrossRef]

52. Sadriev, A.R.; Pratchenko, O.V. Idea management in the system of innovative management. *Mediterr. J. Soc. Sci.* **2014**, *5*, 155. [CrossRef]

53. Kokogawa, T.; Maeda, Y.; Matsui, T.; Itou, J.; Munemori, J. The effect of using photographs in idea generation support system. *J. Inf. Process.* **2013**, *21*, 580–587. [CrossRef]

54. Schlecht, L.; Yang, M. Impact of prototyping resource environments and timing of awareness of constraints on idea generation in product design. *Technovation* **2014**, *34*, 223–231. [CrossRef]

55. Goldschmidt, G.; Sever, A.L. Inspiring design ideas with texts. *Des. Stud.* **2011**, *32*, 139–155. [CrossRef]

56. Yang, M.; Han, C. Stimulating innovation: Managing peer interaction for idea generation on digital innovation platforms. *J. Bus. Res.* **2021**, *125*, 456–465. [CrossRef]

57. Di Guardo, M.C.; Castriotta, M. The challenge and opportunities of crowdsourcing web communities: An Italian case study. *Int. J. Electron. Commer. Stud.* **2014**, *4*, 79–92. [CrossRef]

58. Mikelsone, E.; Spilbergs, A.; Segers, J.P.; Volkova, T.; Liela, E. Idea quality forecasting models for different web-based idea management system types. In Proceedings of the 21st International Scientific Conference Engineering for Rural Development, Jelgava, Latvia, 25–27 May 2022.

59. Cheng, X.; Fu, S.; De Vreede, T.; De Vreede, G.J.; Seeber, I.; Maier, R.; Weber, B. Idea convergence quality in open innovation crowdsourcing: A cognitive load perspective. *J. Manag. Inf. Syst.* **2020**, *37*, 349–376. [CrossRef]

60. Fowler, C.; Jiao, J.; Pitts, M. Frustration and ennui among Amazon MTurk workers. *Behav. Res. Methods* **2022**, 1–17. [CrossRef]
61. O'Donovan, P.; Lībeks, J.; Agarwala, A.; Hertzmann, A. Exploratory font selection using crowdsourced attributes. *ACM Trans. Graph. TOG* **2014**, *33*, 1–9. [CrossRef]
62. Trischler, J.; Pervan, S.J.; Kelly, S.J.; Scott, D.R. The value of codesign: The effect of customer involvement in service design teams. *J. Serv. Res.* **2018**, *21*, 75–100. [CrossRef]
63. Inie, N.; Dalsgaard, P. A typology of design ideas. In Proceedings of the 2017 ACM SIGCHI Conference on Creativity and Cognition, Singapore, 27–30 June 2017; pp. 393–406.
64. Selart, M.; Johansen, S.T. Understanding the role of value-focused thinking in idea management. *Creat. Innov. Manag.* **2011**, *20*, 196–206. [CrossRef]
65. Davies, R.; Harty, C. Initial use of an idea capture app in a UK construction organisation. In Proceedings of the 30th Annual ARCOM Conference, Portsmouth, UK, 1–3 September 2014; pp. 987–996.
66. Taylor, D.W.; Berry, P.C.; Block, C.H. Does group participation when using brainstorming facilitate or inhibit creative thinking? *Adm. Sci. Q.* **1958**, *3*, 23–47. [CrossRef]

MDPI
St. Alban-Anlage 66
4052 Basel
Switzerland
www.mdpi.com

Applied Sciences Editorial Office
E-mail: applsci@mdpi.com
www.mdpi.com/journal/applsci